MATERIALS

AN ENVIRONMENTAL PRIMER

Editors
Hattie Hartman
Joe Jack Williams

RIBA Publishing

Materials - An Environmental Primer
is printed on Cairn Eco Kraft 170gsm
and White Offset 120gsm paper FSC® carbon
balanced by the World Land Trust. Printed on a
high-quality four-colour LE-UV press.

www.carbonbalancedprint.com
CBP2275

Published by RIBA Publishing, 66 Portland Place, London, W1B 1AD

ISBN 978 1 915722 21 8

The rights of Hattie Hartman and Joe Jack Williams to be identified as the
Editors of this Work have been asserted in accordance with the Copyright,
Designs and Patents Act 1988 sections 77 and 78.

British Library Cataloguing-in-Publication Data
A catalogue record for this book is available from the British Library.

Commissioning Editor: Clare Holloway
Production: Marie Doinne
Designed and typeset by Sara Miranda Icaza
Printed and bound by Gomer Press, Llandysul
Cover design: CHK Design

While every effort has been made to check the accuracy and quality of the
information given in this publication, neither the Author nor the Publisher
accept any responsibility for the subsequent use of this information, for any
errors or omissions that it may contain, or for any misunderstandings arising
from it.

www.ribapublishing.com

ACKNOWLEDGEMENTS

A heartfelt thank you to every contributor for generously sharing their knowledge and patiently working with us to find the emphasis and tone for each chapter.

Thank you to the steering group participants who helped frame the content and structure of the book through two enormously helpful online meetings at the outset of the project: Jane Anderson, Will Arnold, Duncan Baker Brown, Simon Corbey, Joe Giddings, Janna Laan, Martha Lewis, and Craig Robertson.

In addition to numerous anonymous peer reviewers, special thanks to the following for providing additional technical peer reviews: Will Arnold, Tab Binding, Joe Giddings, Robert Hairstans, Eva MacNamara, Rachael Owens, Karen Scrivener, and Elaine Toogood.

Thank you to Tom McVeigh, Stephanie Sandall, and Oliver Baldock of FCBStudios for bringing clarity and life to the book's technical content through drawings and diagrams.

And to Kathryn Glendenning for reviewing the manuscript with a finetooth comb in her copyediting.

And finally to the RIBA Publishing team, Clare Holloway, Anna Watson, and Marie Doinne, for getting this book over the line and making it a reality.

We hope it will prove a useful contribution to this rapidly evolving topic.

Hattie and Joe
November 2023

FOREWORD

Material exploration occupies centre stage as architects reframe the challenge of designing in the face of climate and biodiversity emergency. It is now more crucial than ever that everyone involved in design and decision-making in the built environment is equipped to converse in the universal language of 'good choice.'

A holistic understanding of materials must seamlessly inform both our daily design process and the architectural journeys we embark upon. Without this knowledge, we are susceptible to compromise, ultimately causing more harm to the environment and the future of our planet.

Materials: An Environmental Primer serves as a comprehensive compendium to arm designers with a basic understanding of construction materials, the environmental impacts of their manufacturing processes and the relevant questions for project teams to consider as they tackle the planetary crisis.

At Morris+Company, we've confronted the challenge of material selection by establishing a material library and cataloguing system that enhances the environmental literacy of our team. Its focus extends to carbon reduction, pollutants, reusability, and more. Much like our library, this book demystifies the material journey—from manufacture and sourcing to application—providing an invaluable reference for readers.

Instead of relying on a crystal ball for zero-carbon alternatives, this primer equips teams with the knowledge to make informed decisions in real-time, staying attuned to material advancements and innovations that promise an optimistic, low-carbon horizon. Regenerative materials, like timber, are integral to reducing the carbon footprint of the built environment and are well covered in this compendium. For architects working at scale amid tightening regulations, current opportunities for regenerative materials are limited. Buildings often span over a decade from concept to construction, which means that decisions made today may be outdated before practical completion.

Material choices require urgent collaboration, careful negotiation, as well as upskilling and education. Materials are often predetermined before architects even enter the conversation, whether to meet heritage and conservation requirements, design preferences, or, more frequently, due to cost and procurement limitations. As an industry, we must reevaluate the decision-making framework, putting the environment on a par with the traditional hierarchy of time, cost, and quality.

A shift toward prioritising the environment in material selection is evident, but it needs to dramatically accelerate. How can architecture transcend style to place the environment at the apex of the pyramid? What will this new architecture look like? These questions are yet to be defined, and *Materials: An Environmental Primer* should be essential reading for all designers who want to better understand this question and make a difference in shaping the future.

MORRIS+COMPANY
David Storring BSc Arch MArch ARB RIBA is Director Sustainability + Resilience at MORRIS+COMPANY.

INTRODUCTION

Materiality is central to architecture. The materials we build with shape what we build, as well as how buildings look, feel and perform. In the 2020s, the climate emergency has upended mainstream architectural practice, resulting in heightened awareness of the critical importance of material selection in buildings. These ecological concerns have led to intense scrutiny of the impact of conventional materials and a burgeoning interest in regenerative materials.

This increased focus on materials is partly due to a better understanding of the critical role of embodied carbon in the overall environmental footprint of a building. It is also a result of increased public understanding of the impacts of climate change. Designers are pioneering new approaches, and clients are demanding it.

This spike in interest in materials has pervaded the architectural press, made its way into the broadsheets and manifested itself in polemical debates. Complex issues which deserve nuanced consideration are reduced to soundbites, such as '*Concrete is bad*' and '*Timber will save the planet*'.

This is exactly what this book is about, and why it is timely. We hope to dispel prevailing myths about common building materials without demonising individual materials. We are also keen to point out the frequent unintended consequences of material choices: robbing Peter to pay Paul.

Designers must think locally but also understand the global impacts of material choices. Design decisions can affect people, communities and natural environments with unintended consequences across the globe, such as unfair employment practices, deforestation and local resource depletion. The use of a 'harvest map' to survey manufactured, quarried, bio-renewable and reclaimed materials within a given radius of a project should play an increasingly important role in mainstream practice in the future.

The purpose of this book is to help designers navigate the minefield of choosing the most appropriate material for a particular application on a project. Over the last decade, the amount of technical information about sustainable materials has

Figure 0.1 Sands End Arts and Community Centre, Fulham, London, Mae, 2020. This community centre incorporates an Edwardian villa converted into an arts centre and is comprised of a series of newbuild structures knitted into an existing park, with mature trees. Flexible spaces, grouped around an outdoor courtyard, lend themselves easily to changing uses. CLT, glulam and a bespoke brick made from construction waste reduce embodied carbon. Adhesive-free CLT without steel flitch plates was installed with bolted connections to ensure that the building will be easy to dismantle in future. In keeping with the building's lean approach, interior materials are generally self-finishing, and a cradle-to-cradle-certified acoustic lining is used throughout.

exploded, be it in the form of low-carbon route maps from trade organisations or through manufacturers' product literature. There is no shortage of information, yet it can be difficult to find impartial, even-handed guidance.

It should be no surprise that the longest chapters in the book are devoted to the 'big three' structural materials: concrete, steel and timber. While prevailing industry discourse promotes less use of concrete and steel and greater use of timber, we prefer to advocate the right amount of the right material in the right place. Vilifying concrete is counterproductive because it is here to stay for large-scale infrastructure projects and below-grade applications. Likewise, timber is not the silver bullet that many advocates claim.

Numerous industry studies have demonstrated that the most significant embodied carbon savings can be made by optimising substructure and structure on any project, and this is now increasingly understood by designers. Many architectural practices

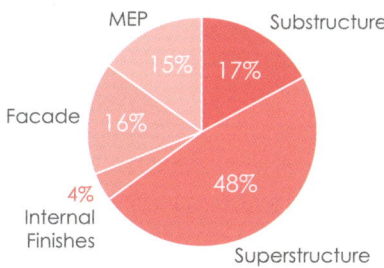

Figure 0.2 LETI (Low Energy Transformation Initiative) embodied carbon primer, 2020. Typical breakdown of life-cycle embodied carbon emissions from an office building.

have developed in-house tools for calculating embodied carbon, and optioneering in early design stages is increasingly common.[1] Buro Happold's useful *Embodied Carbon: Structural Sensitivity Study* (2020) (see the Primer section of this book) graphically presents clear decision trees for structural design in each of the 'big three' materials.[2] The material chapters in this book explore the nuances behind these choices.

Another valuable visualisation tool that has paved the way for better understanding of the environmental impacts of materials is the Material Pyramid developed by CINARK at the Royal Danish Academy (see Figure 0.3).[3] Available since 2020 in a digital version, the pyramid helps visualise which construction materials designers can use abundantly, and which materials need to be used more sparingly due to their carbon intensity.

Another challenge designers face is that assessment of materials quickly gets quite technical, with key concepts buried in complex life-cycle assessment (LCA) modules and jargon. Nonetheless, life-cycle assessment (see the Primer section for more) is a useful tool for understanding material impacts.

To get to grips with the carbon impact of materials, designers must familiarise themselves with the sequence of LCA modules, and for that reason, we have used them to structure the material chapters in this book. Ideally, the modules should be internalised as an intuitive sequence for thinking about materials impacts. Key information then needs to be distilled so that it can easily be communicated and shared with design team colleagues and clients.

Selecting exemplar projects for each chapter has proven tricky because projects that excel in one area, such as low embodied carbon, often do not address another, say, design for disassembly. Image captions highlight why a particular project is featured.

Retrofit, a crucial growth sector, has enormous implications on material choices, particularly for insulation. Choosing the right amount of the right insulation is a complex task. For this reason, we have grouped insulation materials together to facilitate comparison between different alternatives.

THE BUILDING MATERIALS PYRAMID

GWP (KG CO₂EQ /KG MATERIAL)

The pyramid lists building materials with their GWP values (KG CO₂EQ / KG MATERIAL):

- 28242.0 Aluminium sheet
- 26570.0 Roof panel (steel)
- 22923.1 Galvanised steel
- 12433.0 Copper sheet
- 12209.4 Zinc
- Structural steel
- 5733.5 EPDM foil
- 4095.5 Vinyl flooring (PVC)
- 2851.0 Paint, matte
- 1725.3 Ceramic tiles
- 1694.0 Cement-bonded particle board
- 1507.3 Slate
- 1172.7 Aluminium frame window
- 898.2 Brick, red, double-fired
- 781.4 PIR insulation
- 762.6 Wood-Aluminium frame window
- 699.0 Fibre cement boards
- 618.0 Clinker – stoneware
- 565.2 Brick, red, single-fired
- 529.0 Fired clay brick
- 474.1 Wood frame window
- 420.1 Brick roof tiles
- 415.6 Glass pane, triple-glazed
- 407.8 Roofing felt V60
- 375.1 plaster
- 366.1 Concrete roof tiles
- 288.0 Concrete C30/37
- 271.5 PP roofing membrane
- 266.1 Glass pane, double-glazed
- 266.3 PE film (vapour barrier)
- 244.8 Lime sandstone
- 244.2 Foam glass
- 229.0 Concrete C20/25
- 202.3 Lightweight concrete elements
- 190.6 Lime render
- 180.0 Aerated concrete blocks
- 169.6 Gypsum board
- 158.0 Poroton bricks
- 123.3 PUR insulation
- 96.3 XPS insulation
- 93.5 Unfired clay brick
- 93.2 Clay plaster
- 91.2 Gypsum fibre board (paper)
- 83.5 Linoleum
- 70.4 Stone wool
- 60.1 Expanded perlite
- 43.5 EPS insulation Graphite 80
- 19.2 Hemp fleece / PE
- 12.8 Glass wool
- 9.3 Rammed earth wall
- 6.2 Paper wool
- 4.9 Reused brick
- **0**
- -128.2 Straw
- -173.1 Wood fibre insulation
- -182.9 Wood fibre board
- -511.0 Modified wood
- -533.0 Parquet floor, 14 mm
- -610.0 Glulam
- -649.0 Plywood
- -664.0 Cross-laminated-timber CLT
- -667.0 MDF
- -680.0 Construction timber
- -777.5 Spruce
- -1063.0 Oak tree

THINK OF THE QUANTITIES

RECYCLING OR WASTE

REMEMBER LIFESPAN

THINK ABOUT TRANSPORT

THE DETAIL IS ESSENTIAL

CINARK
centre for industrialised architecture

Figure 0.3 Construction Material
Pyramid, by CINARK, 2022. The
Material Pyramid is a digital
visual tool for assessing various
environmental impacts of materials,
including global warming potential,
ozone depletion potential and
others.[4] It provides 'at a glance'
guidance on which materials have
significant environmental impacts,
and therefore their use should be
carefully scrutinised.

Reclaimed materials and waste streams are important growth areas that are mostly beyond the scope of this book, yet we do capture some of the innovation currently underway in this area.

The climate emergency has altered architectural discourse about materials. Exemplary work by the campaign group Architects Climate Action Network (ACAN), the long-standing Alliance of Sustainable Building Products (ASBP), the Frugalité movement in France, research consultancy Material Cultures and others have broadened the palette of material choices.

What was deep green is not yet mainstream, but it is now firmly on the professional radar. The Stirling shortlisting of Matthew Barnett Howland's Cork House in 2019 is one example of much-needed material exploration. Niche explorations of materials are a first step, but as Architects Declare cofounder Michael Pawlyn has repeatedly pointed out, only systems change will bring about the profound disruption of the construction industry that is required.

Every built environment professional plays a part in the industry transformation that lies ahead. Designers must repeatedly ask themselves at what point a material choice becomes an architectural folly at the expense of the environment. It is our hope that this book will help readers answer that question by deciphering the complexity of material choices.

We deliberately selected authors without a strong affiliation to industry trade associations and challenged our contributors to present a balanced and informative view of each material. We would like to express our appreciation for our contributors' patience and willingness to work with us to find the voice for each chapter. Readers will find a range of viewpoints in the book. While it is true that most metals are almost infinitely recyclable, as various authors explain, this does not mean that we can continue to ignore the environmental consequences of mining or the carbon emissions of the associated manufacturing processes of these materials.

Material specification is not a neutral subject, and this is not a neutral book. The time for action is now. The lengthy incubation time of a building from design concept to practical completion means that many buildings in the pipeline now will be completing as

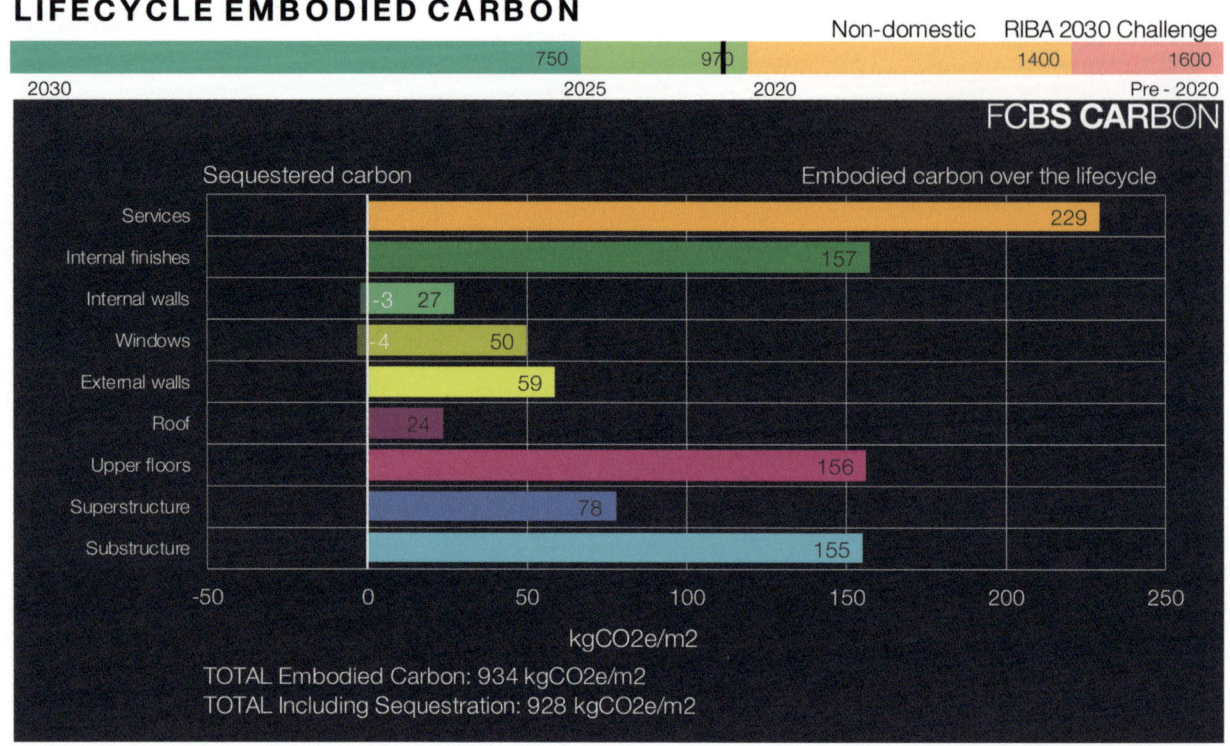

LIFECYCLE EMBODIED CARBON

Figure 0.4 Output from FCBS CARBON, showing embodied carbon of different building elements.[5]

2030 approaches, the year of the UK's legally binding target of 68% reduction in greenhouse gas emissions (over 1990 levels) in order to remain aligned with the Paris Agreement.

Every building design and material choice must be scrutinised to assess whether it is within planetary boundaries.[6] We hope this book will facilitate more informed material choices.

HOW TO USE THIS BOOK

An **introductory primer** draws out key themes that are common to all materials.

Materials are listed alphabetically to make them easy to navigate and to put them on a level playing field. Each chapter, authored by an industry expert (or experts), varies in length roughly corresponding to how widely used the material is in construction. Due to the crucial importance of retrofit, insulation materials are grouped together under 'I' to facilitate comparison, even though insulation *per se* is not a material.

Each chapter loosely follows the life cycle of a material and references within the chapters refer to the life-cycle assessment modules (A1–A5, B1–B5, C1–C4 and D), which are explained in the Primer. Each chapter includes:
- key facts and relevant statistics about each material
- an introductory overview
- construction applications
- a description of key impacts, structured around the LCA modules
- a section on wider sustainability impacts
- design guidance on how to reduce impacts
- industry trends and innovation
- exemplar projects that point the way forward
- key takeaways
- questions designers should ask about each material.

In each chapter there are often carbon emission figures quoted, backed by environmental products declarations or peer-reviewed journals. Unless otherwise noted, the figures do not include any sequestered carbon, putting the focus on immediate carbon reduction opportunities through material choices.

A short chapter on **innovation** looks at current laboratory research and imagines what materials may be on the horizon a decade or more from now.

The **conclusion** is a call to arms, outlining crucial actions for designers to revamp their working practices to design within planetary boundaries.

At the end of the book, a guide to **How to Read an Environmental Product Declaration** has been included, which explains how to get the most from these increasingly common product datasets.

PRIMER

Throughout this book, myriad approaches to using materials in a more responsible way are presented. Some are specific to a particular material. Individual chapters delve into the nitty-gritty of cement replacement, brick firing and straw panel construction. Yet many themes have emerged which apply to all materials. This primer provides an overview of these overarching issues and should provide a base-level of knowledge to meaningfully engage with the content of the material chapters.

EMBODIED CARBON: THE BASICS

Embodied carbon in the built environment is calculated according to BS EN 15978:2011, which sets out life-cycle assessment (LCA) modules that ensure consistency and comparability of reporting. To encourage designers to familiarise themselves with LCA modules, the material chapters in this book are structured around these A to D modules to highlight where in the production process the emissions are for each material. It is important to understand the distinction between life-cycle embodied carbon (modules A–C, excluding operational emissions B6 and B7) and whole-life carbon, which is life-cycle embodied carbon plus the carbon arising from energy and water use.

The LCA modules are split into the following common boundaries, which are further subdivided, as shown in Figure 1.1:

- **Module A** Upfront carbon (cradle to gate): carbon emitted during the creation of a material/product, and the construction of a building, up until it is occupied.
- **Module B** In-use carbon: operational carbon emitted during use and maintenance of a building for 60 years.
- **Module C** End of life: demolition and disposal of the building.
- **Module D** Impacts beyond the building's boundary: notably circular economy impacts.

A simple formula is used to calculate embodied carbon: multiply the quantity of material used by the carbon dioxide equivalent (CO_2e) emissions associated with that product. Most commonly, off-the-shelf software is used to undertake the analysis, with paid-for tools such as OneClick and eTool prevalent in the UK. However, there are many other free tools, including FCBS CARBON,[1] H\B:ERT[2] and EC3,[3] which can be used at various stages of a project.

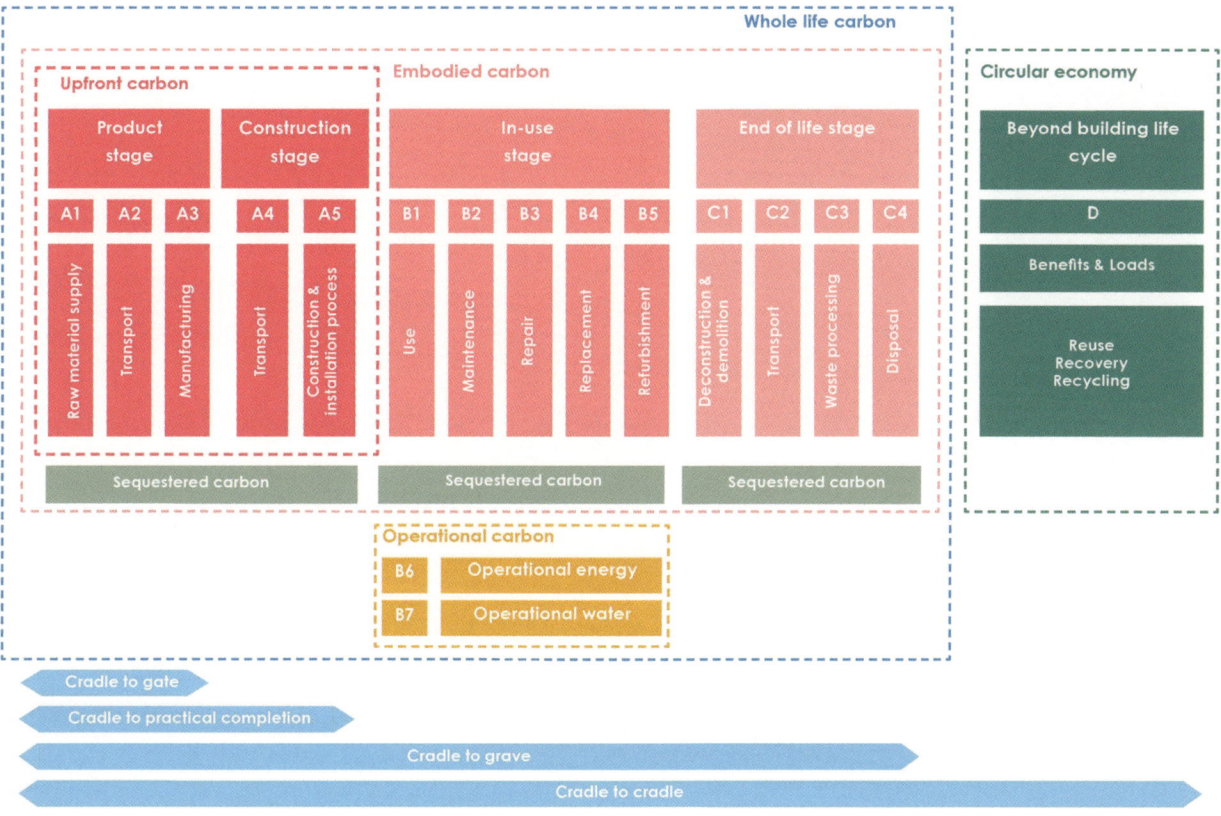

Figure 1.1 Life-cycle assessment modules, from LETI (Low Energy Transformation Initiative) and adapted from BS EN 15978:2011.

Embodied carbon calculations can also be undertaken manually, using the greenhouse gas emissions data that can be found in Environmental Product Declarations (EPDs). EPDs are third-party-verified documents, often prepared by a product manufacturer, which contain data on a product's environmental impact. EPDs provide robust environmental global warming potential data per LCA module for a given quantity of the material, and include other useful metrics such as water use, resource depletion and waste creation. They are often freely available in online databases, or from open databases of generic material emission rates.[4]

When discussing greenhouse gas emissions, it is important to understand the units used. The most common unit is $kgCO_2e$, often expressed per m^3 or kg of a product. Many other gases are greenhouse gases, and their impact is expressed in terms of CO_2e, referred to as the global warming potential (GWP) of a product. For

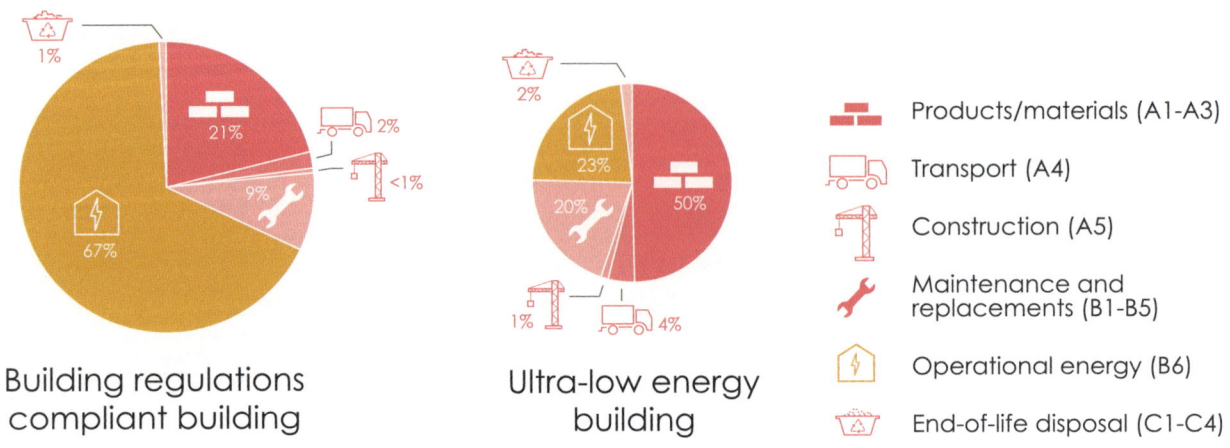

Building regulations compliant building

Ultra-low energy building

Products/materials (A1-A3)

Transport (A4)

Construction (A5)

Maintenance and replacements (B1-B5)

Operational energy (B6)

End-of-life disposal (C1-C4)

Figure 1.2 Emissions arising from different LCA stages for a medium-rise residential building. As a building becomes more efficient in its operational energy use, the relative impact of embodied carbon increases. Source: Low Energy Transformation Initiative

example, 1kg of methane is the equivalent of 29.8kg of CO_2, whereas 1kg of the common refrigerant R410a is equivalent to 2,088kg of CO_2. Even though carbon dioxide is frequently mentioned throughout this book, it is important to be aware that other gas emissions also contribute to global warming.

It's more than just carbon
While carbon is a key element of sustainability, particularly given the climate crisis, it is only one part of the broader sustainability spectrum. **A project must look beyond its site boundaries to fully understand its planetary impacts.** Frameworks such as One Planet Living or the Living Building Challenge include the wider considerations for any project, such as water use, land and nature, waste and community. EPDs provide quantifiable data on some environmental metrics, but do not include these broader metrics that can provide a wider context.

Through thoughtful material choices, designers can influence positive change towards regenerative design practice that can help not only the current project, but also future projects by improving the wider supply chain.

Retrofit is key
The most sustainable design decision is to avoid activity that emits carbon, including any form of construction. However, an ever-evolving need for shelter, for homes and other activities, particularly

- Health and happiness
- Equity and local economy
- Culture and community
- Land and nature
- Sustainable water
- Local and sustainable food
- Travel and transport
- Materials and products
- Zero waste
- Zero carbon energy

Figure 1.3 Focusing solely on carbon emissions can miss the wider environmental and societal impacts of material choices as outlined in Bioregional's One Planning Living framework above.

in the Global South, means that we will continue to build. The clearest method of reducing the construction industry's impact is to build less, encourage the retrofit of buildings, increase their usefulness and improve their efficiency. This is a growing trend, with campaigns such the *Architects' Journal's* RetroFirst pushing the agenda through the industry.[5] With many of the buildings to be used in 2050 already constructed, methods of improving and adapting them will become increasingly important as we move through the 21st century. Some materials are better suited to retrofit, such as breathable insulations or lightweight structural solutions, but all will undoubtedly have a part to play.

Embodied carbon reduction hierarchy

Material use and reduction of embodied carbon is under the spotlight. The most obvious way of reducing embodied carbon is to simply *build less*. The IStructE's Hierarchy of Net Zero Design (see Figure 1.5, adapted from PAS 2080) clearly outlines steps around simply using less, prioritising retrofit and reuse of existing structures.

Once the need for a new building has been agreed, there are still significant opportunities for material reduction. Scrutinising structure and substructure is a good place to start. This is true of any structural material, and Buro Happold's *Embodied Carbon: Structural Sensitivity Study* (2020), published on the IStructE website, demonstrates that savings of up to 40% can be achieved when adopting efficient structural designs.[6] The useful seven-page guide includes clear diagrams which explain the sequence of decisions to optimise structural efficiency for different materials. Some options are less obvious, for instance reducing the structural grid may increase the number of columns, but there will be a significant decrease in beam and floor slab sizes, reducing the amount of material used overall.

Further savings can be made through simple design choices such as having an exposed soffit rather than a ceiling system, or through more complex options, such as reducing the weight of rain screen cladding to decrease the amount of secondary structure required.

It is only once the design has been made highly efficient that alternative, low-carbon material options should be considered. With a competitive materials market, and even with robust specifications and

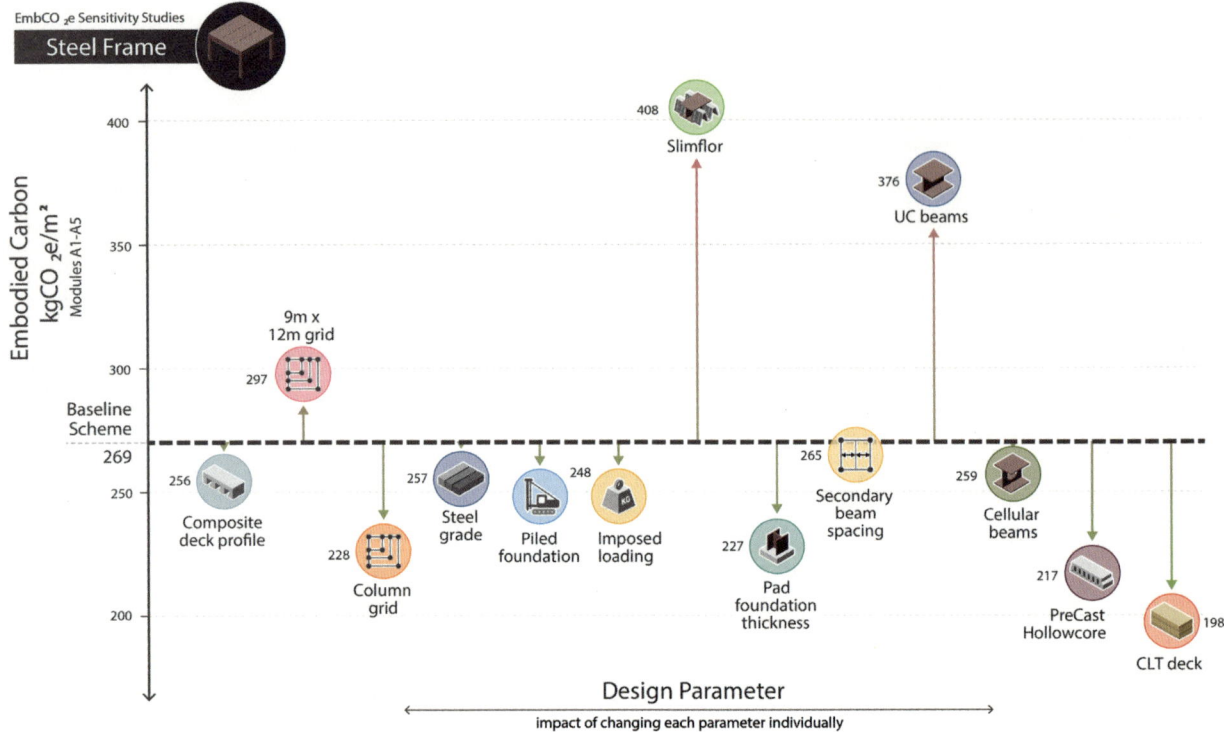

Figure 1.4 Carbon reduction options for a steel frame, from Buro Happold's *Embodied Carbon: Structural Sensitivity Study*, 2020. This chart shows the impact of altering different aspects of a steel framing system on the embodied carbon of the super structure.

contractual documentation, material substitutions can occur during construction, for example changing concrete mixes or different timber suppliers, which will impact the embodied carbon of a building. By designing to an efficient building form in the first instance, the risk associated with material substitutions can be minimised.

CIRCULAR ECONOMY

A knock-on effect of embodied carbon reduction is increasing interest in the circular economy, providing options for material reuse over new materials. Reuse of a material is often incredibly low carbon, with most new emissions arising from transport, and typically significant savings over recycling of materials. Sources of reclaimed materials can be found through an increasing number of services, such as GlobeChain, or directly through suppliers, such as Cleveland Steel and Tubes or Rype Office. Reuse of materials needs to be carefully considered, particularly in situations that require guaranteed performance, such as structural loading or fire resilience, because reused products may need recertifying prior to use in the building. All reuse should be checked with the relevant authorities, including building control and insurance providers.

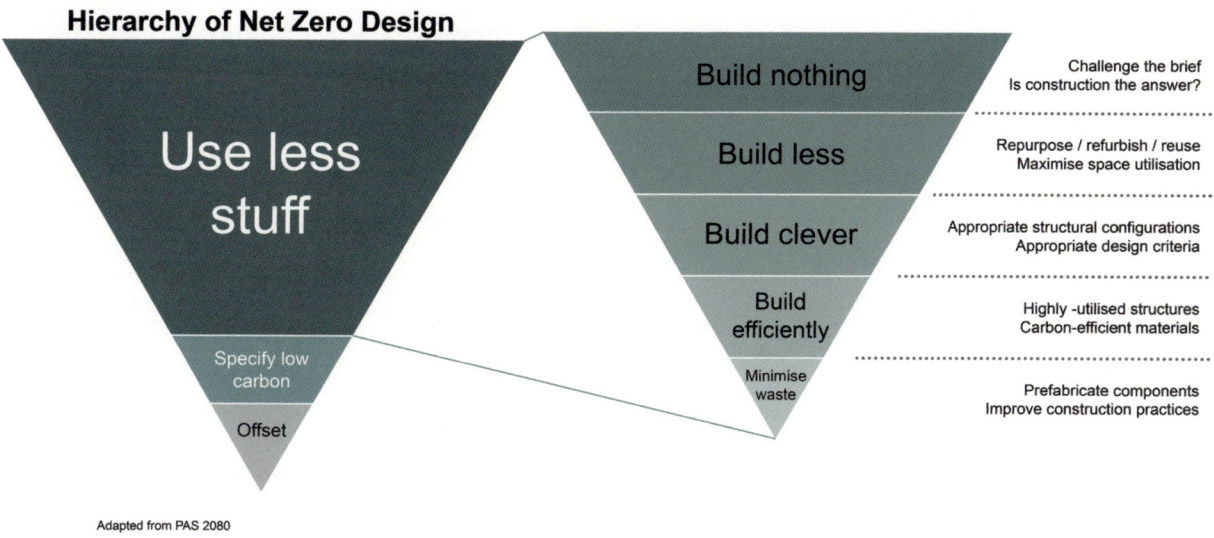

Hierarchy of Net Zero Design

Use less stuff

Specify low carbon

Offset

Build nothing — Challenge the brief / Is construction the answer?

Build less — Repurpose / refurbish / reuse / Maximise space utilisation

Build clever — Appropriate structural configurations / Appropriate design criteria

Build efficiently — Highly-utilised structures / Carbon-efficient materials

Minimise waste — Prefabricate components / Improve construction practices

Adapted from PAS 2080

Figure 1.5 IStructE's Hierarchy of Net Zero Design.

A barrier to the circular economy is that a majority of buildings were not designed to be deconstructed; instead, they require destructive demolition. To enable a future circular economy, buildings should be designed to be dismantled, with the design focus moving away from the materials to the connections between them. Use of mechanical fixings over glues encourages and enables deconstruction, minimising damage and maximising reuse potential.

Some elements will need to be regularly replaced and repaired over the life of the building. These elements should be able to be repaired without causing damage to the wider building, for example by having visible and accessible screw heads. A widespread example of this is unitised cladding, where glazing units often require replacement long before the framing systems.

RENEWABLE MATERIALS

When any product is manufactured, material is converted into a final form to create a functional building element. With many materials, that requires using minerals mined from the earth, which are transformed into a specific product, such as iron ores or limestone. Mineral-based products are a finite resource, and while many are recyclable, their use often contributes to their depletion. While some minerals are plentiful, such as limestone, others are more limited, such as exotic metals, including cadmium. Some materials that seem plentiful can still be under stress, such as the right grade of sand for glass and concrete production.

By contrast, a renewable material is one that can be regenerated, most typically bio-based. Timber can be regrown over a period of 50 to 100 years, whereas hemp can be regrown in less than a year and bamboo in under six years. The use of these materials is often seen as more sustainable because their use today does not deprive future generations of the ability to also use them, and they also draw down atmospheric CO_2 that can be locked away in a material. In the manufacture of bio-based products, there will likely be some use of non-renewable materials, so regardless of the product used, checking the Environmental Product Declaration (EPD) for the amount of resource depletion should be a key check during design (see the Appendices for a guide on how to read an EPD).

WASTE STREAMS AS SUSTAINABLE

With the proliferation of cheap materials and endless fossil-fuelled energy, **the planet is inundated with waste products that need to be dealt with**, typically buried in the ground, dumped in the sea or burnt. Solutions that provide a use, and importantly a value, for a waste stream should be explored, removing that burden off the wider environment. Innovative use of waste streams today may well be the building products of tomorrow.

Keeping materials in use supports the circular economy. But just as waste streams should be examined, ways to improve the original design process and avoid initial wastage should be explored, supporting leaner manufacturing processes and products that produce less waste. Many of the metals within this book feature increased recycled content as a key action in reducing the impact of the materials, captured in module D of an LCA, but this needs to be carefully balanced with the ability to directly reuse a product.

CARBON SEQUESTRATION, IN OR OUT?

When undertaking a carbon assessment for a product or material, carbon sequestration is inevitably a topic for debate. Carbon sequestration is the carbon dioxide stored by a material through its manufacture or use, typically as carbon, effectively locking away the carbon within the material during the product's life. Most commonly this is for bio-based materials, which absorb carbon dioxide during the growth of the plant/tree that makes up the final product. Other non-biological materials also sequester carbon, typically through carbonation, where there is a chemical reaction converting CO_2 into an inert mineral, such as on the surface of untreated concrete. The quantities of carbon locked away vary significantly, but many bio-based materials lock away more carbon dioxide than they emit through their manufacture and transport.

Life cycle emissions of a CLT panel

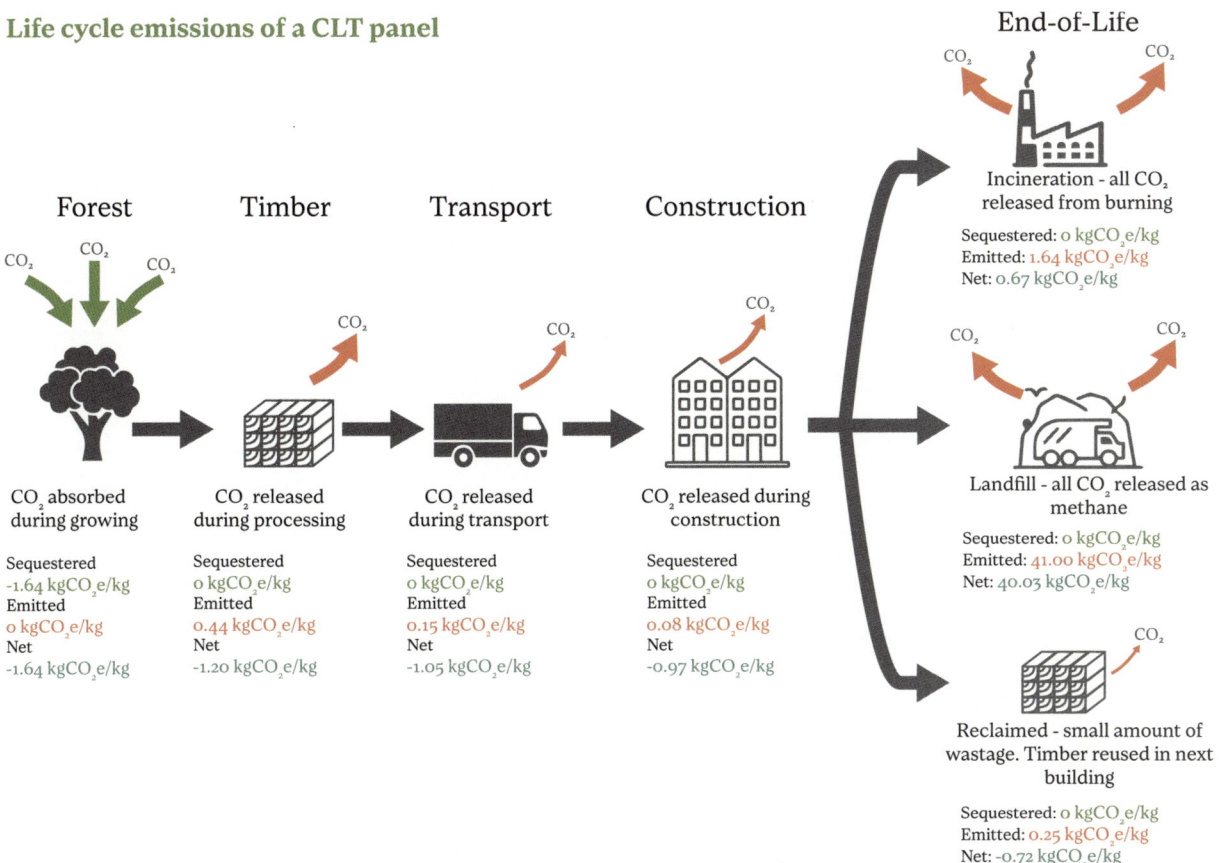

End-of-Life

Forest   **Timber**   **Transport**   **Construction**

CO_2 absorbed during growing

Sequestered
-1.64 kgCO₂e/kg
Emitted
0 kgCO₂e/kg
Net
-1.64 kgCO₂e/kg

CO_2 released during processing

Sequestered
0 kgCO₂e/kg
Emitted
0.44 kgCO₂e/kg
Net
-1.20 kgCO₂e/kg

CO_2 released during transport

Sequestered
0 kgCO₂e/kg
Emitted
0.15 kgCO₂e/kg
Net
-1.05 kgCO₂e/kg

CO_2 released during construction

Sequestered
0 kgCO₂e/kg
Emitted
0.08 kgCO₂e/kg
Net
-0.97 kgCO₂e/kg

Incineration - all CO_2 released from burning

Sequestered: 0 kgCO₂e/kg
Emitted: 1.64 kgCO₂e/kg
Net: 0.67 kgCO₂e/kg

Landfill - all CO_2 released as methane

Sequestered: 0 kgCO₂e/kg
Emitted: 41.00 kgCO₂e/kg
Net: 40.03 kgCO₂e/kg

Reclaimed - small amount of wastage. Timber reused in next building

Sequestered: 0 kgCO₂e/kg
Emitted: 0.25 kgCO₂e/kg
Net: -0.72 kgCO₂e/kg

Figure 1.6 Tracing the carbon dioxide emissions from a tree to a CLT panel, into a building, and finally to the end-of-life scenarios. The CLT remains carbon negative until it is assumed to be burnt or placed in landfill, creating significantly more emissions than it sequestered as a tree. Finding alternative end-of-life scenarios must be a priority.

Accounting for sequestered carbon is a very tricky topic. The RICS Professional Statement on whole-life carbon suggests that sequestered carbon only be included for a full cradle-to-grave assessment for a building or product, not for upfront carbon.[7] However, in practice, due to the life-cycle modelling conventions, any potential benefits of sequestered carbon are negated when the end of life is considered, because the default scenario for any bio-based product is incineration, which releases any sequestered carbon back into the atmosphere. Whether this is a true reflection of the end of life of bio-based materials in 60 years is still being debated, but every opportunity for enabling reuse now should be prioritised.

HEALTHY MATERIALS

Using materials sustainably means not just thinking about their impact on the environment, but also their impact on people. Many regularly used products release chemicals through off-gassing, but also through other processes such as leaching, which impact the indoor air quality (IAQ). Finishes such as glues, varnishes and paints are a common source of chemicals, but so also are many plastics widely used in carpets, furniture, fixings, etc. These chemicals are often volatile organic compounds (VOCs), but there are many chemicals found in building products that can impact occupant health.

Databases of these chemicals are extensive, with Perkins & Will's Transparency website regularly updated with the latest research on toxic substances.[8] The impact does not end with the product itself, but continues through the life of the building, with the cleaning products, and then the end of life of the products.

Until recently, identifying products without harmful chemicals was difficult. However, schemes such as the Living Future Institute's Declare Label[9], NaturePlus certification[10] and the Healthy Product Declarations (HPDs) increase the ability of designers to make informed choices. Often, bio-based products have a reduced likelihood of containing harmful chemicals, with many made using fewer industrial processes and associated chemicals. However, in line with the carbon reduction hierarchy, it is preferable to design out the need for those materials that cannot be shown to be healthy. Is that paint finish really needed? Does the product need to be glued together? etc. To create a healthy environment, it's not just the products that need to improve, but also how they are used in buildings.

It's not just indoor air quality that is a concern, but the product manufacture also directly impacts people. From the extraction of raw materials, transportation across a global supply chain and often multilayered manufacturing processes, every product touches the lives of those involved. These people are those directly employed in the product supply chain, but also the communities that neighbour the many processes. Product manufacturers should be able to demonstrate a fair and equitable supply chain, without issues such as forced and child labour practices that disproportionately affect the Global South. This is an emerging area of interest in the built environment, with clothing manufacturers currently leading the way, but the recently developed Design for Freedom Framework by Grace Farms provides a way for assessing building products.[11]

A GLOBAL SUPPLY CHAIN

Many products rely on a global supply chain. This means that many materials undertake numerous journeys, with some traversing the planet multiple times before the material gets converted into a final product. This can be explicit, such as British sandstone for the façade of Bloomberg's European headquarters in London being shipped to Italy for finishing before being returned to the UK, or more complex, such as the sub-assemblies within a heat pump. For most materials, transport represents a fraction of their carbon footprint, typically around 5%, but this percentage is very sensitive to the modes of transport and the weight of material transported.

Shipping is around 150 times more carbon efficient than flying, and freight trains are the least carbon emissions-intensive land transport method. Many final-mile deliveries are made by a mix of HGVs and vans, which are more flexible at the expense of their carbon efficiency. Grouping deliveries together where possible can save significant fuel and carbon, while simplifying site manoeuvres. Designing to completely fill an HGV or van can also reduce site deliveries and maximise efficiency.

Sourcing local materials is a robust method of reducing the impact of material transport. This can be hyper-local, such as using the clay and earth on site to create rammed earth walls, or supporting local businesses such as brick factories, quarries or timber yards. Using local materials roots a building in its immediate locality, with the finished building reflecting the local vernacular. Any due diligence required of the supply chain is also far simpler, with the working environments easily visitable compared to imported goods.

A BIODIVERSITY CRISIS

Part of the global crisis that threatens the planet is a steep plunge in biodiversity. The materials we use to build can be part of the exploitation of the natural world that is quickening this biodiversity loss, but similarly they can support a regeneration of the habitats that can slow, and potentially reverse, biodiversity loss.

The extraction of raw materials, from the large opencast mines for aluminium ore extraction to the clear cutting of monoculture forests, are obvious examples. In the case of mining, habitat loss is inherent in the process, although reclamation plans to restore the habitat once the mine is closed can mitigate these impacts. The Eden Project in Cornwall is an inspirational example of how nature can recolonise an abandoned quarry in just two decades.

In addition, there are instances of contaminated by-products leaching into the local environment. Standards such as BES 6001 include specific clauses on biodiversity and land use, and are a way for a manufacturer to demonstrate that they have

Figure 1.7 Travel distances to emit 100kgCO₂e when moving 1 tonne of material by different methods. Transport by ship is far less CO₂e-intensive than transport by road. Shipping 1 tonne of material from Scotland to southern England by road emits as much CO₂e as shipping 1 tonne of material by container ship from the Caribbean to the UK.

the policies and checks in place to minimise and mitigate any issues arising from material extraction.

With farmed products, the impact is less obvious; however, monocultures of plants or trees often do not provide the habitats needed to encourage a biodiverse range of flora and fauna. Fertilisers and pesticides can be a significant issue for annual crops, harming local insects and polluting water courses. But there is a growing organic movement, which supports crop growth that is in harmony with nature and significantly limits chemical usage.

The impact of materials does not end with production but continues through their use and end of life, particularly those that contain chemicals that can leach into the environment, such as plastics and paints. These products are best avoided where possible, and if they are necessary, ways of keeping them in-use and out of landfill past the end of life of the building are highly important.

BALANCING MATERIAL CHOICES

Buildings are complex. They must balance conflicting priorities, and that extends to sustainability. Some common design conflicts, each with their own particular performance impacts that often include intangible elements, create ambiguity in decision-making.

Glazing exemplifies this problem, with heat loss through a window being far higher than through a typical wall, yet glazing provides views, daylight and much-needed solar gains in winter. These first two points are tricky to analyse because human need is difficult, if not impossible, to reduce to a simple number.

Another frequently debated issue, which is crucial to the retrofit agenda, is how much insulation is too much? This involves balancing the conflicting drivers of reducing both operational and embodied emissions. With the proliferation of renewable electricity generation in the UK grid, electricity is expected to be nearly zero carbon within the next 15 years, significantly reducing the carbon emissions from buildings arising through energy use. However, there is limited capacity for green energy generation, and the grid can only decarbonise if we reduce the underlying energy use to match the available renewable energy sources.

These performance trade-offs must be resolved by every client and design team. They should be openly discussed, taking in qualitative factors, such as the view from a window, as well as an evidence-based approach, such as dynamic simulation modelling. There is often no perfect answer. It may help to think about the boundaries that a decision influences; is it just the building, the surrounding local environment, a national boundary or a wider global supply chain? No single boundary is more important than another, but considering the boundaries may help to prioritise the impacts of a particular product or material in order to reach a decision on its use.

Every material has benefits and constraints. Throughout the following chapters, these pros and cons are explained, with a focus on how to best use each material to its advantage. **Building design is not a linear process, and it is through an iterative and collaborative process of refinement that each material finds its place.**

MATERIALS
A- Z

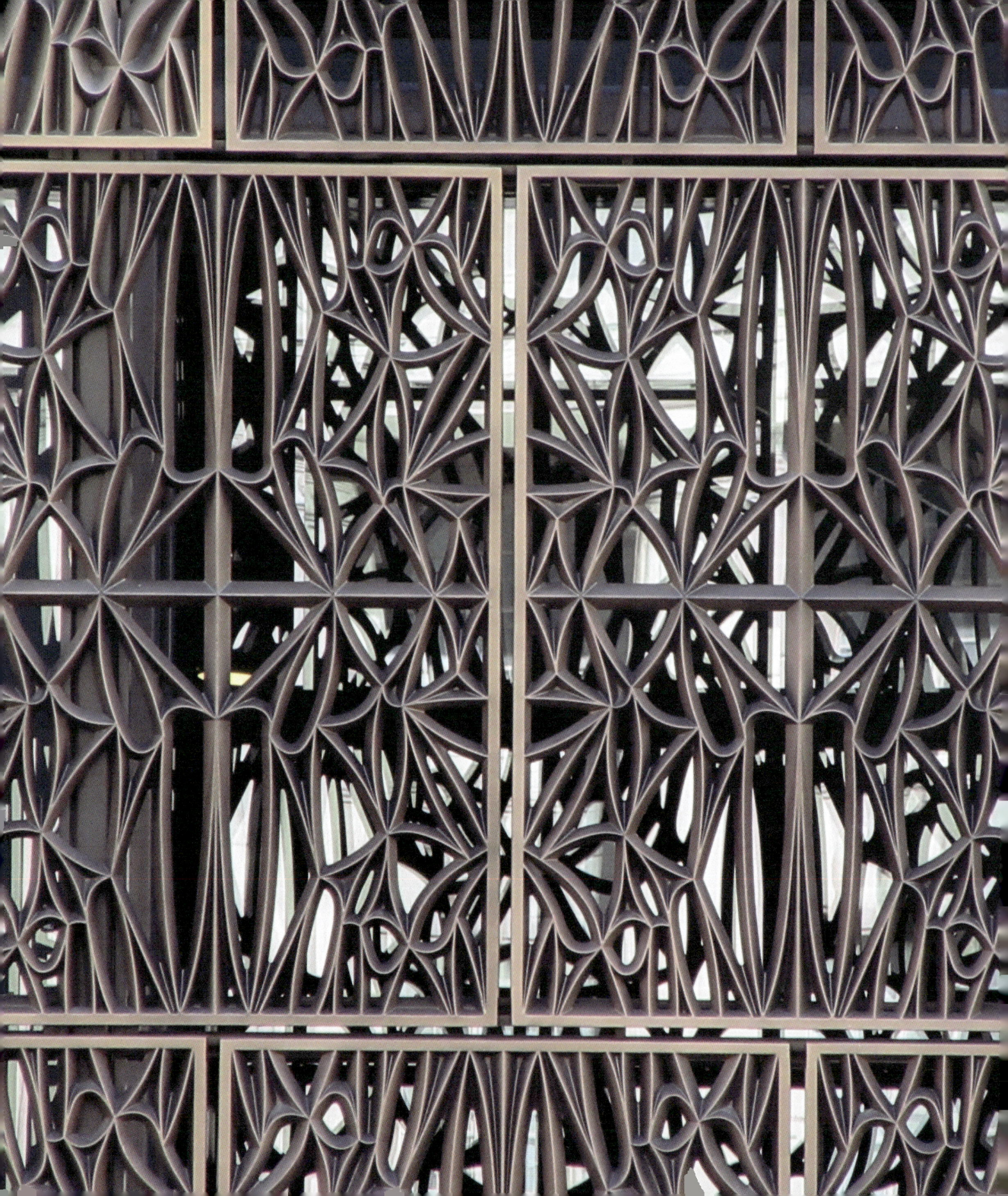

ALUMINIUM

Michael Stacey

Embodied carbon

6,256–19,040

$kgCO_2e/m^3$ [1]

Global emissions from aluminium production, 2022

270 million

tonnes CO_2e [2]

Global production of aluminium

67,243,000

tonnes [3]

UK primary production of aluminium

48,000 tonnes [4]

UK post-consumer recycled aluminium

1 million tonnes

Aluminium and its alloys are a wide class of materials with many applications in contemporary life. Aluminium is lightweight yet durable. It is two thirds the weight of steel, yet 6000 and 7000 series alloys can match the stiffness of mild steel. Some 75% of the aluminium produced since 1886 is either in-use or is available to be used.

Aluminium is found in bauxite and it takes 4 – 6 tonnes of bauxite to make approximately 2 tonnes of alumina, which is then smelted through an energy-intensive process to create about one tonne of aluminium. When considering aluminium for a project, it is important to understand the power mix used to smelt the aluminium which may be renewable (hydroelectric) or fossil fuels. Coal-fired smelters must progressively be phased out.

The three primary methods of finishing aluminium are anodising, polyester powder coating and wet paint systems. All three can be used to coat recycled aluminium. It is most durable when washed by rain so it should be detailed to enable this.

Aluminium is almost infinitely recyclable without any loss in quality and recyclability is the key to sustainability for aluminium. Recycling it uses only 5% of the energy compared to winning it from bauxite. Approximately one-third of global aluminium demand was met by recycled aluminium in 2019.[5] A halt in the rise of primary aluminium production is crucial to limiting global temperature increase to 1.5 by 2050. [6]

APPLICATIONS

- Cladding and curtain walling
- Window and doors
- Ceilings
- Rainwater goods
- Ironmongery
- Solar panels
- Structures, including bridges

PROS

- Low maintenance and durable
- Very low weight
- Highly recyclable, with good established recycling streams
- Can be produced using renewable electricity to create low-carbon aluminium

CONS

- High embodied carbon for primary aluminium

QUALITIES OF ALUMINIUM

Aluminium has six primary qualities that make it ideal for use in applications within architecture and the built environment. It is:

- durable
- recyclable
- flexible
- light and strong
- economical
- sympathetic and powerful.

Aluminium, like many materials and technologies, is very dependent on design – the skills and sensitivity with which it is used – thus in the hands of a good design team it can be considered sympathetic and powerful: contributing to health and wellbeing in the form of a durable built environment.

EXTRACTION, REFINING, SMELTING AND MANUFACTURING

Aluminium is a light, ductile and highly corrosion-resistant metal and is, in itself, unchanged since it was identified by Sir Humphry Davy in 1808, as a constituent of alumina.[7] **Aluminium is the third most abundant material in the earth's crust** and the most abundant metal. It is typically found in the form of bauxite. The key stages for the production of primary aluminium are the mining of bauxite, the refining of alumina and the smelting of alumina to produce aluminium. It takes 4–6 tonnes of bauxite to make approximately 2 tonnes of alumina, which is smelted to about 1 tonne of aluminium. The major bauxite-producing countries are Australia, China and Guinea (production mass in the order stated).

The Bayer process for converting bauxite into alumina was invented by Carl Josef Bayer in 1888. The cost of producing aluminium was very significantly reduced by the Hall-Héroult electrolysis process, which was simultaneously invented in the USA and France in 1886 and named after its three inventors: Charles Martin Hall with his sister Julia Brainerd Hall and Paul Héroult.[8] Combined with the reduction of the cost of producing electricity at the beginning of the 20th century, the cost of aluminium reduced significantly. In the 1850s, the cost of one kilogram of aluminium produced by chemical reduction was over \$70, reducing to \$33 by 1886 and only 40 cents in 1915.[9]

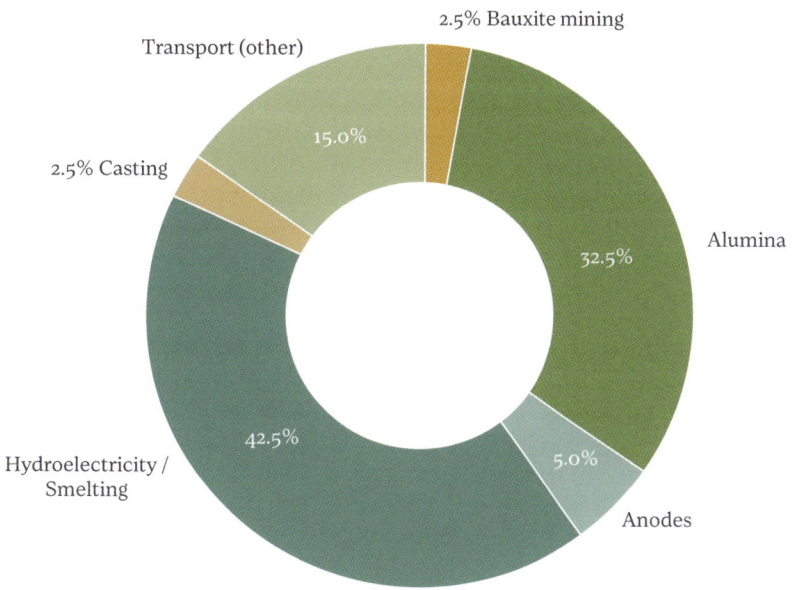

Figure 2.1 The embodied carbon breakdown for the stages of the production of primary aluminium using hydroelectricity, source Norsk Hydro.

2.5% Bauxite mining

Transport (other)

15.0%

2.5% Casting

Alumina

32.5%

Hydroelectricity / Smelting

42.5%

5.0%

Anodes

Smelting

Smelting is an energy-intensive process which requires a continuous supply of electricity, using approximately 48% of the required primary energy, and the anodes are consumed in this process. The energy required to produce primary aluminium has reduced significantly over the past 20 years, with a 15% reduction in electrical energy used in the smelting process.[10] Similarly, the energy required by the Bayer process, used to extract alumina from bauxite ores, has also seen a significant reduction of 25% between 2010 and 2020.[11] The significant alternative route for life-cycle assessment (LCA) modules A1–3 (see Figure 1.1) is recycling post-consumer aluminium (see below).

In the UK, the only remaining smelter is at Fort William, run by Alvance. It produces aluminium using hydroelectricity, therefore the embodied carbon for UK-produced primary aluminium is 9,248 to 10,880kgCO_2e/m³ compared to a European average of 19,040kgCO_2e/m³.[12] Aluminium is a lightweight material, therefore an embodied-carbon-per-component figure is more appropriate. Annual primary aluminium production in the UK is 48,000 tonnes, all at Fort William, compared to a global production of primary aluminium in 2021 of 67,243,000 tonnes.[13]

Aluminium is also imported into the UK in the form of semi-finished products, known as semis, in the form of billets or cast slabs for roll forming. They may be smelted with a range of power sources: hydroelectricity, other renewable, nuclear, natural gas, oil and coal. Many UK producers of aluminium window and curtain walling systems extrude their own sections. System sections are also imported into the UK from countries including Turkey and China, where the aluminium may be smelted using coal, with a much higher carbon footprint, therefore the power mix used is of vital importance.

POWER MIX

When evaluating aluminium for a current project, it is important to consider the power mix used to smelt the aluminium. Aluminium was first smelted using hydroelectricity in 1886 using the potential energy of Niagara Falls. When produced using hydroelectricity, it has a minimal global warming potential, but there is some fossil fuel depletion (anodes and transport). Taking Canada as an example, the production of aluminium is based on 96% renewable energy, primarily hydroelectricity (however, smelters can be used as grid batteries). **Specifying the power mix used to produce aluminium and its alloys should be part of project specifications.** The means of production is not inspectable on site, so certificates of conformity to evidence the power mix used are essential, in the form of an Environmental Product Declaration (EPD) or Aluminium Stewardship Initaitive (ASi) certification.

Figure 2.2 Global power mix used to produce aluminium, reported between 1980 and 2020 (GWh).[14] Note the use of renewables beyond hydroelectricity since 2016 and the continued dominance of coal, especially in China.

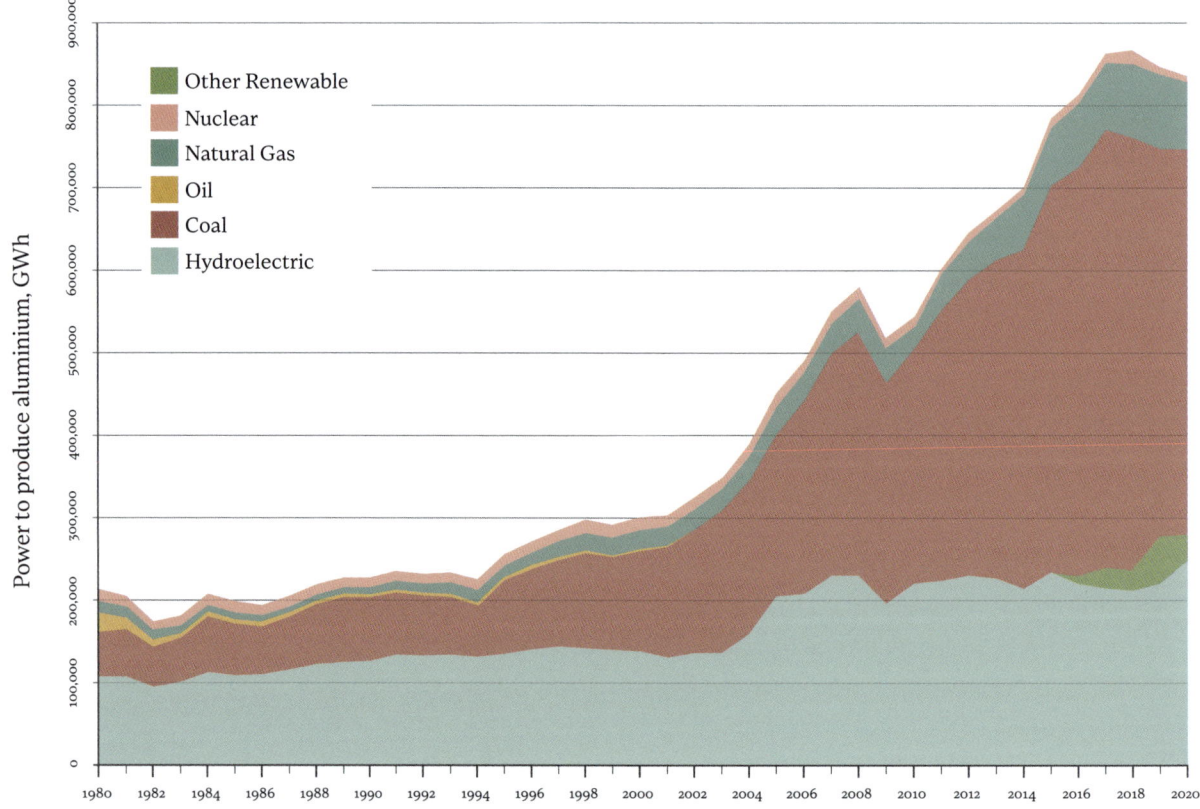

Aluminium (pure) key data	
Density	2,720kg/m³
Coefficient of thermal expansion	2.3 x 10⁻⁵/°C
Corrosion resistance	Excellent
Durability	80 to 120 years
Melting point	660°C
Recyclability	Infinite, with no loss of material quality
Embodied carbon of 75% post-consumer recycled aluminium	6,256 kgCO₂e/m³
Embodied carbon of hydroelectrically smelted aluminium	9,248–10,880 kgCO₂e/m³
Embodied carbon of aluminium – European average	19,040 kgCO₂e/m³

APPLICATIONS

Primary aluminium can be cast, extruded, roll-formed, press-moulded, spun, digitally printed and more; it is inherently flexible by design. Many of the forming processes exploit the inherent ductility of aluminium.[15] The form of aluminium components is in the hands of architects, designers and engineers.

Aluminium and its alloys form a wide class of materials with many applications in contemporary life, from aluminium foil, an iPhone case, a coffee pot or an Airbus A380 to windows, curtain walling and solar shading. Aluminium is used in its pure form (99% pure) only in aluminium foils, known as 1000 series, used in buildings as a vapour check layer for their vapour resistance. The two *Voyager* spaceships launched in 1977 are now the most distant human-made artefacts, having entered interstellar space, and are made from 92% aluminium and silicone, and the scientific equipment onboard was earthed with 1000 series aluminium kitchen foil.[16]

Aluminium alloys are set out in *International Alloy Designations and Chemical Composition Limits for Wrought Aluminum and Wrought Aluminum Alloys*, issued by the Aluminum Association of the USA. This is a four-digit system in which the first digit, from 1 to 9, indicates the principal alloying element. This system is used in British and European Standards, for example BS EN 575:1996 Aluminium and aluminium alloys.

For further information, including heat treatment and temper of aluminium alloys, see *Aluminium Flexible and Light: Towards Sustainable Cities*.[17] Structural application of aluminium alloys should be designed to Eurocode 9: Design of aluminium structures (2007).[18]

Cast-aluminium cladding and solar shading

Research has found a wide adoption of cast aluminium as cladding and solar shading. Examples include the 92% recycled cast-aluminium solar shading of Heelis National Trust Headquarters, Swindon, by Feilden Clegg Bradley Studios, 2005. This was produced by Novacast in Melksham, only 28 miles away.[19] The Smithsonian's National Museum of African American History and Culture (NMAAHC) was built on the last remaining site on the National Mall in Washington, DC, completed in 2016, by a team led by David Adjaye.[20] Above the glazed ground floor, all four façades are clad with cast-aluminium panels that provide solar shading, using six patterns that range in opacity from 65 to 90%, coated in a bronze-coloured PVDF (polyvinylidene fluoride) wet paint system by Dura Industries. The aluminium panels of NMAAHC were cast by Morel Industries using recycled aluminium; however, they travelled over 2,700 miles to Washington, DC.

Overcladding

Guy's Hospital Tower was designed by Watkins Gray Architects and opened in 1974. During 2012 and 2014, the tower was overclad and reglazed with curtain walling, designed by Penoyre & Prasad, working with Arup Facades. To inform the design development of the overcladding, a life-cycle assessment (LCA) was conducted using GaBi software.[21] The carbon payback period for this overcladding and curtain walling is 12.5 years. The new building envelope will save over 22,100 tonnes of CO_2e over its 80-year predicted life. This is a successful and non-combustible overcladding project that will extend the use of this hospital for at least a further 80 years.

DURABILITY AND FINISHES

The durability of aluminium and its finishes are two of the primary reasons it is specified for architecture and infrastructure projects. Mill-finish aluminium will rapidly develop a coating of aluminium oxide on exposure to air; this forms a protective grey oxide coating on exposed aluminium components of about 4μm, which will patinate in time. It is a durable finish in urban and rural locations. Aluminium is more durable when washed by rain and therefore architects and engineers should detail buildings, where possible, to facilitate the washing of the surface by rain.

There are three main methods of finishing aluminium components: anodising, polyester powder coating and wet paint systems including PVDF. Guarantees of up to 40 years are available for super-durable polyester powder coating and PVDF, and 80 years for anodising; however, such guarantees are dependent on periodic cleaning. All three systems can be used to coat recycled aluminium.

Led by the author, together with Philadelphia-based architects KieranTimberlake, the Towards Sustainable Cities (TSC) team reviewed and inspected more than 50 projects that feature the use of aluminium.[22] Three projects were selected for non-destructive testing, including the New Bodleian Library in Oxford.[23] The aluminium windows of the library were tested during its refurbishment, led by WilkinsonEyre, reopening as the Weston Library in 2015. When tested, the anodising on all but one of the windows (then 75 years old) of the New Bodleian Library would not meet coating thickness requirements for new installations by current standards (BS/EN/ISO 7599:2010). However, based on the durability of the windows, the design team at WilkinsonEyre concluded that the finish is satisfactory for another 60 years and so the original

Figure 2.3 Tinhouse, Isle of Skye by Rural Design, 2016. Referencing the corrugated metal sheeting typically found on rural buildings across the Highlands, the walls and roof of Tinhouse are clad in mill-finish aluminium. Designed and built by Rural Design's founders, Tinhouse was a self-build project and aluminium ws selected because it could be erected by a single person. The mill finish has weathered extremely well in its exposed coastal setting.

Location or typology	Cleaning cycle
Normal environment	Every 12 months
Marine and/ or industrial environment	Every 3 months
Swimming and leisure pools	Every 3 months

Table 2.1 Qualicoat guidance on cleaning intervals for finishes on aluminium.

windows have been retained following just cleaning and reglazing in the refurbished library.[24] The TSC research revealed that finishes, both PPC and anodising on aluminium, have outlived the guarantee period provided at the time of construction. Well-maintained external aluminium components should have a service life of at least 80 years and well-maintained internal aluminium components, 120 years or longer.

Regular cleaning is much better than repainting, offering a lower carbon footprint during the service life of building and key to the warrantees provided on finishes on aluminium.[25] Qualicoat – a global organisation committed to maintaining the quality of coatings on aluminium – provides recommended cleaning frequencies dependent on the location of a project.[26] Coated aluminium should be cleaned with warm soapy water using a detergent that is pH neutral, using a soft microfibre cloth. Then wipe off with a dry, soft microfibre cloth. The relevant cleaning cycle should be included in the in-use section of a project's LCA (module B2).

Infrared light is invisible to the human eye; however, it can account for 49% of solar gain on a sunny roof or wall. High-reflectance powder coatings reduce the surface temperature of aluminium components when subject to solar radiation. This can reduce the energy required to cool a building by limiting solar gains and reduce the risk of overheating in fabric-first buildings, which have a

Figure 2.4 Weston Library, Oxford, Sir Giles Gilbert Scott, 1940; retrofit by WilkinsonEyre, 2015. The anodised windows of this library are now more than 80 years old. When tested, the anodising on the majority of existing windows sampled did not meet coating thickness requirements for new installations by current standards (BS/EN/ISO 7599:2010). However, based on the durability of the windows, the design team concluded that the finish was satisfactory for another 60 years or more, so the original windows were retained after cleaning and reglazing.

QUESTIONS
FOR PROJECT TEAMS AND SUPPLIERS

☐ Is a certificate of conformity for the power mix used to produce or recycle the aluminium available, either as an Environmental Product Declaration (EPD) or via Aluminium Stewardship Initiative (ASi) certification?

☐ What is the certfied recycled aluminium percentage in a particular product?

☐ What is the percentage of renewable energy used in the production of both primary and recycled aluminium?

☐ Think about recycled and primary aluminium on a regional basis (for example the UK or Europe). Consider supporting indigenous manufacturing and a circular economy.

well-insulated external envelope. It is now possible to specify powder coat that appears black (RAL 9005) to the human eye yet provides a reflectance of infrared light that is 25% higher than the standard formulation of black (RAL 9005) polyester powder coatings.[27] Therefore an aluminium roof may be the appropriate specification for future projects, as reflective coating can also reduce the heat-island effect experienced in cities.[28]

FIRE

Aluminium is non-combustible, in the form of, say, a 3mm-thick cladding panel, and will melt at over 660°C and not burn. Under EN 13501-1, anodised aluminium and PVDF (polyvinylidene fluoride) coating is classed as A1, the highest level of non-combustibility. Polyester powder coating, which is typically at least 40 microns thick, does add a very small fire load and is classified under EN 13501-1 as A2-S1, Do (where S1 equates to little or no smoke and Do represents no flaming droplets). Therefore, all three of these very durable coatings on aluminium can currently be used on buildings of over 18m in England and Wales and 11m in Scotland under Part B of the building regulations, but should always be checked with an appropriate professional. In three large-scale holistic tests undertaken since the Grenfell Tower fire, 3mm-thick aluminium-clad façade systems, all using non-combustible mineral wool insulation, have been successfully tested to BS 8414.[29]

A CIRCULAR ECONOMY: END OF LIFE, RECYCLABLE AND RECYCLED

The recyclability of aluminium is a fundamental quality of this light metal, and it has been consistently recycled since global flows were first recorded in the 1950s. It is a cornerstone of a circular economy. More than one billion tonnes of aluminium have been produced since 1886, the year the Hall-Héroult process was invented. Some 75% of the aluminium produced since then is either in-use or is available to be used by humankind. It is a resource that can be considered a material and energy bank for humankind today and in the future. **Aluminium is almost infinitely recyclable, with no loss of material qualities.**[30] Based on an 80-year life span for an aluminium product, with a collection rate of 95% and 1 to 2% losses during recycling processes, the aluminium in a building could still be in use in 4,000 years. **Recycling aluminium requires only 5% of the energy used to produce aluminium from bauxite.**[31]

Anodising aluminium does not inhibit recycling, as this is simply a thicker oxide coating that occurs on mill-finish aluminium. Nor does polyester powder coating present a problem during recycling; however, flue extract precautions may be necessary. Each tonne of aluminium recycled saves approximately 15 tonnes of CO_2e.[32] The embodied carbon of 75% post-consumer recycled aluminium is 6,256kgCO_2e/m³.[33]

Globally in 2019 around 33 million tonnes, or 34% of global aluminium demand, is met from recycled aluminium.[34] That year in the UK, a million tonnes of recycled aluminium was available – potentially a supply of over 95% recycled content if all the aluminium for your next project is sourced from the UK. **It is the recyclability of aluminium that is key to its sustainability**, and this recycling should always be assumed for the end-of-life scenario in an LCA.[35] Recycling rates of aluminium from buildings in Europe and the UK are between 92 and 98%.[36]

In Germany during 1994, five aluminium system suppliers founded an association with a closed-loop recycling scheme for aluminium curtain walling, doors and windows.[37] A|U|F is a not-for-profit aluminium recycling scheme and is an exemplar of closed-loop recycling on a local basis, with more than 200 members across Germany, including 11 recycling centres. A|U|F's aim is for other countries to set up national schemes to minimise the components miles and retain the resources in each region. In 2019, the Council for Aluminium in Building (CAB) set up a closed-loop recycling scheme for the UK. A pilot project included the recycling of fire-damaged, 11-year-old curtain walling from a hotel in Bournemouth by a senior aluminium systems fabricator, as part of the façade replacement contract. As the aluminium components in 20th-century buildings reach their life expectancy and are disassembled, and as recycling processes improve, it is likely that humankind has reached peak primary aluminium production. **A reduction in primary aluminium production is crucial to limiting global temperature increase to 1.5°C by 2050.**[38]

The Hive by Wolfgang Buttress is a project based on ecological concerns and an exemplar of Design for Disassembly (DfD) using aluminium components. It was first assembled in Milan in 2015 and reassembled at Kew Gardens in 2016. Deployable aluminium bridges designed by MAADI Group are another demonstration of DfD.

WIDER SUSTAINABILITY ISSUES
Members of the International Aluminium Institute (IAI) are miners who agree to mine responsibly. This includes maintaining the biodiversity of the mining site, encompassing: pre- and post-mining flora and fauna surveys and ongoing monitoring; retention of the original topsoil; establishing a reserve or nursery for native plant species; and leaving strips or islands of native vegetation within mining areas as wildlife reserves and corridors. IAI members also commit to being socially responsible

employers.[39] Processing bauxite uses water – the IAI considers the world average input to be 0.5m³ of freshwater to extract and prepare one tonne of bauxite. Tailing dams are used to store the unused fraction of the bauxite and are typically earth-based embankment dams used to permanently store waste from mining operations. Aluminium and energy company Hydro has developed a dry backfill process in Brazil that avoids the need for new tailing dams. Tailings are dried and used to backfill areas previously mined so that rehabilitation can begin and biodiversity can be reinstated.[40]Another route to ensure that aluminium products are produced by an environmentally, socially and ethically responsible supply chain is certification by the Aluminium Stewardship Initiative (ASi).

As a lightweight metal, aluminium can readily be transported, typically by sea. At Sophos Operational Headquarters, Abingdon, Oxfordshire, in 2003, architect Bennetts Associates and the author successfully specified roof lights fabricated solely from aluminium produced by hydroelectricity within the context of a commercial building contract. The aluminium was sourced from Canada – thus successfully lowering the embodied impacts of this project. To maintain this saving in embodied impacts, it was of vital importance to source the double- or triple-glazing locally.

FUTURE TRENDS
- Long-span, lightweight aluminium roofs with aluminium cladding/roofing.
- Foamed aluminium as insulation, structure and acoustic absorbency.
- Phasing-out of coal-fired smelters and the increased use of renewable energy sources.
- Increased availability of post-consumer recycled aluminium.

KEY TAKEAWAYS

- Aluminium is infinitely recyclable. The key to sustainable aluminium specification is ensuring high (close to 100%) recycled content.

- A reduction in primary aluminium production is crucial to limiting global temperature increase to 1.5°C by 2050.

- Recycled and primary aluminium sourcing should be considered on a regional basis, from the UK or Europe, to support local manufacturing and a circular economy.

- The production of aluminium is dependent on electricity, which is becoming increasingly decarbonised. Understanding the power mix of aluminium production behind a product is an important element of its sustainability.

- Used wisely, well-detailed and carefully executed aluminium can play a role in a sustainable built environment.

BAMBOO

Hector Archila, David Trujillo and Edwin Zea Escamilla

Embodied carbon for culms (stems)

17-47 $kgCO_2e/m^3$ [1]

Embodied carbon for engineered bamboo products (EBPs)

120-267 $kgCO_2e/m^3$ [2]

Bamboo products imported into the UK in 2021

around **8,000** tonnes[3]

China's global share of production of EBPs

80% as of 2019[4]

Time to grow bamboo

3-6 years

Transport emissions to Europe from China

15-130 $kgCO_2e/m^3$ [5]

Bamboo is a fast-growing 'giant' grass; it does not undergo secondary growth like trees. It attains full height and diameter within six months (although full strength is not achieved until maturity at between three and six years), with the root network surviving harvest for new growth. Historically used in tropical and subtropical construction, it is now increasingly seen as a potential construction material in Europe as well. There are more than 1,600 bamboo species, but only a few – generally those with a diameter over 50mm – are used for construction due to their high compressive and bending strength along the fibres. Like wood, bamboos have moderate shear strength and low tensile strength perpendicular to the fibres.

While bamboo lacks natural durability due to the absence of tannins and oils, it can be protected from degradation by appropriate detailing and the use of boron-based solutions. Engineered bamboo products (EBPs) are made from smaller pieces of bamboo glued together and display higher strengths than softwood. Bamboo's limited fire resistance necessitates additional safety measures, while EBPs follow similar protection procedures to timber products.

Bamboo's growth rate, simple silviculture, along with its ecosystem services and mechanical properties make it a valuable resource for a climate-neutral built environment. Nevertheless, its irregular shape and labour-intensive processing present challenges. EBPs can offset these issues, though structural use is still limited. As of 2019, China produced 80% of global EBPs, used mainly for flooring, with the UK showing a growing trend in EBP usage. Plantations are currently being established in southern Europe, primarily Portugal, Greece and Italy.

APPLICATIONS

- Structure
- Interior panels
- Flooring
- Window framing
- Furniture

PROS

- Fast-growing
- Carbon sink during cultivation and use
- Versatile
- As strong as a hardwood

CONS

- Limited durability if exposed to the elements
- Nascent supply chain
- Limited knowledge about fire performance

Bamboo culms – or stems – emerge from the ground at the diameter they will always be and reach their full height in less than six months. They do not become taller or wider over time. What follows is a process of internal consolidation of tissues. **When a culm is harvested, the root network remains alive and retains its potential to produce more culms.** This is a significant advantage over trees. Bamboo has been used as a vernacular construction material for millennia in the tropics and subtropics where it natively grows. In light of the climate emergency, it is increasingly perceived as an engineering material with applicability in Europe.

While there are more than 1,600 identified species of bamboo, less than 100 are used in construction. Giant bamboo species such as *Guadua angustifolia* (Guadua) and *Phyllostachys pubescens* (Moso) – used in the building in Figure 3.2 – tend to have larger diameters (>50mm), very high tensile and compressive strengths when loaded parallel to their fibres (higher than most timber species), high bending strength and stiffness (as high as some tropical hardwoods), but only moderate shear strength (still higher than softwood), and low tensile strength perpendicular to fibres (similar to softwood). Their shape (tapered and normally a hollow cylinder) reinforces these strengths and weaknesses, making bamboo an excellent column, prop and arch, but a mediocre beam.

The term engineered bamboo product (EBP) is a catch-all term used to describe products mainly made from glued bamboo strips. Though there are a range of board and beam products, the two most commonly available products are glued laminated bamboo and bamboo scrimber. Most EBPs display higher strengths and hardness than softwood, but similar stiffness properties.

Unlike wood, bamboo does not accumulate tannins and oils over time, hence does not develop natural durability. This means that if left untreated it is susceptible to biological degradation by insects and fungi. Bamboo culms are normally protected from insect attack by impregnation with boron-based solutions, and from fungal damage by keeping the stems dry, drained and well ventilated through durability-by-design practice. EBPs are protected through diverse procedures; it is important to refer to a manufacturer's literature and guidance.

Bamboo culms have very limited fire resistance due to their hollow shape. Suitable fire performance needs to be attained by other means, including minimising the hazard or encapsulating the culms in fire-resistant materials (e.g. boards). EBPs have a performance more akin with timber products. Some EBPs have been classified as B_{fl}-s1 and B-s1, d0 to EN 13501-1. Currently, EPBs are rarely used for structural applications. Their scope has generally been limited to semi-structural use or internal and external finishes.

HARVESTING

Maturity in bamboos occurs faster than in most wood species, which makes bamboo plantations an attractive option for high biomass yield and rapid CO_2 fixation compared to fast-growing softwood species. Depending on the species, bamboo culms take between three and six years to reach maturity. The harvesting process of bamboo is mainly manual and labour intensive. Naturally occurring bamboo forests are usually within dense forest sites, on steep hills, close to rivers and/or in areas of difficult access.

As is the case for wood, biogenic CO_2 sequestration in bamboo depends on the annual yield in cubic metres per hectare (m^3/ha) of the specific species and varies with harvesting practices and forest management. Estimates report a yield between 45 and 190m^3/ha for tropical bamboo species *Guadua angustifolia* Kunth. Furthermore, as seen in Table 3.1, embodied carbon of bamboo products also varies depending on processing practices and transport distances.

APPLICATIONS

While there is a widespread conception of bamboo as a material being used for temporary structures, due to its versatility and attractive natural look, there are more and more examples of the material being used appropriately for long-lasting applications in buildings.

Bamboo product	A1 - A2 - A3	A4	A5
Bamboo culms	16.7 – 47 kgCO$_2$e/m^3*	14.5 kgCO$_2$e/m^3* - China to Europe	Waste rates ranging from 5 to 25%
Laminated bamboo	120 – 267 kgCO$_2$e/ m^3*	128 kgCO$_2$e/m^3* - China to Europe	Waste rates likely to be similar to engineered timber.

* Calculated using Ecoinvent 3.8 and IPCC2013. The results are sensitive to the carbon intensity of the electricity grid of the country of origin. Lower range corresponds to countries with a high component of renewable energies.

Table 3.1 Embodied carbon of bamboo products.

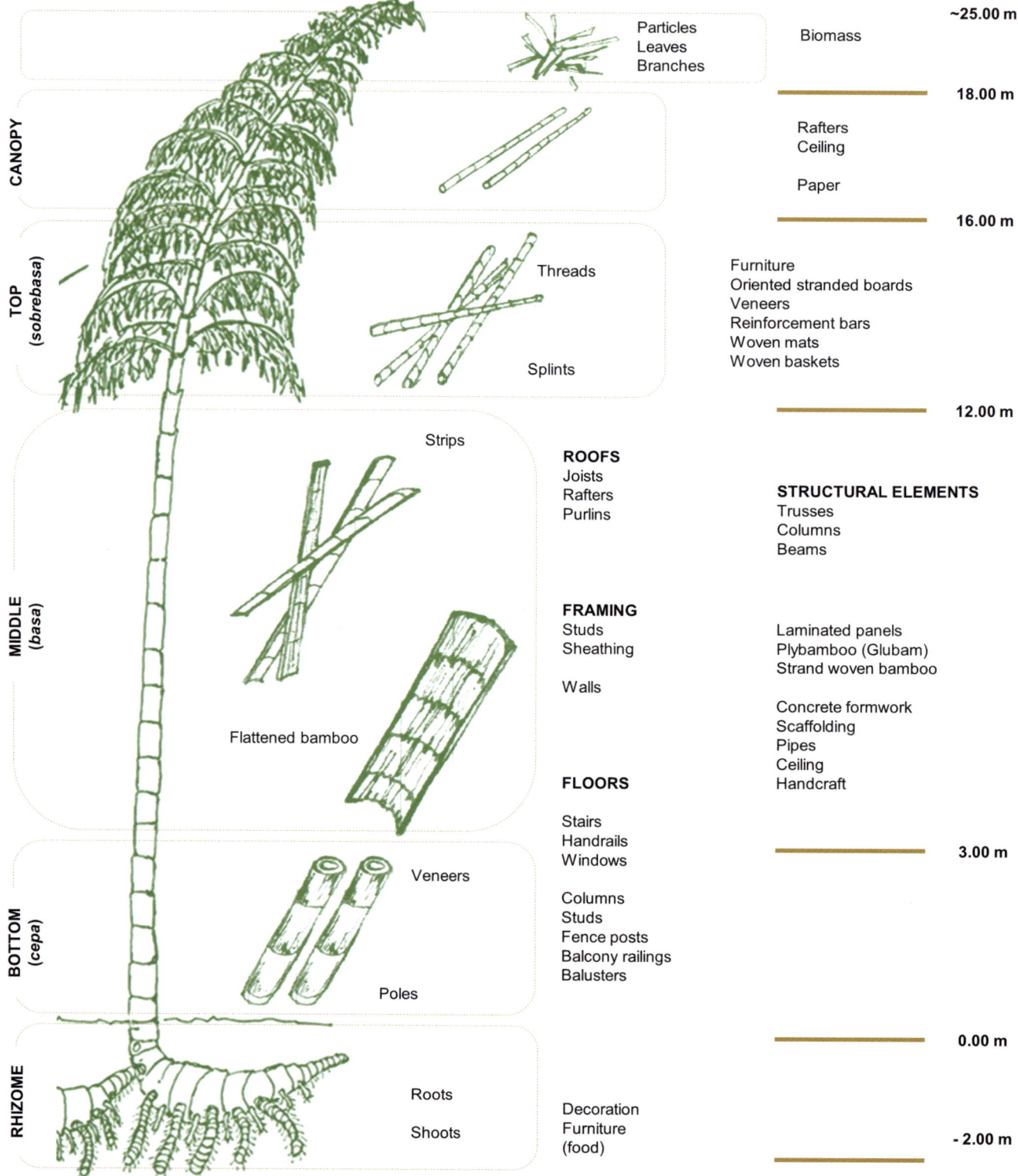

CANOPY

Particles
Leaves
Branches

Biomass

~25.00 m

18.00 m

Rafters
Ceiling

Paper

TOP
(sobrebasa)

Threads

Splints

16.00 m

Furniture
Oriented stranded boards
Veneers
Reinforcement bars
Woven mats
Woven baskets

12.00 m

Strips

ROOFS
Joists
Rafters
Purlins

STRUCTURAL ELEMENTS
Trusses
Columns
Beams

MIDDLE
(basa)

FRAMING
Studs
Sheathing

Walls

Laminated panels
Plybamboo (Glubam)
Strand woven bamboo

Concrete formwork
Scaffolding
Pipes
Ceiling
Handcraft

Flattened bamboo

FLOORS

Stairs
Handrails
Windows

3.00 m

BOTTOM
(cepa)

Veneers

Poles

Columns
Studs
Fence posts
Balcony railings
Balusters

RHIZOME

Roots

Shoots

0.00 m

Decoration
Furniture
(food)

- 2.00 m

Figure 3.1 Construction applications of different parts of the giant bamboo species *Guadua*.

Figure 3.2 INBAR Pavilion, Horticultural Expo, Beijing, Studio Cardenas, 2019. Bamboo culms are used in a climate with marked seasonal changes (Beijing). Bamboo has been used for the arches, purlins, ceiling and flooring.

To achieve durability of bamboo in buildings, some basic design considerations include:
- appropriate grading, drying and preservation according to international standards (see the standards list in the Further Reading section)
- appropriate preservation and protection of building elements, depending on the intended use; round bamboo should not be used in contact with the ground, in-ground or water-logged, except in structures having a design life of less than five years
- appropriate design for serviceability, structural integrity, fire, assembly and maintenance; for instance, EBPs are more appropriate for doors and windows, whereas round bamboo should be avoided for these applications
- avoidance of embedding bamboo in concrete: this is neither environmentally friendly nor structurally sound; use of bamboo as steel replacement in reinforced concrete is an ill-conceived practice that should always be dissuaded.

Figure 3.3 Jakarta Hotel, Amsterdam, SeARCH 2018. More than 35,000m² of bamboo slats, 7700m² of bamboo veneer and 2750m² of bamboo flooring were used in the hotel interior. Engineered bamboo can play an important role in numerous applications in the Global North.

Traditionally, in tropical areas, bamboo has been used in temporary preconstruction works (e.g. scaffolding and formwork). However, the use of bamboo for specialist enabling works and most substructure applications is not recommended due to the harsh environment it would be subjected to (e.g. high moisture, ground contact, concrete alkalinity, etc.).

Fabric

Lime and/or cement renders are used in composite bamboo shear walls (also known as light cement bamboo frame construction) in external and internal walls, to offer weather and fire protection. As is normal practice in wood construction, bamboo exposed in intermediate floors, internal and external walls should generally be protected against fire, weather (e.g. behind mortar render or plasterboard) and condensation.

Hybrid materials

Rather than a supplementary material to wood, concrete and other conventional building materials, **bamboo should be considered a complementary material for building applications** which can significantly reduce the overall carbon footprint of a building. An effective and sustainable use of bamboo entails the appropriate use of a combination of materials for the intended application. Hybridised uses of bamboo with timber in structural applications and other bio-based materials (e.g. insulation using hemp, straw, cellulose, wool, cotton and flax) are a way forward to reduce embodied carbon in the built environment.

REDUCTION OF WASTE

While most waste occurs during the A1 to A3 life-cycle assessment modules – through selection (e.g. straight culms), drying of bamboo and the manufacture of EBPs – actual figures for the waste produced during construction of bamboo buildings are hard to come by. However, carbon storage in bamboo forests and buildings largely surpasses the CO_2 from product waste.[6] Cascading and 'circularity' approaches should therefore be favoured to make full use of the material in long-lasting applications, such as buildings.

In traditional construction with bamboo culms, offcuts generally become waste. Nevertheless, recent DfMA (Design for Manufacture and Assembly) practices implemented in affordable housing projects led by BASE Bahay in the Philippines demonstrated exemplary waste reduction practices from the product stage through to construction.[6] Rigorous quality control at source (e.g. selection of straight and uncracked dried and preserved culms), ordering pre-dimensioned culms, planning for use of offcuts, along with the setting up of flying factories next to the site (temporary facilities established for the duration of a project), significantly contributed to reduced waste in modules A4 and A5.

DfMA and premanufacturing approaches can be easily implemented when utilising EBPs in buildings as these products are manufactured to conform with the straight-edged nature of modern construction. Aspects to consider include the wear of cutting and drilling tools due to the higher density of bamboo and the slender sections/thickness of commercially available products.

END OF LIFE

In common with other bio-based materials, recyclability is not a real option (EBPs cannot become culms). Instead, reuse, downcycling (cascading), energy gain (combustion as a fuel) and composting are the most viable alternatives. Design for Deconstruction (DfD) should be encouraged as the most sustainable end-of-life practice. This mindset is still uncommon in bamboo projects.

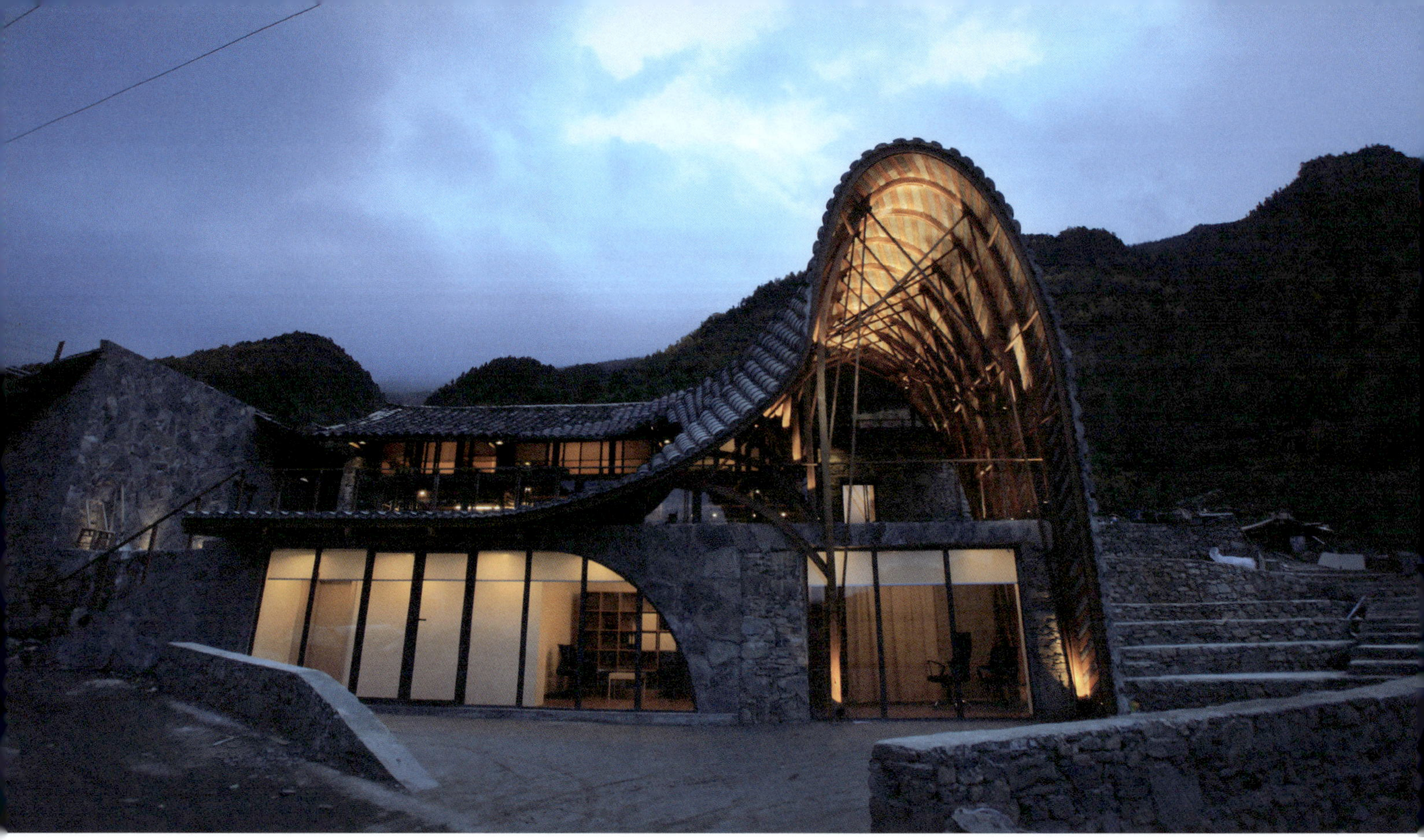

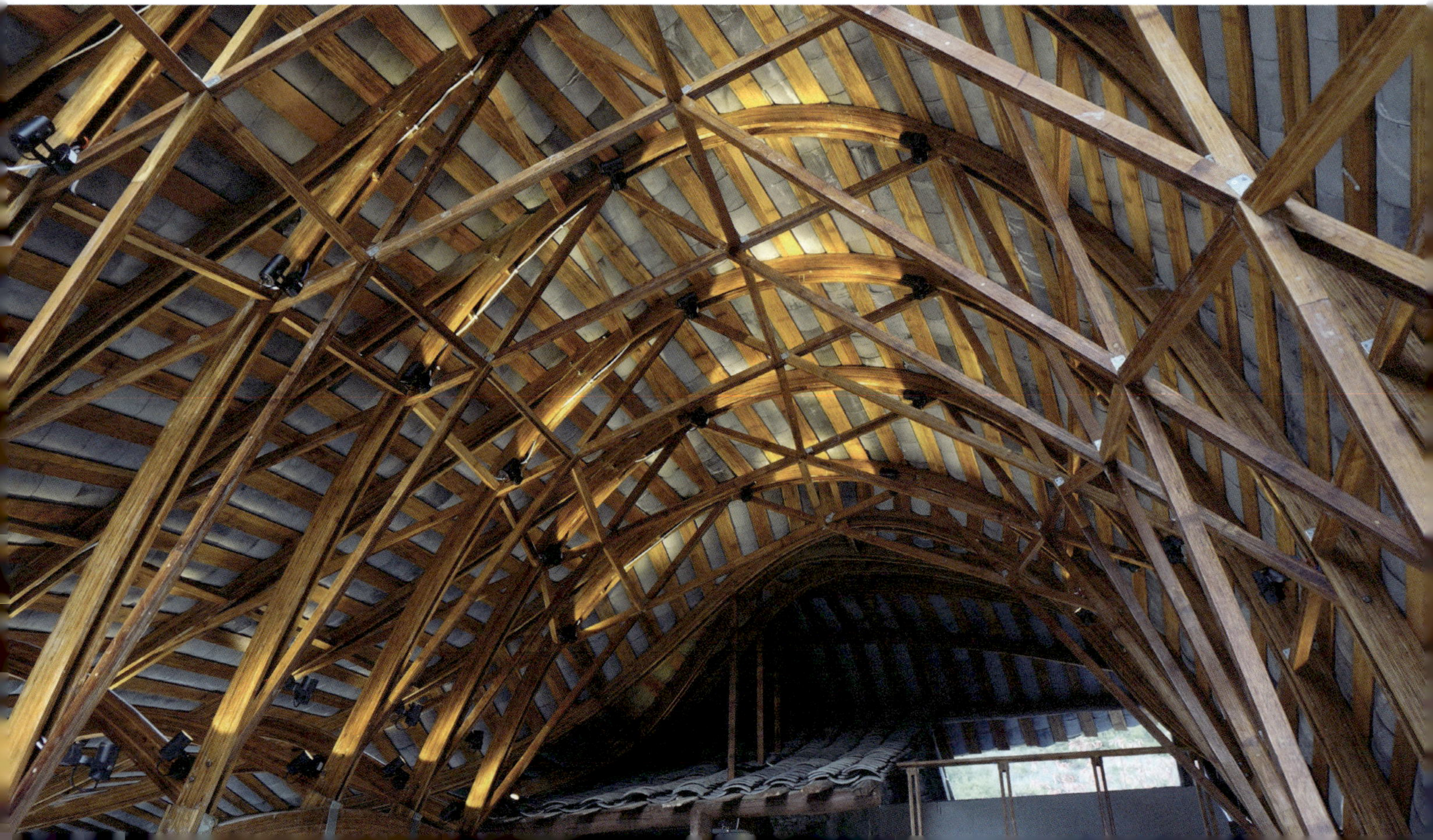

Figures 3.4a and b Cattle Back Mountain Volunteer House, Sichuan Province, China, dEEP Architects, 2015, exterior and interior view. The building is an example of using engineered bamboo products (scrimber), combining curved and straight elements, and the integration of traditional building techniques with digital design.

DESIGNING FOR LOWER IMPACTS

When used close to source, **bamboo culms have a very low embodied carbon and are an inexpensive resource.** Growing and harvesting bamboo has a minimal environmental impact, and plays a positive role in terms of carbon sequestration and other ecosystem services. **The greatest potential environmental impact of using bamboo culms comes from the unavoidable use of chemical preservatives.** It is necessary to strike a balance between effectiveness and toxicity. Boron-based preservatives seem to offer an optimal compromise, though it is important to note that they are water-soluble and therefore can leach out if the stems are exposed to regular wetting. Therefore, very exposed locations, such as cladding, should be avoided.

Criticisms of EBPs include bamboo scrimber's very high content of resins (10 to 25%) and high wastage while manufacturing laminated bamboo (40 to 70%). Currently, common EBPs tend to be costlier than engineered timber, and have only marginally better stiffness, and therefore are unlikely to be an effective substitute in all structural applications. However, EBPs perform particularly well when strength (particularly cross-grain strength) and hardness are needed but stiffness is less important, for example ceilings and flooring. Some EBPs can perform well in environmentally exposed locations such as cladding and decking.

WIDER SUSTAINABILITY IMPACTS

The use of bamboo brings benefits beyond its low carbon footprint or the storage of biogenic carbon. One of the most important benefits is the protection of biodiversity in regions producing it. Thanks to its heterogenous growth, a bamboo forest is never clear cut, thus maintaining vital ecosystems while providing valuable resources to the communities around them.

One matter of concern with bamboo culms is the safe use and environmentally sound disposal of chemical preservatives. It is common for producers to recommend more toxic preservatives than boron-based compounds on the grounds that the latter are not effective. However, 'poor effectivity' claims may be a consequence of poor preservation procedures or poor detailing. Another matter of concern is exposure to volatile organic compounds (VOC) that might be present in adhesives used in EBPs; hence, it is important to check

QUESTIONS
FOR PROJECT TEAMS
AND SUPPLIERS

☐ Are the parties aware of, and complying with, relevant ISO standards (where applicable)?

☐ Is the environmental exposure appropriate for the product and the intended design life?

☐ Has the material been cut after reaching maturity (three to six years), dried and preserved appropriately?

☐ Is the preservation process adequate for the environmental exposure?

☐ Are the preservatives used effective, legal and benign?

☐ Is any exposed bamboo being protected against weathering and fire?

☐ Do you know the species, age, moisture content and origin of the material?

☐ Is the material coming from a well-managed plantation?

☐ Is the product/species well researched and documented?

☐ Is the supplier reputable and does it adopt quality-control procedures?

with the supplier that the product complies with minimum and safe levels of formaldehyde emissions (i.e. formaldehyde classes for boards E1 <0.1ppm (parts per million) and E2 between 0.1ppm and 0.3ppm).

FUTURE TRENDS

Bamboo culms Given its strong environmental credentials and efficient shape, the use of bamboo culms is likely to increase in the future, if its current limitations can be overcome. Plantations are being established in southern Europe (e.g. Portugal, Italy and Spain) to shorten transportation distances; however, it should be noted that the carbon footprint of bamboo is smaller than most conventional materials, even when transported from another continent. Digital fabrication systems using 3D scanning and CNC (computer numerical controlled) cutting should help to overcome the labour-intensiveness of making joints.

Connections/joints As with any other construction material, connections are crucial to ensure structural stability. Traditional construction with bamboo culms often used mortar grout to create stiffer connections between elements, but the nature of the material (a hollow, tapered cylinder) is the biggest challenge in bamboo construction. Recent innovations can tend to be costly and at times overengineered. Undeniably, connections for EBPs are more straightforward and can utilise timber connection technology.

Engineered bamboo products Other trends include the combination of EBPs with timber to reduce the manufacturing costs associated with glue and energy intensity. For similar reasons, low-processing or minimally transformed EBPs are also being developed commercially. These aim to bring bamboo elements closer to a prismatic section through less intensive and less wasteful processes than currently used with EBPs.

KEY TAKEAWAYS

- Bamboo grows very quickly, reaching its full height in six months. Then, bamboo does not grow any taller; what follows after six months and until reaching maturity (three to six years) is a process of consolidation/strengthening of the internal cells.

- Sustainably managed exploitation causes minimal impact to bamboo forests, because the root system remains alive, generating new stems.

- Bamboo for construction comes in two forms: as a culm (in the round) and as engineered bamboo products (EBP).

- Bamboo makes an excellent column or arch, but a mediocre beam.

- Bamboo used in the round MUST be preservative-treated and MUST always be shaded and protected from wetting and moisture.

- If a project is located close to the source (<500km), bamboo culms have a very low embodied carbon. If greater transportation distances are required (>1,000km), EBPs are likely to have lower embodied carbon than conventional construction materials transported over the same distances. Always look at whole-life carbon emissions and consider bamboo's advantages in terms of wider sustainability impacts.

- Currently most bamboo production is non-European. However, due to bamboo's qualities and potential, this is rapidly changing.

- Designers of bamboo structures should make use of updated British and international standards that provide guidance for structural use of bamboo.

BRICK

David Watson

Embodied carbon

404 kgCO$_2$e/m³ [1]

UK usage per year

2,370 million bricks (2018) [2]

UK imports per year

370 million bricks (2018) [3]

Global production per year

1,500 billion bricks (estimated) [4]

Largest global producer

China at up to **1,000** billion bricks per year

Bricks are small building units most commonly made from fired clay but also made in smaller volumes from calcium silicate, concrete or natural stone. Larger units known as blocks are generally made from concrete. Brick units are most often coursed with cement-based mortar to form a composite material.

Embodied carbon emissions result mainly from the fuel used to heat the brick and cement kilns, process emissions from clay extraction and transport, and direct emissions from calcination of the binder materials used in mortar. To reduce emissions, alternative fuels, binders and mineral sequestration are being explored, together with opportunities for increased reuse. Local production is still most common, given the weight and low value of the feedstock. Durability is achieved by suitable material choice and appropriate detailing. Brick buildings are often extremely long-lived as a result of brick's inherent durability and predominant use in residential structures. Most bricks used in the UK are of the modern British standard size (w215 x d102.5 x h65mm), assembled with a 10mm mortar joint. The units often incorporate a few internal voids or depressions, which assist with even firing and reduce overall material use. Today, bricks are most commonly used in the load-bearing external walls of low-rise domestic properties. Increasingly, we see bricks used as non-load-bearing cladding for high-rise residential developments, an echo of the past vernacular.

APPLICATIONS

- Foundations
- Walls and partitions (structural and non-structural)
- Cladding
- Arches
- Chimneys

PROS

- Highly durable with correct specification and detailing
- Plentiful raw materials, available locally in the UK
- Assembled easily by hand
- High thermal mass
- Self-finishing
- Inherent fire resistance
- Good acoustic performance
- Ability to help regulate humidity
- Non-toxic and largely inert
- Potential for reuse (particularly when laid with lime mortar)

CONS

- Carbon emissions from high-temperature firing of clay units
- Carbon emissions from the production of cement and lime used in mortars
- High volume of quarried and mined non-renewable materials
- High embodied carbon compared with other common building materials when normalised for structural strength or stiffness

INTRODUCTION

The use of small building units in the construction of large buildings has a long history, from the earliest application of natural stone and earth bricks, through to the development of fired clay bricks.

In the UK, the use of fired clay bricks increased rapidly from the beginning of the 18th century. Small-scale local production gave way to large-scale manufacturing during the Industrial Revolution.

This chapter focuses principally on fired clay bricks, since these are the dominant type in both the UK and globally. Mortar shares many similarities with concrete but warrants separate treatment here.

MASONRY CONSTRUCTION

Today, as in the past, bricks are typically built up into masonry construction, a composite of different materials, which includes the following:

- **Units** - bricks or blocks generally of regular size, that are coursed and bonded.
- **Mortar** - comparatively thinner horizontal bed joints and vertical perpends. Mortar absorbs the deviation in accuracy of the units to avoid stress concentration, tolerate movement and close gaps to enhance durability, acoustics and fire resistance.
- **Accessories** - in load-bearing construction, accessories include ties between leaves of cavity walls, straps for connecting floors and roofs to walls, lintels over openings and padstones for load introduction. In non-load-bearing cladding, they also include shelf angles, restraint ties and wind posts, which help resist lateral loads. In external walls, the use of stainless steel is common for durability, with reinforced or prestressed concrete used for lintels.

Masonry may also be reinforced, by means of bed-joint reinforcement or grouted rebar within voids. Prestressing is also possible, though limited in application due to masonry's comparatively low compressive strength and its tendency to creep under load.

STRUCTURAL BEHAVIOUR

The mortar used between bricks is generally weaker than the unit itself. The difference in strength and stiffness between unit and mortar sets up a complex stress state. It is advisable to limit the difference in strength between the unit and mortar. High-strength during firing, production has dropped and accounts for less than 5% today.

Lime mortar A low-strength and low-stiffness, slow-setting mortar made using natural hydraulic lime or less commonly non-hydraulic lime as a binder instead of ordinary Portland cement; generally found in older masonry construction pre-1900, with later construction adopting quicker-setting cement-based mortar.

London stock A yellow brick originally handmade using a soft-clay process, typically using locally excavated earth from construction sites in the capital, mixed with small amounts of town ash and fired in a temporary, intermittent kiln called a clamp. Today, replica bricks are produced and there is a market for reclaimed original bricks.

Pointing Finishing of mortar joint; different forms are used, which can help to improve the weathering resistance of the wall while creating different visual effects.

Timbrel vault Otherwise known as a Catalan vault, a materially efficient form of masonry shell construction using slim ceramic bricks or tiles set in mortar and arranged to be self-supporting during construction.

Vitrification The process of turning a substance into a glass-like non-crystalline amorphous solid. In the case of clay, this requires high-temperature firing.

bricks generally result in higher emissions and the benefit can be lost when paired with a low-strength mortar.

The typically low strength of masonry means that concrete padstones or steel spreader beams are required at points of concentrated load introduction.

Masonry is susceptible to reasonably high movement due to moisture change and temperature effects. For clay masonry, joints are typically provided to allow at least one millimetre of movement per metre run of wall. Guidance on spacing of joints and associated detailing is provided in national design standards and literature. Joints can be minimised through use of more flexible mortars (e.g. lime mortars) and bed-joint reinforcement, though it is generally more economic to provide joints. The provision of joints is more important in lightly loaded and non-load-bearing walls.

UNITS

Brick units can be classified according to whether they are naturally derived or man-made, with a further distinction between ceramic and mineral-bound composition. The focus in this chapter is on fired clay bricks. A brief description of the other main types is included below.

Concrete blocks

Conventional concrete blocks are made with either normal-weight or lightweight aggregate, pressed and cured in specially controlled chambers. Block dimensions are typically 440mm in length by 215mm high, with thickness ranging between 75mm and 215mm.

Aerated autoclaved concrete blocks are made with only quartz sand as aggregate and incorporate aluminium powder, which reacts with calcium hydroxide in an autoclave to aerate the block.

The mineral binder is typically ordinary Portland cement, often in combination with supplementary cementitious materials such as PFA (pulverised fuel ash) or GGBS (ground granulated blast-furnace slag). Only very small amounts of secondary materials, such as coal fines, slags and crushed brick, are currently used within new clay brick manufacture. Mineral-bound units such as concrete blocks offer the potential to increase the use of secondary materials as aggregates.

Natural	Stone		
Man-made	Ceramic	Fired clay bricks	Common brick (11.5% of UK demand)[5]
			Facing brick (81% of UK demand)[6]
			Engineering brick (7.5% of UK demand)[7]
			Perforated bricks (not common in the UK)
	Mineral bound	Concrete	Normal weight
			Lightweight
			Aerated autoclaved concrete (AAC)
		Calcium silicate	
	Unfired clay		

Table 4.1 Common units used in masonry construction.

Annual demand for concrete blocks in the UK is around 7.5 million cubic metres, more than double the volume of fired clay bricks.[8] Concrete blocks are most commonly used for load-bearing and non-load-bearing walls and partitions.

Calcium silicate bricks

Calcium silicate bricks are made by combining appropriate silica sand and lime with water, before reacting, pressing and setting under steam heat and pressure in an autoclave.

Suitable sand is not common in the UK and as a result there is currently no domestic manufacture of calcium silicate bricks. They remain popular in Europe, and estimates suggest they may represent up to 12.5% of the brick market.[9]

MORTARS

Mortars represent around 15% of a total masonry wall by volume. The use of thin-bed mortars offers a way to reduce this but requires higher accuracy in the manufacture of units and may offer less tolerance of movement.

In modern construction, cement-based mortars are most common due to widespread availability of ordinary Portland cement (OPC) binder, offering rapid strength gain and good adhesion. Historically, lime-based mortars were most common and there has been a recent increase in their use to permit larger joint spacing and facilitate brick reuse. Mortar is generally pointed to enhance durability and achieve a given aesthetic. Less durable mortar in exposed conditions may need to be raked-out and repointed over time.

Bricks absorb moisture from the mortar. A brick that is too absorbent can prevent proper setting of hydraulic binders such as OPC and hydraulic lime, which require the presence of water to set. Both the absorbency of unfired bricks and their comparatively low strength means they are incompatible with OPC or hydraulic lime-based mortar. Non-standard mortars are the subject of current research.

EMBODIED CARBON

Environmental Product Declarations (EPDs) are now available for some clay bricks, concrete blocks and calcium silicate bricks. These include blended industry averages (available from industry trade associations, e.g. BRE) and product-specific EPDs (available from manufacturers). Reported values vary considerably due to differing energy efficiency, fuel types, firing times and cycles, extraction processes, manufacturing and transport.

Group	Type	Benefits	Challenges
Cement-based	Ordinary Portland cement (OPC, also referred to as CEM I)	Fast setting, good early strength; high strength possible; widely available	Difficult to remove; reduced permeability to moisture
	Blended cement (including PFA, GGBS, etc.)	Can enhance durability; can reduce embodied carbon	As for OPC mortars; availability
Lime-based	Hydraulic lime	Easy to remove to facilitate reuse; increased tolerance to movement, allowing larger joint spacing; can offer lower embodied carbon	Slow setting with low early and final strength, with non-hydraulic lime being slowest; lack of availability
	Non-hydraulic lime		
Thin-bed mortar	Polymer modified cement	Reduced mortar volume offering improved material efficiency	Requires higher-precision blocks; reduced tolerance to movement
	Polymer based		
Novel mortar	E.g. sodium silicate for use with unfired clay bricks	Shows promise for use with bricks with high moisture absorption	Lack of precedent and standards coverage; further research and development needed
No mortar	Traditional dry rubble, rough coursed, coursed and ashlar walls	Reduced mortar volume	Reduced structural, acoustic and thermal performance; difficult to construct
	Novel forms including high-accuracy blocks with mechanical interlock	Reduced mortar volume; improved reuse potential; speed of construction	Lack of availability and precedent; generally proprietary technologies, which can impede access and uptake

Table 4.2 Mortars used in masonry construction.

It is also instructive to consider the embodied carbon of masonry compared with other typical building materials when normalised to structural capacity. **Masonry does not compare well purely on this metric but can bring other benefits, being largely maintenance free and avoiding the need for add-on materials.**

RAW MATERIALS EXTRACTION AND SUPPLY
Brick clay
The weight of raw materials needed for brick manufacture means that production tends to be local, to avoid unnecessary transport. In the past, this was particularly important, with brickworks generally established next to quarries, leading to large-scale brick manufacturing in the UK in places such as Bedfordshire and Leicestershire.

Suitable deposits of brick clay are found in all countries of the UK but are most abundant in England, which accounts for more than 90% of production. Recent

Variable	Discussion	Impact on embodied carbon
Fuel type	Natural gas is the most common fuel in Europe. Biogas is being used as an alternative and will generally reduce the reported embodied carbon. Coal firing is, however, still used by some smaller-volume European producers to achieve pleasing variations in face colour, with increased emissions compared with natural gas. Note that self-firing clays with high organic content (e.g. Lower Oxford Clays) will reduce input energy but will not reduce overall emissions, given that the organic matter is combusted. Indeed, emissions of other harmful substances will generally increase.	Higher
Firing cycles	To achieve darker visual finishes, a second firing can be undertaken under low-oxygen conditions (reduction firing). Fired bricks may also be engobed in a bath of clay slip or cement wash before a second firing to modify surface finish. Second firings can significantly increase embodied carbon, particularly when not undertaken as part of a continuous kiln process.	
Firing temperature	Higher temperatures result in higher embodied carbon. For a given type of clay, higher firing temperature will be associated with darker colour, lower absorption/porosity and higher strength. Given the use of different clays, it cannot be said, however, that darker bricks generally have higher embodied carbon!	
Energy efficiency of kilns	Modern continuous kilns offer best efficiency, including heat reclamation. Standard bricks produced in large volumes facilitate efficient kiln use.	
Additions	Oxides of iron and manganese are often added to alter colour. The impact on embodied carbon is generally low.	Lower

Table 4.3 Factors related to manufacturing affecting embodied carbon of fired clay bricks.

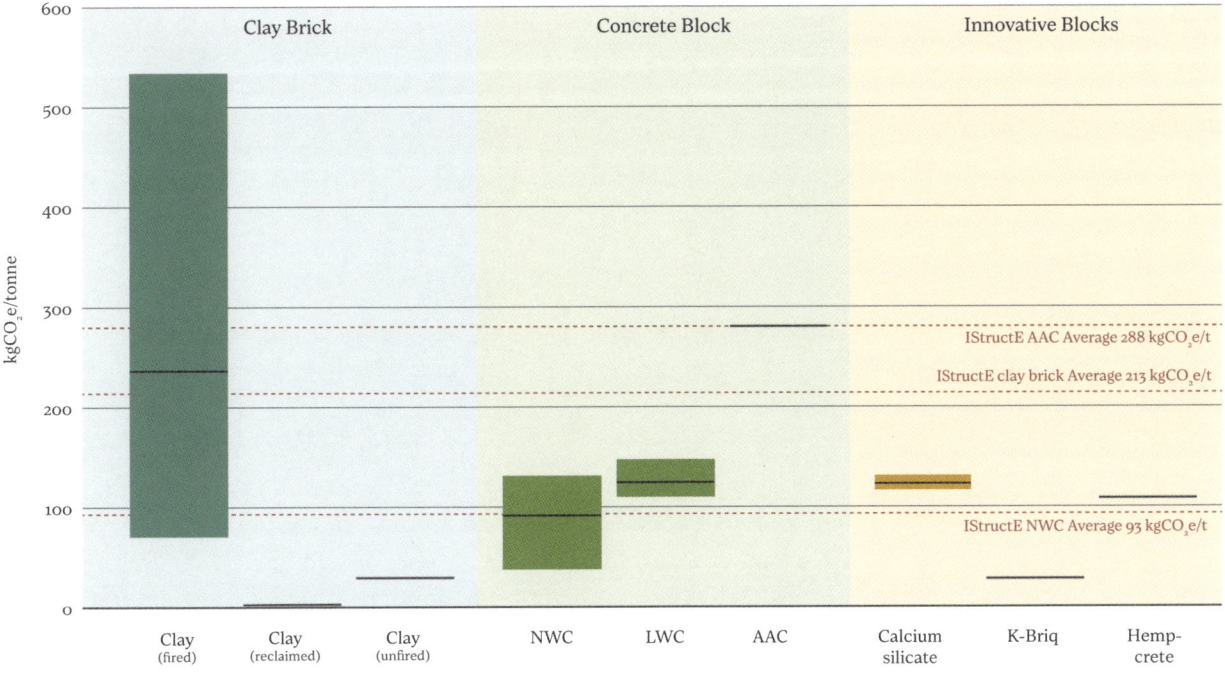

Figure 4.1 Comparative embodied carbon of masonry units.[10]

Notes:
NWC – Normal-weight concrete
LWC – Lightweight concrete
AAC – Aerated autoclaved concrete
K-Briq – Example of proprietary brick using waste products, embodied carbon estimated, based on manufacturer's data and not yet a full EPD

studies by the British Geological Survey have suggested that there are sufficient current reserves for UK brick manufacture to at least 2030 and likely 2040, but permissions on new resources must be sought. Compliance with the National Planning Policy Framework requirement for producers to maintain a 25-year land bank is found to be limited, suggesting difficulty in securing future reserves.[11]

In the UK, brick clay is extracted using open-pit methods. Clay has low value before processing (3 to 6% of the value of the construction products). With increased blending of clays from different sites, the cost of clay for brickmakers can now be higher due to increased transport.

Mineral binders

Mineral binders are used in the production of mortar. Most mortars used today are bound with ordinary Portland cement. The principal input for OPC is limestone (calcium carbonate) which is heated with smaller amounts of other materials in kilns to produce reactive oxides, releasing CO_2 in the process. Slower-setting, weaker lime may also be used as a binder, using a similar process.

The resulting binders are then combined with sand and small amounts of admixture to produce mortars.

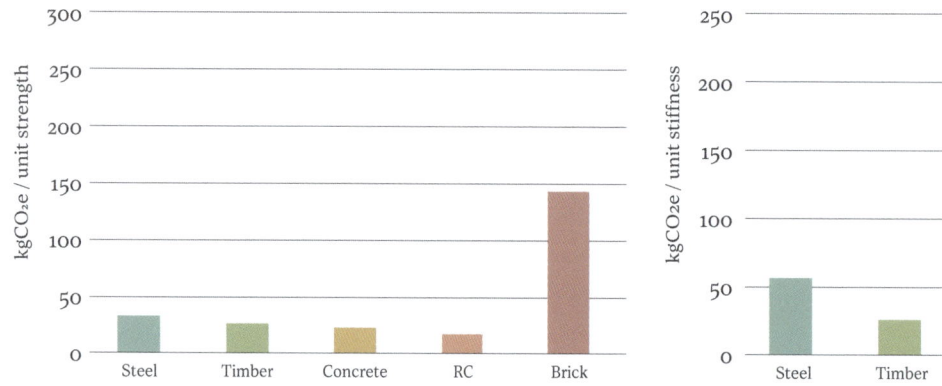
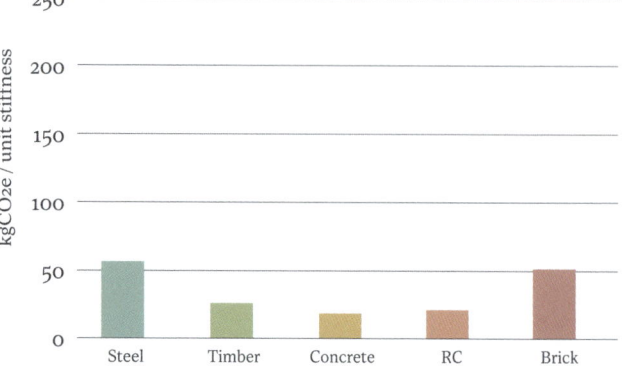

Figure 4.2 Embodied carbon normalised for strength and stiffness for common building materials (RC = reinforced concrete).

Fuel

Clay bricks require drying, followed by carefully controlled firing over a number of days at temperatures between 900 and 1,200°C, which is responsible for the majority of emissions. In the UK and Europe, these high temperatures are most commonly achieved using natural gas as the main fuel source. Biogas is also used, which will typically reduce the embodied carbon reported, given the backwards sequestration assumed under BS EN 15804. Elsewhere in the world, coal firing is still most common, though this is also used in Europe by small-volume producers aiming to achieve special visual finishes.

Trials are ongoing using hydrogen as a fuel and research continues into the development of electric kilns, though challenges remain in achieving the high temperatures required.

MANUFACTURE
Clay bricks

Clay brick manufacturing involves the following steps:
- Extraction of clay followed by ageing/weathering
- Preparation of clay prior to milling and the addition of water to increase workability
- Moulding – either by soft clay process (40% of production) or extrusion (60% of production) with small quantities pressed or hand-made[12]
- Drying – water removal (initially 25% water content for soft clay process, 12% for extrusion) to control shrinkage and cracking; this takes place between 80 and 120°C and lasts 18 to 40 hours
- Kilning – generally using a modern tunnel kiln with variable firing temperatures, typically up to 1,050°C, over a number of days; higher-strength bricks are fired at higher temperatures.

Binder type	Input feedstock	Availability	UK annual demand
Natural hydraulic lime (up to 1,200°C kiln temperature)	Argillaceous or siliceous limestones (containing alumina and silica)	Reserves exist but currently no production in the UK; all imported, mainly from France	Around 1,500 tonnes[13]
Quicklime (non-hydraulic lime) (up to 1,000°C kiln temperature)	Limestones and chalk	Produced in the UK; Tunstead Quarry is the largest producer.	Around 2.5 million tonnes but used across many industries beyond construction[14]
Ordinary Portland cement (1,400 to 1,500°C kiln temperature; however, kilns are generally more efficient than for lime production)	Carboniferous and Jurassic limestones and chalk (87%) with clay and shale (13%)	Mainly domestic production (>85%) from UK quarrying	Around 10 million tonnes, principally used in concrete[15]

Table 4.4 Binders typically used in mortar.

In the UK, different types of continuous kilns are used, either tunnel kilns (with bricks moving through different temperature zones) or Hoffmann kilns (with a moving fire and stationary bricks). When fired by natural gas, these are energy efficient in comparison to the basic intermittent or batch kilns used historically.

MATERIAL SOURCING

The annual UK demand for bricks and blocks can be summarised as follows:[16]

Fired clay bricks	UK produced	2.86 million m³	(2018)
	Imported	0.53 million m³	(2018)
	Total demand	3.39 million m³	(2018)

= 2.37 billion bricks total or 35 bricks per person

Production in the UK is currently dominated by four manufacturers (Forterra, Ibstock, Michelmersh and Wienerberger) which together account for greater than 95% of the bricks made.[17] Following consolidation, around 60 brickworks still operate, with capacities ranging from 30 million to over 100 million bricks per year. Following reductions coinciding with the 2008 recession, domestic production has not increased in line with demand, meaning that imports have increased in recent years, both of standard bricks for general housebuilding and specialist architectural bricks.

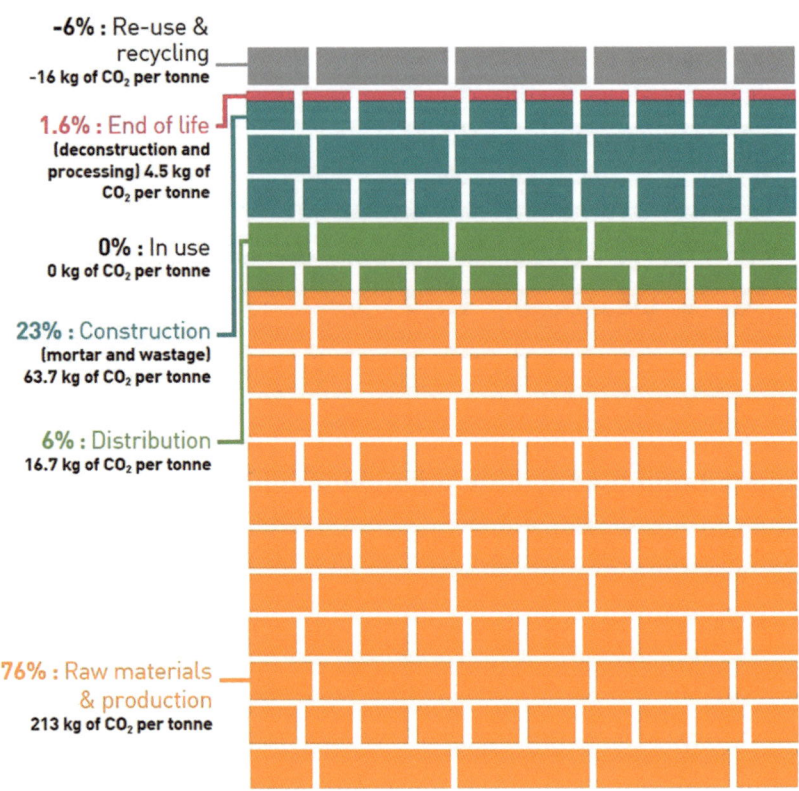

Figure 4.3 Breakdown of CO₂e emissions through the life cycle of fired clay brick.[18]

-6% : Re-use & recycling
-16 kg of CO₂ per tonne

1.6% : End of life (deconstruction and processing) 4.5 kg of CO₂ per tonne

0% : In use
0 kg of CO₂ per tonne

23% : Construction (mortar and wastage) 63.7 kg of CO₂ per tonne

6% : Distribution
16.7 kg of CO₂ per tonne

76% : Raw materials & production
213 kg of CO₂ per tonne

Specialist bricks from Danish, Belgian and Dutch producers are sought for their visual appearance. These are, however, often coal-fired or double-fired to achieve the desired finish, significantly increasing embodied carbon.

APPLICATIONS

Bricks in the UK are predominantly used to construct the external load-bearing walls of low-rise domestic properties. They are increasingly used to form non-load-bearing cladding to reinforced concrete and timber-framed multiple-occupancy residential structures. More than 70% of all dwellings constructed in England between 2002 and 2010 were completed using load-bearing masonry.[19] Each year, residential buildings account for 83% of the market demand for bricks in the UK.[20]

While bricks do have comparatively high embodied carbon, use in long-lived residential structures requiring little maintenance means they can compare favourably to other less durable alternatives.

Historically, bricks were used more widely to construct pavement vaults, basement walls, foundations and vaulted floor structures. Bricks were also widely used in infrastructure including sewers and masonry arch bridges and piers. In these applications, the inherent compressive strength of masonry is utilised, and horizontal spans are achieved via arching. When using lime mortars, tension cannot be accommodated and so elements must be sized to ensure geometric stability. Despite this limitation, structures of great elegance and economy of material are possible when architectural and structural concerns are synthesised.

Cladding

Non-load-bearing cladding panels typically rely heavily on metal accessories to provide structural support and restraint. For a typical hand-laid masonry façade, the total weight of stainless-steel accessories may approach 5kg/m². Given the high embodied carbon of stainless steel, the emissions associated with these accessories can be similar to the brick itself. Considering the lack of EPDs or even easily accessible weights for accessories, combined with the fact that these elements are often quantified only by the contractor, it is likely that these embodied carbon emissions are not currently being counted accurately.

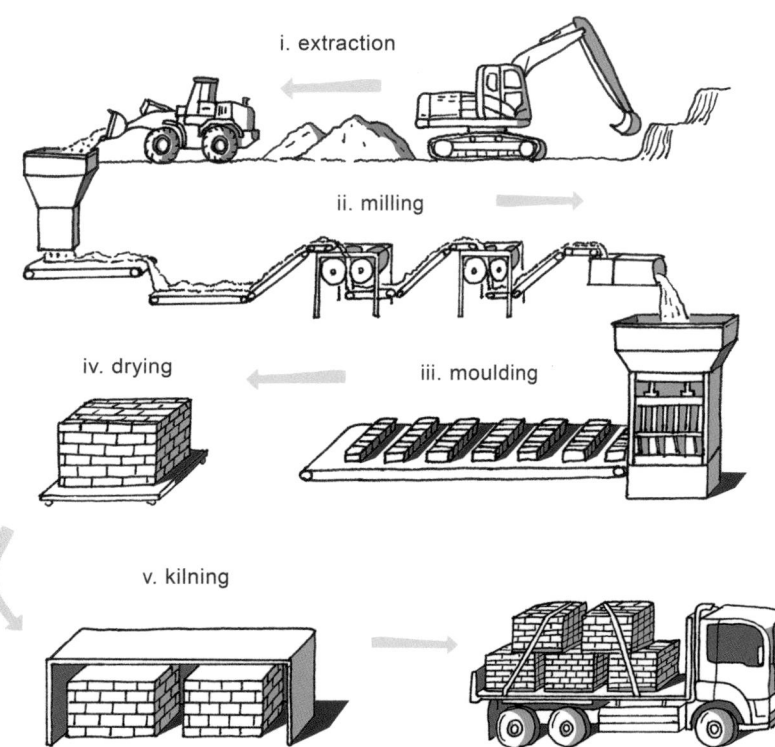

i. extraction

ii. milling

iv. drying

iii. moulding

v. kilning

Figure 4.4 An overview of the brick manufacturing process.

It is possible to reinforce masonry using grouted steel reinforcement or prestressing rods. These techniques provide enhanced bending resistance akin to reinforced concrete. Adopting this approach can reduce the need for high embodied carbon steel wind posts and is particularly relevant when designing for seismic loads.

The heavyweight nature of masonry cladding adds to the load carried by the structural frame, increasing a design's overall carbon footprint. For typical single-skin masonry and blockwork cavity construction supported floor by floor, cladding can represent 20% of the load carried by the perimeter structure, rising to more than 30% in the case of timber-framed construction.

Alternatively, masonry cladding can be stacked to support its own weight but this presents challenges with respect to differential movement and is rarely seen, though examples do exist, in project such as the Aga Khan Academy, Dhaka, by Feilden Clegg Bradley Studios and AKTII, 2022.

REDUCTION OF WASTE

Given variability in the clay, drying and firing processes, some bricks fail to meet necessary manufacturing standards. Clay bricks are reasonably fragile and breakage occurs during transit and placing. With extra care, wastage rates can be reduced but some residual losses are inevitable given the material. Waste bricks are typically diverted to use as general fill.

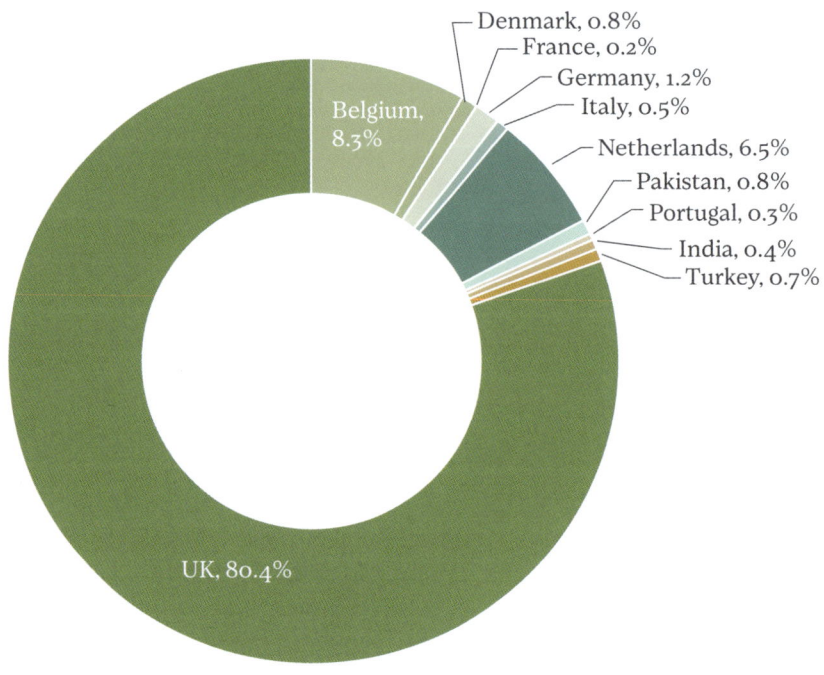

Figure 4.5 Geographic distribution of UK brick imports in 2021.[21]

Figure 4.6 York House, London, dMFK Architects, 2019. This extension and retrofit of a 1980s building includes a welcoming new entrance with a self-supporting brick screen comprised of a folded lattice that provides additional stiffness and reduces the amount of supports required, using only half the amount of bricks of a typical wall.

Figures from the circular economy NGO Waste and Resources Action Programme (WRAP) suggest that typical on-site waste rates may be as high as 20%, dropping to 5 to 10% with best practice.[22] Careful setting out of the façade based on the brick dimensions can minimise the number of bricks that need to be cut, further reducing waste.

LONGEVITY

When used appropriately, **fired clay bricks can be extremely durable, with a reasonable lifespan measured in centuries**, as evidenced by many examples of historic buildings and infrastructure. Durability is achieved principally through:

- **Appropriate choice of unit for a given exposure** – consideration of freeze-thaw risk from driving rain and climate, together with assessment of the aggressiveness of the environment.

Figure 4.7 Shrewsbury Flaxmill Maltings, Shrewsbury, Feilden Clegg Bradley Studios, 2022 (originally built 1796). Notice the many layers of repair in the external wall, shown by the varying colouration. Bricks used for repair were specially manufactured to conform to the larger, historic 'great brick' dimension.

- **Appropriate detailing** – water must be shed away from the face of the brick. Traditional detailing incorporates cornices, overhangs, sills, drips, flashings and chimney pans. Damp-proof courses and/or non-porous engineering brick at the base of walls provide protection from low-level moisture.

Generally, more durable bricks require better control of clay preparation and manufacture, and importantly, longer, hotter firing. This means higher emissions, so **it is important to avoid over-specification which can needlessly increase impact**.

Masonry structures are readily repairable using both traditional and modern proprietary techniques and this contributes to their longevity. Matching lime mortar is important when repairing historic structures since the use of cement-based mortars can cause structural and durability issues.

When considering the use of novel units not currently covered by standards, testing is required to demonstrate appropriate structural capacity and longevity. A suite of test standards (BS EN 772) addresses matters of principal concern, including:

QUESTIONS
FOR PROJECT TEAMS
AND SUPPLIERS

☐ Are product-specific EPDs available for the bricks being specified?

☐ What emissions are associated with the surface finish chosen?

☐ Are there options to reduce emissions using alternative fuels?

☐ What level of resistance to exposure is actually required for the application in question?

☐ Is the detailing appropriate to optimise durability?

☐ Does the cladding design achieve the maximum service lifespan?

☐ Is it possible to design the façade as self-supporting to eliminate stainless-steel support accessories?

☐ For novel bricks, what extent of testing is required with reference to BS EN 772 and has this been undertaken? Alternatively, are the products covered by ETA/UKTA approval or similar?

☐ Can reclaimed bricks be used and, if so, what is the source of the bricks being considered; is the risk of damage or deterioration compatible with the proposed use?

☐ Is there potential to support pathfinder technology on low-risk applications such as internal non-load-bearing walls?

- Compressive strength
- Water absorption
- Density
- Soluble salts content
- Moisture expansion
- Frost resistance.

END OF LIFE

It is difficult to ascertain accurate figures for the amount of bricks arising from demolished buildings each year. The Environment Agency's Waste Data Interrogator for England identifies the following quantities:[23]
Brick waste = 0.84 million tonnes
Mixed demolition bricks, concrete, tiles and ceramics = 7.72 million tonnes.

It is unclear what proportion of the mixed arisings bricks comprise and this highlights the challenge in separating bricks for reuse. Figures from the now defunct Construction Resources and Waste Platform, based on 2007 data, suggest the total amount of ceramic materials from demolition could be as high as 2.08 million tonnes.[24] Other authors have suggested an even higher estimate, at around 6 million tonnes, approximately equal to annual UK demand. Whatever the figure, **it is clear that at present very few bricks are reused, perhaps less than 5%, with most being crushed for use as general fill**.

Reclaimed bricks
One challenge to more widespread reclamation is the difficulty in removing cement-based mortars. This is currently done mechanically and is slow, costly and prone to damage the unit. A return to low-strength, low-adhesion lime mortars in future could address this, but comes with the downsides of reduced masonry strength and slower setting times. This may be less of an issue in non-load-bearing applications.

As with many reclaimed building products, there are questions surrounding the route to compliance with the EU and UK Construction Products Regulation (CPR). Compliance with this regulation permits a product to be CE or UKCA marked, respectively, which is a requirement for all new construction products placed on the market.

EMERGING INNOVATIONS

Reduction in emissions from firing

Use of biogas and, in the longer-term, green hydrogen as alternatives to fossil fuel for firing.

Kiln electrification to allow use of renewable energy.

Development of new lower-energy ways of achieving desired visual finishes to replace coal-firing and double-firing techniques.

Alternatives to fired clay

Bricks made with alternative binders such as lime or gypsum incorporating demolition, construction and other industrial waste as aggregate in mineral-bound blocks.

Use of unfired clay bricks, potentially using clay extracted from local basement excavation. Associated development of mortars optimised for this use.

Carbon-negative bricks produced using products resulting from mineral sequestration of CO_2 such as Seratech magnesium carbonate bricks.

Bricks and tiles produced using calcifying bacteria.

Reduced material demand

Increased use of more easily removed mortars, e.g. lime mortars to facilitate reuse of bricks.

Development of reusable cladding panel systems.

Novel proprietary systems such as interlocking blocks, avoiding the need for mortar, and blocks with voids for grouted reinforcement, avoiding the need for stainless-steel restraint systems.

Most products achieve this by conforming to relevant designated standards relating to manufacture and performance. However, the manufacturing process for reclaimed bricks is generally not known and may have occurred prior to these standards existing.

Two alternative routes have been proposed and are the subject of ongoing discussion.

- **Exemption route:** It is possible for products to be exempt from compliance if they are defined as individually manufactured or custom-made for a given declared use. This has been the view adopted most regularly in the UK to date, where reclaimed bricks have been used in conservation and heritage projects. It does generally mean that conservative assumptions on strength and durability are adopted.

- **European or UK Technical Assessment (ETA/UKTA) route:** This route sees a parcel of bricks tested for compliance with the harmonised European Assessment Document (EAD) 170005-00-0305. The tests are similar to those used for new bricks and allow a specific declaration of performance to be provided by the supplier. There are companies in countries such as Denmark and Sweden that offer reclaimed bricks covered by such an ETA and it is expected that others will follow. This route does place greater burden on the supplier.

It is advisable to engage with local authorities and insurers at an early stage to agree the approach to be adopted on a specific project.

Reclaimed panels

Another approach is to extract panels of conventional hand-laid masonry intact. This presents difficulties with respect to cutting, cranage and transport. In future, we may see masonry cladding panels designed from the outset for reuse, though care must be taken to ensure backing materials do not needlessly add to the carbon footprint.

DESIGNING FOR LOWER IMPACTS

As for all materials, we should aim first to simply use less through design efficiency and longevity. We should ask whether the useful service life of our current cladding designs is unnecessarily limited by the support systems we adopt. Is the use of brick as a facing to precast concrete panels an appropriate use of material? Further research is warranted.

There are also opportunities to reduce impacts through product specification and a switch to lower-emissions products. These changes can be encouraged through demand, though there may well be a cost premium, at least in the short run. Examples of best practice exist, which can point to future opportunities.

WIDER SUSTAINABILITY IMPACTS

As with any product made from high-volume, extracted materials, there are issues with respect to land use and environmental impact, including the potential loss of natural habitat. In the UK, any new extraction proposal requires submission of a planning application and an associated Environmental Impact Assessment. Mineral extraction companies are responsible for the restoration of sites following use.

Carbon-reduction potential	Measure	Example	What to do now to enable it
High benefit	Reuse in place.	Maintenance and adaptive reuse of existing masonry structures.	Prioritise retrofit where possible. Detail for longevity in new structures. Encourage the keeping of construction records.
	Reuse elements in new location.	Reuse of brick cladding panels.	Design brick cladding to be reused but avoid compromising longevity.
	Reclaim materials for high-grade reuse.	Use of reclaimed bricks as individual units or as cut panels.	Demand reclaimed bricks to help drive improvements in cementitious mortar removal, which currently inhibits reuse. Consider use of easier-to-remove mortars in new design, e.g. lime mortars. Use of cut panels is currently impeded by practical difficulties in extraction, transport and erection.
	Reduce emissions from production.	Substitution of lower-emissions alternatives in place of fired clay bricks.	Explore use of brick units cut from natural stone, particularly waste stone. Consider unfired clay bricks for indoor use.
		Use of lowest embodied carbon brick options.	Ask for product-specific EPDs and prefer products with lower emissions to drive change. Prefer single-fired bricks from efficient continuous kilns. Avoid over-specifying strength or durability, which may require higher firing temperatures.
Minimal benefit	Reclaim materials for low-grade reuse.	Use of crushed masonry as hardcore and general fill. Potential use of crushed fired clay brick in the production of binder additions for cement.	Use as hardcore and general fill currently common.

Table 4.5 Carbon-reduction measures.

Figure 4.8 A visualisation of Gent Design Museum, Ghent, Carmody Groarke, due 2026. The new building utilises a bespoke novel brick and white mortar made from local demolition waste bound with lime. Cured rather than fired, the bricks contain approximately one third of the CO$_2$e of conventional Belgian bricks. Note the detailing of window reveals and parapets to protect the façade from rainwater.

Flue gas from kilns includes small amounts of noxious gases, including nitrogen oxides, fluorides and chlorides, though these can be scrubbed to some extent. If released into the atmosphere, these compounds can be harmful to human health and so permitted emissions are limited by law in the UK. Elsewhere in the world, kilns fired with coal are a major source of black carbon, accounting for an estimated 20% of the total emission.

Brick clays also generally include sulphur compounds, which are released during firing and contribute to the formation of acid rain.

Significant amounts of water are consumed in brickmaking, though marginally less than concrete and far less than steel. Best practice can reduce water demand to less than 2l per brick (0.8l/kg) but the water footprint could be much higher than this overall.[25]

FUTURE TRENDS

UK brick demand per capita is far lower than its peak following World War II and throughout the Industrial Revolution. Based on current best estimates, average global demand per capita is comparable to the historic peak demand in the UK and rising, with current consumption in China estimated to be more than three times this.[26]

In the UK, demand is projected to rise in line with UK government plans for increased housebuilding. For low-rise, load-bearing construction in particular, masonry will remain a popular option, hence minimising the embodied carbon in manufacturing is critical in order to reduce emissions. The largest savings will come from the reduction of emissions from firing or through alternative production that avoids the need for firing altogether.

KEY TAKEAWAYS

- Fired clay bricks are capable of extremely long service life when designed for longevity. Look to historic precedent for good examples of detailing, maintenance and adaptation.

- Make use of the non-structural benefits of brick, given that embodied carbon is high when considering structural performance alone compared to alternatives.

- When selecting bricks, ask manufacturers for EPDs and prefer low-carbon options to drive innovation and support pathfinder projects. Be aware of the large variation in embodied carbon and that specific visual finishes can involve high emissions processes.

- Be aware of the embodied carbon in the restraint and support structures used in masonry cladding. Ensure that the intended service life of the elements in the system is compatible.

- Be open to new types of bricks offering reductions in emissions, opportunities for local sourcing and the use of waste products. Become familiar with the testing required to demonstrate adequate performance for the intended application.

CONCRETE

Helen McGarry, Bruce Martin and Eva MacNamara

Embodied carbon

30–600 $kgCO_2e/m^3$ [1]

UK concrete production per year

90 million tonnes [2]

Annual UK greenhouse gas emissions from concrete

10 million tonnes CO_2e/year (9 million tonnes CO_2e/year from cement production) [3]

Proportion of UK construction emissions from concrete production

25% [4]

Global emissions from concrete

Approximately **8%** (4–5% from cement production) [5]

Concrete is made from a mixture of aggregate (broken stone, gravel, sand), cement and water that is spread or poured into moulds and then hardens to form a mass resembling stone. The cement (also referred to as the 'binder') undergoes a chemical reaction that results in it hardening and binding all the ingredients together. Currently, most concretes rely primarily on Portland cement. Portland cement typically contributes over 70% of the embodied carbon of a given mix despite only accounting for approximately 10% of its volume. 'Concrete' is a general term, and every mix will have different properties depending on the nature and quality of the ingredients, the proportions of the mix, its preparation and the construction processes. Concrete may also contain admixtures which are used to modify concrete properties during construction and/or in the long term.

Reinforced concrete (RC) refers to concrete used in combination with another material (often steel bars). This primarily increases its tensile strength, Typically, the amount of steel reinforcement is 1 to 2% of the concrete volume.

The most effective way to reduce the Co2e of concrete is to reduce the overall volume used. Another approach, widely used in the UK and abroad, is the introduction of cementitious materials, such as GGBS, to 'replace' some, or all, of the Portland cement in a concrete mix. GGBS must be used with caution due to both national and global supply shortages. Much current R&D is focused on developing low carbon alternatives to Portland cement.

APPLICATIONS

- Infrastructure assets such as roads, viaducts and bridges
- Foundations for buildings
- Basements
- Retaining walls
- Superstructure frames (best used in elements that work mostly in compression, such as columns and walls)

PROS

- Strong in compression
- Durable if constructed and maintained properly
- Economic, widely available, easy to work with, industry established
- High thermal mass
- Vibration and acoustic benefits
- Low combustibility

CONS

- Weak in tension
- Can crack and degrade from within due to reinforcement corrosion
- Biggest contributor to greenhouse gas emissions from any single building material
- Currently mostly downcycled at end of life

GLOSSARY

Alkali-activated cementitious materials (AACMs) and geopolymer concretes
These concretes usually do not contain any Portland cement; instead, their cement is made of cementitious materials plus an alkali activating agent. These cements gain their strength, and other properties, via a chemical reaction between a source of alkali and aluminate-rich cementitious materials. The latter could be ground granulated blast-furnace slag (GGBS), fly ash, pozzolana, municipal solid waste incinerator ash (MSWIA) or other alumina-rich materials. These concretes can have much lower embodied carbon than an equivalent Portland cement-based concrete.

Cement content The amount of cement in a concrete mix (measured in kg/m³ of concrete or kg/kg of concrete). It significantly affects a concrete's strength and it also contributes to its durability. Higher cement content generally correlates with greater strength properties.

Ground granulated blast-furnace slag (GGBS) GGBS is the most extensively used SCM in the UK today. It is a by-product of the steel industry, whose production is rapidly decreasing, and therefore a finite resource. Use of GGBS to replace a high percentage of Portland cement often requires an overall increase in the total cement content of a concrete mix. It is currently drastically overused to reduce embodied carbon in concrete mixes.

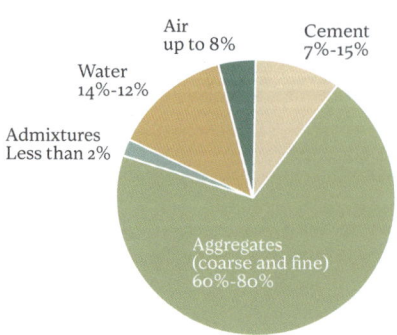

Figure 5.1a The components of a 'typical' concrete, proportioned by unit volume.

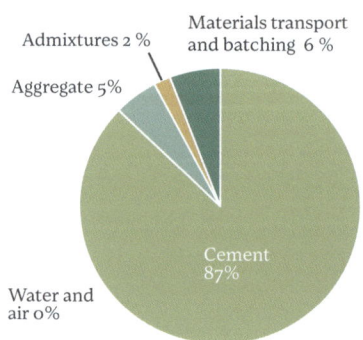

Figure 5.1b Breakdown of CO_2e in a m³ of a typical concrete, including life-cycle assessment (LCA) stages A1 to A3.[6]

Concrete is highly durable and long-lasting if used wisely. The Pantheon in Rome has boasted the world's largest unreinforced concrete dome since the second century AD, making it more than 1,800 years old! Often the limiting factor for the longevity of reinforced concrete is degradation of steel reinforcement.

Direct reuse and adaptive reuse of concrete structures is a viable circular mode; reuse of individual concrete elements is possible but usually more complicated (because the joints are often grouted in or cast in situ). Concrete material recycling has historically been a downcycling process (most commonly, concrete is crushed and used for road base). Other potential uses, for example as aggregate for concrete, are often limited by current methods of demolition, which do not sufficiently separate concrete waste from other construction waste materials.

Concretes are strong in compression (and relatively weak in tension). As a construction material, **concrete is economic, widely available, easy to work with, industry established and, if used properly, durable and resilient.** General-use concretes have good (i.e. low) combustibility ratings and do not support the spread of flames, making them relatively fire resilient. Concrete is a high-thermal-mass material (i.e. it is good at absorbing and storing heat). Exposed concrete floors and walls can be used to help passively cool spaces, preventing excessive temperature rises, reducing the risk of overheating and reliance on mechanical cooling. Concrete can be utilised for its vibration control and acoustic isolation benefits.

Portland cement (PC) This is the most common type of cement used in the UK today (also referred to as ordinary Portland cement (OPC)). It is made by heating limestone (calcium carbonate, $CaCO_3$) and clay in a kiln to form a clinker (nodular material produced when limestone is heated to high temperatures) which is then ground to a fine powder and mixed with gypsum. During heating in the kiln, the $CaCO_3$ breaks down to release CO_2, limiting the reduction in CO_2e which can be achieved.

Secondary (or supplementary) cementitious materials (SCMs) Other cementitious materials which can be used to replace some of the Portland cement. SCMs included in the British Standards include ground granulated blast-furnace slag (GGBS), fly ash (FA), silica fume, limestone and 'natural pozzolana' (such as volcanic ash and calcined clay). Variations in the blend of SCMs and Portland cement alter the properties of the cement, and therefore the concrete.

What's the outlook for concrete?

There are instances where it is possible to use a lower-carbon material than concrete, and this substitution should be made if appropriate to the use and performance requirements. However, on a larger scale, concrete's importance as a construction material over the next 30 years and beyond will not disappear as there is no alternative so widely and abundantly available. Concrete is indispensable in major infrastructure projects across the globe, such as roads, bridges and railways, as well as below-grade applications in buildings. The trick is to reduce how much concrete is needed and use it wisely.

There are existing and emerging techniques to reduce the carbon footprint of concrete by replacing Portland cement with other cementitious materials. Several such products are already available in the UK market, such as Cemfree, EFC, ECOPACT, Virtua Abs Zero and LoCem. These products all rely heavily on GGBS which is in limited supply. Availability and use of alternative SCMs, such as limestone, calcined clay and volcanic ash, is increasing. Such trends are indicative of a shift away from wholly relying on GGBS as the de facto Portland cement replacement material. With appropriate market signals, industry will likely invest in reducing the carbon footprint of new concretes.

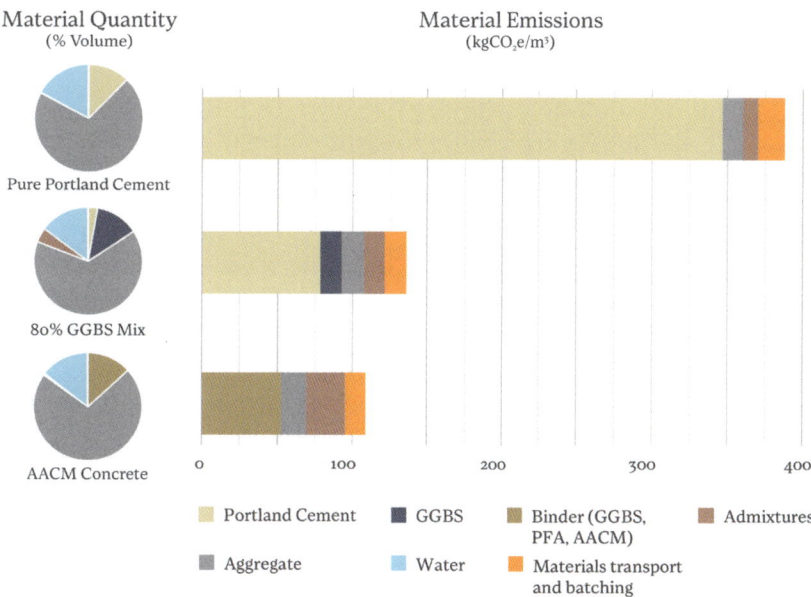

Figure 5.2 Typical concrete mixes and carbon intensities in 2022.

LOW CARBON CONCRETE ROUTEMAP AND BENCHMARK RATINGS

The Green Construction Board's (GCB) *Low Carbon Concrete Routemap*, published by the Institution of Civil Engineers (ICE) in 2022, produced its first benchmark ratings for embodied carbon for standard concretes in a range of strength classes. This is intended to be updated annually and the current version is freely available online for download and use.[7] The ratings, from A to F, relate to concrete recently available in the UK market and summarise the distribution of embodied carbon for various strength grades.

Concretes are usually referenced by their compressive strength, using the designation 'X'/'Y' which refers to the compressive 'cylinder strength' (X) and compressive 'cube strength' (Y). As well as determining the final compressive strength of a concrete, cement content influences how quickly a concrete mix gains strength. The

Figure 5.3 Embodied carbon benchmark ratings graph (LCA modules A1 to A3). In general, the higher the concrete strength class, the higher the carbon, but there is a huge range in potential embodied carbon for a given strength class. Adapted from the *Low Carbon Concrete Routemap*.

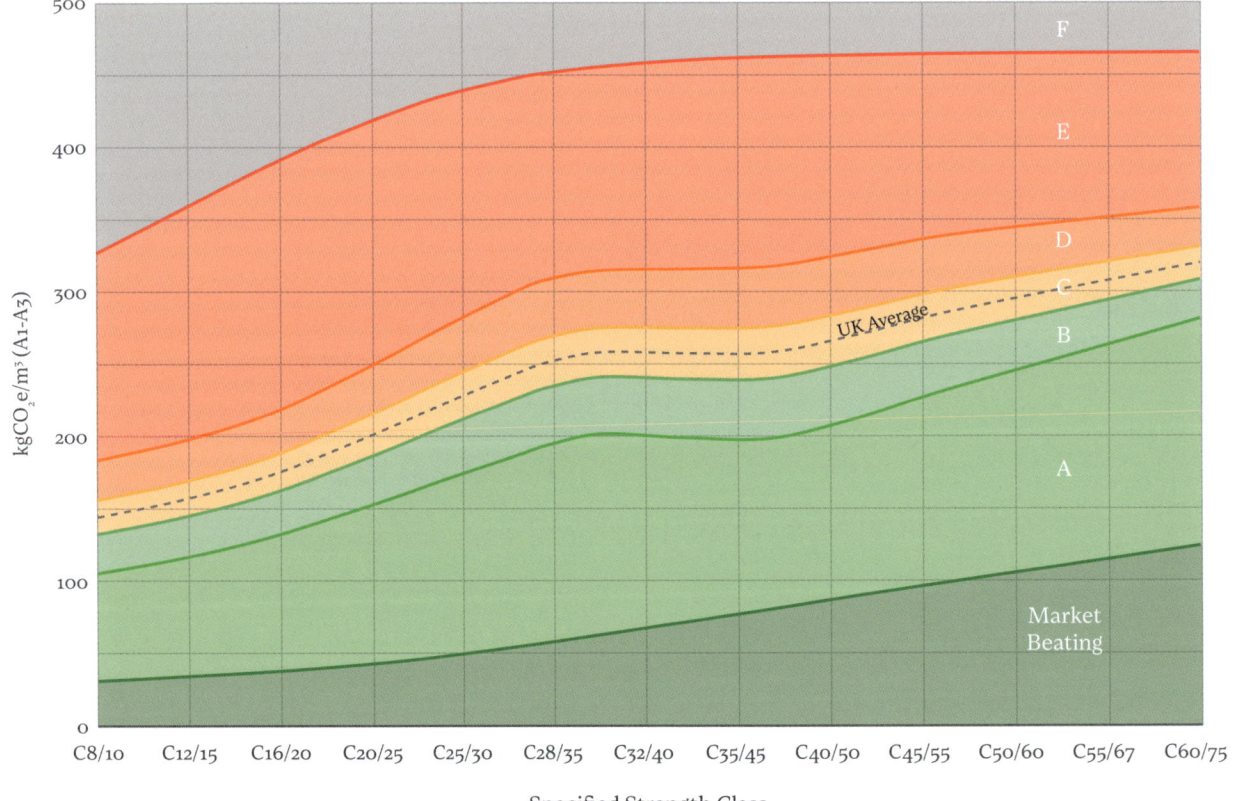

End use	Strength class
Trench fill, mass fill, kerbing, small garden foundations	C8/10
Blinding	C8/10 to C16/20
Mass concrete foundations	C8/10 to C20/25
Domestic garage floors	C20/25 to C25/30
Paving	C25/30 to C30/37
Beams and slabs	C25/30 to C32/40
Walls, piles, pile caps, pad foundations (reinforced)	C28/35 to C32/40
Columns	C28/35 to C40/50

Table 5.1 Typical uses for different strength classes of concrete.

cement content also influences a concrete's durability, technically referred to as consistency. Higher cement contents are associated with faster strength gain, its durability, and increased workability. The cement content also influences a concrete's durability. This often leads to more cement being included in a concrete than is required to achieve the typically specified 28-day strength.[8]

Non-structural applications can generally use lower concrete strengths than structural applications. Typically vertical, load-bearing elements, such as columns, will require higher compressive strengths than horizontal, primarily bending elements, such as beams or slabs.

LIFE-CYCLE ASSESSMENT (LCA) STAGES

In most cases, the whole-life CO_2e of concrete is dominated by the CO_2e arising from its manufacture (i.e. that associated with LCA module A).

Extraction of raw materials

Aggregate is the largest raw material by volume, making up 60 to 80% of the concrete; however, aggregates account for only a small amount of the overall embodied carbon in concrete. Traditional aggregates (crushed stone, gravel, sand) are generally sourced locally. Crushed concrete aggregate and recycled aggregate can also be used.

Cement is the second-largest constituent material of concrete by volume. Although considered a raw material in concrete batching, the production of most cementitious materials requires significant processing, unless they are industrial by-products (like GGBS or FA). At present, the process emissions during production of Portland cement form most of the embodied carbon of most concretes.

Admixtures form a relatively small fraction by volume of the raw materials of concrete and usually make up a very small part of the embodied carbon, but can reduce embodied carbon by reducing the cement content needed for a required strength.

Transport (and storage)

The materials for concrete production are transported to the batching plant or precasting factory in advance of production and stored separately at the site in large quantities. Aggregates are

Table 5.2 Distribution of CO_2e in LCA stages – current snapshot for typical structural concrete.[9] These values vary nationally and between projects and manufacturers, and will evolve over time.

*CO_2e due to maintenance can be significant in some environments, particularly those with exposure to sea salts or de-icing salts.

LCA Stage		Typical proportion of whole-life CO_2e
A	Before use	
A1 to A3	Cradle to factory gate	75%
A4 and A5	Transport and construction	15%
B	In use	Minimal *
C	End of life	10%
D	Subsequent benefits and loads	Varies

typically sourced locally to the batching plant; Portland cement is primarily sourced from national supplies, GGBS is mostly imported and FA is currently mostly nationally sourced. The number of different concrete mixes available from a plant may be defined by the plant's capacity to store multiple types of cementitious materials.

Cast in situ: manufacturing (or batching)

Manufacturing of the concrete material occurs at a batching plant and in the ready-mix wagon. Ingredients are measured in accordance with the mix specification. At a 'wet mix' plant, the ingredients are mixed before being loading into the ready-mix wagon. At a 'dry mix' plant, the ingredients, including water, are placed unmixed into the ready-mix wagon, where they are mixed.

Cast in situ: transport

Typically, ready-mix concrete must be placed on site within two hours of being mixed to control loss of workability and setting of concrete. This requires batching plants to be relatively local to site (or on site). Admixtures can be used to extend the working life of fresh concrete to several hours.

Cast in situ: construction

Most of the site works associated with a concrete pour must be carried out prior to the arrival of the concrete on site. This includes the support and erection of the formwork and fixing of reinforcement.

After the concrete has been placed, it needs to cure long enough to gain enough strength so that the formwork and props can be removed.[10] Additionally, the new concrete building element may be used to support the props and formwork for higher floors.

Industry expectations include highly workable concrete, enabling pumped placement on site and minimising requirements for vibration. Although admixtures are used to control the workability of a concrete, increasing the cement content is a common method.

Quick removal of formwork is also preferred, which requires rapid early strength gain. Similarly, this often leads to concretes with more cement than is required for the permanent works. At present, the most economic construction requires simple geometry that avoids changes in profile of the concrete (steps, voids, coffers, etc.). Formwork is typically made from plywood, which has limited reusability.

Precast: manufacturing (or batching) and casting

Concrete batching for precast elements typically happens at the precasting factory. A 'wet mix' process is used – the ingredients are mixed before the wet concrete is pumped into forms or moulds and left to cure. Precast elements may also be created by slip-form or extruding techniques, after which the final products are cut to the required length (e.g. hollowcore planks).

Highly workable concrete is preferred to enable easy factory placement with rapid early strength gain to enable quick turnaround and reuse of moulds.

There are several working methods available to the precast industry that enable reductions in Portland cement, but at present they are not widely used. Such methods include carefully designed 'blended cements', low-energy heating systems to accelerate early strength gain, alternative methods of demoulding and extension of the time before demoulding.

Precast: storage and transport

Once an element has been demoulded, it can be stored at the factory to gain further strength before being transported to site. The transportability and installation of precast elements needs to be considered in their design.

Transport can be a significant part of the embodied carbon of precast elements that have been optimised to minimise CO_2e.

Precast: site works

Site works for precast elements can be simpler and quicker than cast in situ construction. Connections between elements are generally formed either by bolting together using steel connectors that were cast at the factory, or by grouting or concreting around loops of reinforcement left protruding from the units.

LONGEVITY OF MATERIALS (DURABILITY, MAINTENANCE AND REPAIR)

In many cases, the CO_2e due to maintenance during service is minimal. Exceptions do occur and CO_2e due to repair can be significant, particularly in environments that include exposure to sea salts or de-icing salts. Degradation of steel reinforcement is often the limiting factor in reinforced concrete structures.

END OF LIFE DISMANTLING, REUSE, RECYCLING AND (RE-)CARBONATION

Concrete buildings are generally strong, enduring structures, and a trend towards their direct or adaptive reuse for new projects instead of demolition and rebuilding is growing. While reuse of precast elements is feasible, in practice it is often difficult to disassemble in a manner that enables economic reuse. Mechanical connections (such as bolting) facilitate deconstruction at the end of an element's life but require high precision for fit, whereas grouting is more tolerant of misalignment but requires wet concrete on site and forms a connection which is more difficult to deconstruct.

Although steel reinforcement is widely reclaimed and recycled, concrete has historically been downcycled (typically crushed and used for road base). Crushed concrete aggregate (CCA) and recycled aggregate (RA) can be used as aggregate in new concrete. RA is aggregate from the reprocessing of inorganic material previously used in construction, while CCA principally comprises crushed concrete. The effect of RA and CCA on a concrete mix may be an increase in cement content. Due to the angularity of these aggregates, more water is required in the concrete mix; this means more cement is required to ensure a suitable water/cement ratio. Technologies are emerging which improve the quality of materials recovered from demolition.[11]

If suitably exposed, a concrete structure will reabsorb carbon over its whole life – this process is known as 're-carbonation' and the rate varies considerably with concrete type and location. Re-carbonation during a structure's life may reach about 5 to 10% of its upfront emissions in 20 to 30 years;[12] however, recent studies have concluded that it may be closer to 4%[13] and that the majority will not happen until after the building is demolished. Such contributions will not address the immediate need to severely reduce emissions in the next five years and avoid the initial release of CO_2.

DESIGNING FOR LOWER IMPACTS

Most of the carbon in concrete is in LCA module A. To really tackle these stages, it's critical to reduce the amount of concrete used and use materials with reduced CO_2e (i.e. use less Portland cement). Reducing carbon in LCA modules B, C and D is also important (see Table 5.2).

Figure 5.4 Burntwood School, Wandsworth, London, AHMM, 2014. This campus-wide retrofit of a south London school used prefabricated precast concrete elements to great architectural effect, utilising the full depth of the façade for solar shading. Windows were factory-installed to the rear of the cladding panels, contributing to a low airtightness figure of 2.53m³/h.m² at 50PA. The thermal mass of the precast façade elements also reduces energy loads, supporting a mixed-mode ventilation strategy which includes some operable windows. Factory assembly minimised waste in the production process.

USE LESS CONCRETE – TACKLING LCA MODULE A
Make strategic design choices

It is important to reuse existing concrete structures whenever possible. Choosing to redevelop an older concrete building is a strategic design decision which can mean huge carbon savings compared with demolishing and building from scratch.

For any project, the design stage has significant impacts on the quantity of concrete that will be used, so it is important to choose approaches where materials play to their strengths.[14] Concrete should be used in elements which are primarily under compression and avoided in highly tensioned elements where a lot of reinforcement will be required.

The implications of early design decisions such as material, grid sizes and imposed loading should be thoroughly discussed by the design team with the client. Longer spans are generally associated with higher carbon. Where possible, concrete should be saved for heavy load-bearing structures such as foundations and timber

Figure 5.5 IStructE Headquarters, London, Hugh Broughton Architects, 2015. A 1960s office building in central London was transformed into flexible 21st-century workspace by retaining the primary concrete structure, opening up the entrance into a double-height foyer with a dramatic new staircase and fabric improvements throughout. Structural interventions focused on specific areas to limit material use and embodied carbon, as well as to control costs and shorten the construction period. A large opening was cut through the first floor to create an atrium and allow a feature staircase to be introduced. In order to do this, the existing precast floor planks were strengthened with carbon-fibre strips. The adaptive reuse project saved 95% embodied carbon versus a new build.

should be considered for small- and medium-span superstructures. Higher-strength concretes offer potential material savings in vertical structures with high compressive forces.

Promote a marginal gains approach

The full potential of concrete's material properties should be utilised, including varying section thicknesses to closely match internal stresses, to minimise the volume needed (rather than using a constant section size by default). Marginal gains can come from reducing cover requirements to avoid excess material or considering whether element depths/bar spacing can be rounded to the nearest 5mm rather than the nearest 25mm. Using refinements from the building codes can also provide marginal gains, for example to reduce lap/anchorage lengths.

Consider alternatives to RC flat slabs

Over 50% of the embodied carbon in an office building is in the structure, most of which is in the floor slabs. Reinforced concrete (RC) flat slabs are widely used for floors as they offer shallow floor depths, minimising floor-to-floor and maximising floor-to-ceiling heights. Flat slabs are (currently) economic but rarely a carbon efficient use of concrete. Alternative RC options such as coffered slabs, waffle slabs and beam and slab systems are lighter, deeper solutions which can significantly reduce concrete volume, in some cases by more than 50%.

Use of punching shear heads (local thickening of the slab around the top of a column) with a flat slab system helps control punching shear and deflection, which are typically the governing criteria for a flat slab. Their use can mean a shallower slab for the same span, thereby enabling a reduction in overall volume. Designers should consider the range of concrete structural systems available and their potential carbon benefits when selecting the appropriate solution for a project; this may involve an increased floor zone to accommodate more efficient structural forms.[15]

Detail out volume

If high-quality workmanship can be ensured when fixing and casting, the concrete cover to reinforcement can be reduced, potentially resulting in thinner sections. Voids, coffers, ribs and non-structural fill should be used to reduce the total volume of concrete where possible. The use of pre- and post-tensioning in beams and slabs can also reduce concrete volume. Detailing decisions which appear to be small but are repeated throughout a building can have a significant impact on the amount of concrete used.

Consider constructability

There may be implications associated with casting elements with intricate geometries (such as formwork inefficiencies, increased costs and/or construction time) so an open dialogue between client, design team and contractor is essential to determine what can be achieved and the trade-offs of decisions. Complex forms can be created using intelligent, non-wasteful solutions, such as lightweight able-net and fabric formwork systems.[16]

Change the mix – use less Portland cement

Reducing the CO_2e intensity of materials is the same as saying, 'Use lower-carbon materials.' In the UK, producing 1kg of Portland cement results in approximately $0.86kgCO_2e$.[17] By comparison, a high-blend cement (i.e. one with a low percentage of Portland cement and a high percentage of SCMs) can have an embodied carbon figure as low as $0.25kgCO_2e/kg$.[18] Portland cement use should be minimised and, working with a concrete technologist, projects should use sensible and optimised concrete mixes that are lower carbon.

Be aware: availability of secondary cementitious materials

GGBS and, to a lesser extent, FA are currently the most widely used SCMs in the UK; however, there is an unavoidable problem with supply. They are both by-products of high-carbon industries which are changing rapidly – GGBS is a by-product of the steel industry and FA is produced from burning coal in electricity power stations. The production of steel from ore will be phased out as the industry moves away from creating primary steel in blast furnaces and therefore the supply of GGBS is forecast to drastically reduce. Likewise, as the use of coal in electricity production is phased out, production of new FA is declining. The amount of GGBS available globally is enough for only 5 to 10% of cement production, with a similar figure for FA.[19] There are large quantities of FA stockpiled in landfill throughout the UK, however, it is not yet clear whether this will be a viable source of SCMs.

Avoid over-reliance on GGBS

GGBS, in particular, is used extensively and in high volume in concrete mixes across the UK today. As well as being produced in the UK, it is currently imported from France, the Netherlands, Germany, Spain and Japan, among others. Its use as a cement replacement reduces the carbon footprint of an individual mix at project level. However, since the global supply is fully utilised, unless there is a particular technical benefit from using GGBS, its use in any one mix will not result in an overall reduction in global greenhouse gas emissions. To most effectively reduce greenhouse gas emissions, it is recommended to use GGBS efficiently, in particular for applications where there is a specific technical reason (such as its ability to extend concrete setting times). GGBS should not be used solely to reduce the carbon footprint if it results in an increase in the concrete's total cement content. The Institute of Structural Engineers has recently published a guidance note on its efficient use.[20]

Start using alternative concretes for low-risk applications

Alternative concretes can be described as those made using an alternative cement that is not one of the standard combinations defined in BS EN 197-1 and BS EN 197-5. Alkali-activated cementitious materials (AACMs) and geopolymers are examples of alternative concretes that contain little or no Portland cement. Some of the emerging alternative concretes sequester CO_2 in their production, allowing them to promote their products as zero carbon or even carbon negative.[21]

UK codes permit the use of alternative concretes; however, additional testing may be needed to demonstrate code compliance. Uses of alternative concrete are likely to be in lower-risk items and/or unreinforced concrete or concrete reinforced with non-corrosive fibres or bars such as fibre-reinforced polymer (FRP) bars or stainless steel. Items which may be suitable initially include non-structural blockwork,

Figure 5.6 HiLo rib-stiffened funicular floor system, Zurich Block Research Group, ETH Zurich, 2021. This funicular floor system was inspired by traditional vaulted masonry structures, which transfer loads to supports through arch action – using geometry to achieve strength and efficiency. The concrete arch works in compression and generates a horizontal thrust at its base which is taken out by steel ties. By placing material only where it is structurally needed, following the flow of compression and tension forces, and by keeping all materials separate, the HiLo floor system saves more than 50% of concrete and 90% of reinforcement steel compared to standard concrete flat slabs and allows for easy recycling at end of life.

kerbs, blinding concrete, fencing posts/bases, concrete fill around manholes, drainage runs, buried structures and temporary works.[22]

Don't make concrete specifications too prescriptive

Concrete is a unique material in that it is possible to directly adjust its constituent parts to produce a mix with a minimised carbon footprint that also meets the project performance requirements.[23] Concrete specifications that are too prescriptive can limit the mixes that can be offered by a concrete supplier/contractor. Keep concrete specifications performance-based where possible, affording the concrete supplier maximum flexibility to provide a mix which satisfies project requirements while minimising carbon.

Build-in carbon goals

If a specification is clear but flexible and carbon goals are built into it as a requirement, the concrete supplier will have a clear direction of what needs to be achieved and they will have enough scope to propose something workable. Specify concrete carbon targets ($kgCO_2e/m^3$) and/or a rating to be achieved in accordance with the latest version of the LCCG Benchmark as part of the tender documents, to be discussed and developed with the concrete supplier.

☐ Have concrete carbon target ratings in accordance with the LCCG (Low Carbon Concrete Group) benchmark for all concrete used on your project been added into the contract requirements?

☐ Will the contractor report concrete benchmark ratings during design development as part of the materials approvals criteria?

☐ Will the contractor/concrete supplier accompany every mix design certificate with a specific embodied carbon value, and give clarity on how it was calculated?

☐ What are the implications of grids, imposed loading and floor systems on the sizes of concrete elements? (Discuss this with the structural engineer at the scheme stage, with a view to minimising material volume.)

☐ Will the contractor engage early with the concrete supplier so that there is time to iterate to an optimised concrete mix?

☐ What are the implications of the construction programme and methodology on the embodied carbon of the final concrete mix? (Discuss this with the contractor.)

☐ What are the implications of conflicting project objectives, such as trying to deliver a low-carbon building while under the pressure of very tight construction programmes (which typically drive higher-carbon solutions)? (Discuss this with your client.)

Engage with the concrete supplier and contractor

Change is difficult if only one element of the chain is innovating. For example, it is not enough for the concrete supplier to offer a high-SCM concrete mix, as the contractor may need to allow for increased curing times. Designers should allow for actual mix properties in their design, rather than basing it on overly conservative generic assumptions. It is important to engage with concrete suppliers and contractors as early as possible to initiate a dialogue about project aims and the practicalities of material supply. If a project is part of a masterplan or long-term development, there will be greater opportunity for engagement and innovation.

Communicate with the precaster

Precasters rely on quantities of elements ordered to keep their factories busy. If a project needs a lot of an element, a precaster may be more likely to modify their mode of working to accommodate new materials or geometry. Communicate with the precaster to work out what is possible for a specific project. For elements with a small production run, the cost of making the mould may be a large part of the cost of each element. For larger run elements, it may be possible to modify the demoulding process to allow longer for strength gain if storage is available. Low-energy heating systems, such as heating pads or using hot water in the concrete mix can be used to accelerate early strength gain.

MINIMISE CARBON IN USE – TACKLING LCA MODULE B

As mentioned above, degradation of steel reinforcement is often the limiting factory for the longevity of concrete elements. This can be controlled by protecting the reinforcement appropriately, by using alternative types of reinforcement (such as stainless steel or FRP bars) or by designing to minimise tension (and therefore reinforcement). In harsh environments (such as marine environments and those exposed to de-icing salts) particular care should be taken, perhaps by use of protective barrier layers and proactively planned inspection and maintenance.

MAKE CHOICES NOW TO ENABLE CIRCULARITY IN THE FUTURE – TACKLING LCA MODULE C

Table 5.3 gives examples of things to think about in a new building which can enable circularity in its end of life.

WIDER SUSTAINABILITY IMPACTS

Material extraction of the constituents of concrete (limestone, aggregate and sand) is a temporary land use which has the potential to effect biodiversity negatively through habitat loss and disruption.[23] Historically, extraction of raw materials has caused significant harm to biodiversity and in many parts of the world this is still the case. There is a global sand shortage due to increasing demand and extraction locations are shifting to more fragile environments such as rivers, coastlines and oceans.[24] Poorly executed extraction of sand and marine aggregate can result in riverbank collapse, deepening of riverbeds and coastal erosion, as well as biodiversity loss.

In Europe and the UK, biodiversity damage from raw material extraction has been largely mitigated through effective regulation. Mineral planning authorities provide planning consent, and control and monitor quarrying in the UK. This, in conjunction with regulation by the Environment Agency and implementation by the industry, ensures that development and operation of quarries does not damage and, where possible, enhances the environment. New quarries are planned to avoid areas of biodiversity value, and during and following extraction must demonstrate a net gain in biodiversity. In Europe and the UK, concrete at its end of life is effectively completely diverted from landfill and recycled in various forms. Concrete production and construction do use a lot of water, with the use of rainwater a more sustainable source than mains water.[25]

Carbon reduction potential	Measure	Example	What to do now to enable it
Extremely beneficial	Reuse in place.	Adaptive reuse of a concrete frame building and foundations.	Ensure proper management of the project and as-built information, e.g. BIM data including reinforcement and concrete material certificates.
	Reuse elements in a new location.	Relocate precast planks from a building at the end of its service life to a new structure.	Think about deconstructability, especially in the connection details.
	Reclaim materials for high-grade reuse.	Recycle steel reinforcement, separate high-quality sand, stone and hydrated cement.	Demolition specifications and waste management standards should prioritise circularity; demolition materials should be traceable so that reuse can be maximised, and transport distances should be minimised.
	Reclaim materials for medium-grade reuse.	Recycle steel reinforcement, crush and expose concrete to maximise re-carbonation.	
Minimal benefit	Reclaim materials for low-grade reuse.	Use crushed concrete for road base.	

Table 5.3 Measures that can be incorporated on a project to facilitate future circularity.

At project level, it is important that material sources are local to minimise transportation impacts. Specifiers should also require compliance with BES 6001 Responsible Sourcing of Construction Products and ISO 14001 Environmental Management (or equivalent environmental management systems). Compliance to these standards is typical for the major UK concrete industry players, although smaller suppliers may not always conform.

Specifiers can include requirements for a project- or site-specific Biodiversity Action Plan, which sets out the project's biodiversity aims, and how this will be achieved. Specifiers can also include a requirement for the batching plant or precasting factory to have an independently assessed Biodiversity Enhancement Plan for their facility.

If used insensitively and in unsuitable environments, concrete can create sterile hard-surfaced 'biodiversity deserts'. The 'urban heat island' effect is a phenomenon occurring in cities with overuse of concrete pavement, buildings and other surfaces that absorb and retain heat. This effect increases energy costs and pollution levels and is harmful to the city's human, animal and plant life. Sensitive use of concrete can help to create habitats, such as support for intensive green roofs, as well as innovative forms of bio-receptive concrete, which is a hyperporous concrete that can support plant and animal life.

FUTURE TRENDS
Emerging innovations
- **Emerging carbon-negative concretes** Such as Seratech, Karbonite and Cambridge Electric Cement.[26]
- **Graphene concrete** The addition of tiny amounts of graphene to concrete can strengthen it by around 30% compared to standard concrete, meaning less may be needed to achieve the equivalent structural performance.[27]
- **3D-printed concrete** 3D-printing technologies have potential carbon savings implications by being able to reduce waste and produce efficient concrete forms that follow the stresses of the element. They are best suited to unreinforced or externally reinforced concrete.[28]
- **Lightweight cable-net and fabric formwork** Flexible and reusable formwork systems which can intelligently create structural forms that have been digitally optimised for material efficiency.[29]
- **3D-printed formwork** A formwork technology using 3D-printing foam that can create geometrically complex concrete elements which have been optimised for structural/material efficiency.[30]
- **Smart-crushing** A technology that can return end-of-life concrete to its original sand, cement and gravel components. After recovery, the original raw materials can be fully processed into new concrete.[31]
- **Bio-receptive concrete** Concrete that supports and stimulates plant and insect life.[32]

Upcoming trends

- Adaptive reuse of concrete-frame buildings.
- More structurally efficient concrete elements (coffered slabs, beams and slabs, punching shear heads versus flat slabs etc.)
- Increased use of voids, non-structural fill and other concrete volume substitution measures.
- Upper limits on $kgCO_2e/m^3$ for concrete specified on projects, plus public reporting against the *Low Carbon Concrete Routemap* benchmark.
- Low-energy heating systems, such as heating pads, or using hot water in the concrete mix used to accelerate early strength gain (without the need to increase cement content) in precast production.
- Increased availability and use of limestone, calcined clay and volcanic ash to replace Portland cement.
- Increased availability and use of alternative concretes (such as alkali-activated cementitious materials (AACMs) and geopolymers).
- Better use of admixtures to control workability and strength gain (and thereby enable reductions in cement content).
- Increased use of non-corrodible reinforcement such as GFRP (glass-fibre reinforced polymer) and BFRP (basalt-fibre reinforced polymer).

KEY TAKEAWAYS

- Concrete is indispensable to major infrastructure projects and in below-grade applications. Its importance as a construction material will not disappear; the trick is to reduce how much is needed and use it wisely.

- Often the limiting factor for the longevity of reinforced concrete is degradation of steel reinforcement. Concrete should be used primarily in compression and avoided in highly tensioned elements where a lot of reinforcement is required.

- Over 50% of the embodied carbon in an office building is in the structure, primarily in the floor slabs. Flat slabs are (currently) economic but are rarely carbon-efficient. Coffered slabs, waffle slabs and beam and slab systems are lighter, deeper solutions that can significantly reduce concrete volume.

- Over-reliance on GGBS as a cement substitute should be avoided. GGBS supply in the UK is finite and its global supply is limited so specifying GGBS or PFA should be capped to locally available supply chains. Alternative SCMs that have abundant local supply, such as limestone fill or calcined clay, should be considered.

COPPER

Nick Hodges

Embodied carbon of facade cladding

5,000-10,000
$kgCO_2e/m^3$ [1]

Embodied carbon of rod products

10,000-35,000
$kgCO_2e/m^3$ [2]

Density of copper

8,960 kg/m³

Melting point

1,085°C

Durability

60–250+ years

Copper has a long history as a building material and remains in use today as façade cladding, roofs, rainwater goods and in electrical wiring. Historically, it was prized for its colour and robustness, weathering with recognisable colour and patination to a distinctive verdigris green that is commonly seen on significant buildings such as cathedrals. It is naturally malleable and has excellent thermal resistance and durability, with examples of installations that have lasted more than 200 years. It can be used with many other materials, such as stone and brick, but care should be taken when placing it in close relationship with other metals, particularly aluminium and zinc, where bimetallic corrosion can occur.

As a metal, it has high energy consumption in mining and processing, and wider sustainability impacts relating to chemical extraction of copper from ores. It has excellent circularity – around 40% of copper is recycled and some construction products contain 90–100% recycled source material.[3] Using recycled material can save more than 50% of carbon emissions related to extraction and manufacture, and copper can be recycled repeatedly with no loss of performance.

APPLICATIONS

- Cladding
- Roofing
- Rainwater goods
- Some pipework
- Wiring

PROS

- Malleable
- Extremely long life
- High recycled content
- Excellent potential for recycling at end of life

CONS

- High energy use in extraction, processing and production
- Broader sustainability impacts of extraction
- Bimetallic corrosion

HISTORY

Copper is one of the earliest metals identified and used by humankind. As a native metal (occurring in a directly usable metallic state) it was first used around 8000 BC. It was the first metal to be smelted, cast into a mould and intentionally alloyed, with tin, to form bronze. It formed the key part of other alloys, such as brass, which were known to the Greek and Roman civilisations. Copper was used as some of the earliest coinage, and as an alloy it is still in use in US, Euro and UK coins.

Copper has been recycled and repurposed for many centuries; the dome of the Pantheon in Rome was once clad in copper both inside and out and the material was stripped for reuse over the intervening centuries. The mass production of copper was a significant part of the 19th-century Industrial Revolution in the UK, but today copper extraction and processing are global, with much coming from South America and China.

CHARACTERISTICS

Copper is a highly malleable and ductile metal, with high thermal and electrical conductivity that makes it a vital part of electrical cabling systems. For similar reasons, copper was popular for cooking utensils such as kettles, and continues to be popular for pipework installations, particularly domestically. It has excellent longevity and corrosion resistance when maintained well, with frequent examples of lifespans of over 100 years, and many over 250 years.

Newly manufactured copper is highly reflective and almost pink in tone, but this quickly patinates through oxidisation to first a warm brown, black and finally a distinctive green. The transition of copper to its iconic green colour is highly dependent on atmospheric conditions and is more likely to form in coastal and urban areas now that atmospheric pollutants such as sulphur dioxide have reduced.

Brasses and coppers offer a very high risk of bimetallic or galvanic corrosion and care should be taken with copper, and copper alloys, when used with other metals. This is particularly the case when metals are in contact but can also occur from water run-off from copper, and affects many common metals including aluminium, zinc, steel and lead, and can lead to their damage. As a result of copper's relative softness, fixing clips are often made from stainless steel, which offers rigidity, corrosion resistance and is stable when in contact with copper.

Copper and bronze are recognised for their natural antimicrobial potential, and in recent years there has been increased interest in how this could be used to naturally reduce contact transmission of diseases, including viruses and fungi. Reduced proportions of pure copper content impact the effectiveness, so not all alloys may perform to the same standards.

EXTRACTION AND MANUFACTURE

Copper is abundantly available; however, **extraction and processing are highly energy intensive processes that have the potential to be environmentally damaging**. Copper has a high residual value and capacity for being recycled, resulting in almost all post-consumer copper products being reclaimed.

Copper is mined globally, with Chile, Peru and China currently contributing over 50% of global production. Similarly, 50% of copper smelting production is concentrated in China, Chile and Japan.[4] The process of smelting is energy intensive, requiring high amounts of heat, and forms raw copper blister in the form of ingots or 'pigs'. The blister copper is around 98% purity and can be processed to form rolls and sheets.

The blister can also be further refined to higher purity either through methods requiring further heating or electrolysis. Much copper used in construction is sourced from recycled material, which has benefits of reduced ecosystem impacts and carbon from production and processing. Recycled copper is not free of environmental impacts or carbon resulting from heat energy for re-smelting, and individual suppliers should be consulted.

While copper has excellent recyclability, post-consumer sources currently do not meet supply; around 30% of copper in use worldwide is from recycled sources. Approximately 70% of copper is recycled at end of life, and in construction products the proportion is often much greater, with some products being from 90–100% recycled sources.

MATERIAL SOURCING

Copper used in construction is often sourced close to the manufacturers of raw material. Following manufacture into raw copper rolls, it is then transported for any further processing or finishing. In its simplest form, copper rolls can be cut to sheet sizes or smaller shingles and applied with little finishing other than cutting and pressing processes.

Sheets and rolls are typically between 0.7mm and 1mm thick, although can be sourced thicker if required. The thicker the material, the more difficult it will be to press, cut and work precisely.

These smaller formats are then supplied to product manufacturers for processing into specific copper products, or supplier subcontractors for specific work. This can include cutting primary sheets or rolls into elements unique to specific projects, or inclusion of clips or other fixings.

Figure 6.1 70 Broadwick Street, London, Buckley Grey Yeoman, 2022. This copper-clad extension provides nine residential units as part of a mixed-use retrofit in Soho. Copper was selected as a lightweight, non-combustible cladding to fit within the constraints of the existing building's structure; also because of its service life of more than 100 years and because it is 100% recyclable. Compared with other lightweight cladding materials such as aluminium, copper could be thinner (1.5mm for façades compared with 3mm for aluminium), reducing both material use and embodied carbon. Perforated panels enabled concealed integration of MVHR (mechanical ventilation with heat recovery) vents through the façade.

Discussions with suppliers and subcontractors can also identify the most efficient way to use material when cut down from larger rolls to reduce wastage – although any wastage created is easily converted back into raw material.

For projects where a specific patina tone of copper is desired, then it is possible to select pre-patinated copper that can be fixed in tone or allowed to continue to weather. These are available in colours such as coated warm brown, oxidised dark brown and the classic verdigris tone. Additional chemical treatments are used to create these patinas, but have limited impact on the recyclability of the material at end of life.

WIDER SUSTAINABILITY IMPACTS

Modern copper ores in primary production have under 1% proportion of copper present in the mined material, and so it is necessary to extract copper from the surrounding 'uneconomic' material. These processes make use of chemical reactions and agents that can remain in the waste extraction product, known as tailings.

Copper tailings can include residual ground rocks, water and chemical reagents, including heavy metals and even radioactive material. They can be a dangerous source of contamination and require careful management to reduce risks to local ecosystems and biodiversity.

Rainwater run-off from roofs and facades can be a source of contamination for local water courses, so careful control through copper coatings or sustainable drainage systems should be included in the design.[5]

When considering copper for internal use, particularly wiring, there may be carbon savings in considering alternatives such as fibre or extensive use of WiFi. Circularity should be considered in this so that the benefits of recycling material at end of life can be compared.

COPPER IN CONSTRUCTION

Similar to zinc and other metals, copper is typically available as sheets or shingles which are cut from larger rolls. When installed as sheets, it is typically installed as a standing seam 'tray' between 300mm and 600mm wide and up to 3m long, which is secured to the substrate with copper or stainless-steel clips and the trays interlocked through a folded seam. Timber is the most frequently used substrate and should be ventilated to prevent the formation of damaging condensation. It is usually recommended that copper is laid on a separating geotextile or breather membrane to encourage moisture transfer and reduce the effect of rain drumming noise.

It is possible to introduce seams and welded joints to copper finishes but care should be taken to allow the necessary thermal expansion or there are risks of splits and breaks forming. Typically, copper joints rely on bends, folds and pleats that interlock panels with each other to create a weathertight form.

When used in a smaller format, such as interlocking singles or sheets, fixings are smaller stainless-steel clips, usually in four or six locations on the upper edges. Sheets are installed from bottom to top, so setting out should account for the module of the shingle and how it meets at eaves and openings. Corner junctions can be folded as a continuous part of the façade or joined with a vertical 'seam'.

QUESTIONS
FOR PROJECT TEAMS AND SUPPLIERS

☐ What is the recycled content of the copper being used?

☐ Where was the copper originally sourced and processed?

☐ What are the optimum sheet or roll sizes for the most efficient use of material?

☐ How should the material be fitted in a way that will last and reduce risks of substrate failure or bimetallic corrosion?

Copper in electrical systems remains common and the impact of MEP (mechanical, electrical and plumbing) services on the embodied carbon of buildings is still being understood. Detailed analysis of cable use and longevity may make alternative types of distribution worth consideration.

DESIGNING FOR LOWER IMPACTS

Compared to many other materials, **copper has a relatively high embodied carbon per kilogram produced**; this is principally the result of chemical energy used for extraction and heat energy in processing. The most effective way to reduce this is to specify 100% recycled material, ideally from as local a source as possible to the site. Although current recycled copper cannot meet global demand, specifying recycled sources supports and encourages a circular supply chain.

Given its longevity, copper should be used to maximise design and building life through robust details with well-ventilated substrates. Site waste can be minimised by working with suppliers and subcontractors to identify the most efficient shingle and roll sizes. Where possible, copper should be avoided as part of composite elements that are hard to separate as this will impact recycling at end of life. This might include fixing types and treatments.

CIRCULARITY AND RECYCLING – END OF LIFE

Copper is highly recyclable, and as it is self-finished with no surface coatings to address it is a relatively straightforward material to recycle. It also retains a high residual value, which means it is not uncommon for copper to be a target of opportunist theft. In construction materials, much, or even all, of the copper used in sheets and tiles derives from recycled sources and most suppliers are able to provide details on the specific proportions in each alloy or patina type.

To improve recyclability at the end of the building life, it is important to specify fixings that can be easily removed. Standing-seam and tile installations should be easily demountable with the stainless-steel clips and fixings able to be separated. Composite material use of copper should be avoided where possible to maximise recycling potential.

FUTURE TRENDS

Alloys and new technologies In recent years, there have been evolutions in the use of copper alloys to offer further ranges of colour tones and textures in copper, including brass and 'gold'. These alloys are relatively new to construction and their innate characteristics should be explored when used. Their experience of patination may be relatively unknown, for example, and they are likely to be stiffer in bending or cutting, which could influence how they are most effectively used in a building project.

Reduced-carbon copper Copper producers are transitioning the primary energy needs for smelting and processing to lower-carbon and renewable sources, such as solar, wind power and even hydrogen. It may be that some producers are able to offer reduced-carbon copper to the market as these processes mature.

KEY TAKEAWAYS

- Copper is a metal with exceptional longevity, evidenced by its long use in building and examples of buildings more than 200 years old.

- Copper has a recognisable character, with a bright shine when newly installed that turns to a warm brown and eventually a distinctive verdigris green.

- It has a relatively high embodied carbon, and its mining and processing involves chemical waste products that have potential for environmental impacts.

- Copper has great potential for circularity and when used in construction can be sourced as 100% recycled material.

- Using copper efficiently and from recycled sources can mitigate some of its carbon impacts, and it should be installed in a way that is easily demountable to support recycling at the end of the building's life.

CORK

Oliver Wilton

Area of cork oak landscapes in Mediterranean Basin

2 million hectares

Largest global manufacturer

Portugal

Average density

150 kg/m³ [1]

Embodied Carbon for Insulation

83 kgCO$_2$e/m³ [2]

Embodied Carbon for Flooring

210-420 kgCO$_2$e/m³ [3]

Sequestered carbon

200-1,000 kgCO$_2$e/m³ [4]

Cork is the bark of the cork oak tree, typically harvested once every nine years in a process that does not harm the tree. This harvesting process increases biogenic carbon storage without the need to fell the tree or disrupt the landscape. Cork oak trees can live for up to 200 years and grow around the Mediterranean basin, often as part of traditional, biodiverse, cultural landscapes.

Cork has been used in construction for millennia, with early uses including roof tiles and wall linings. Cork is currently utilised in a broad range of construction products, including pure cork products and a range of engineered products that include a binder and sometimes other materials such as timber. The two main material types are pure expanded cork, a 100% plant-based material used to make thermal insulation and façade cladding boards, and agglomerated cork with polyurethane or another binder, used to make floor and wall tiles and composite flooring.

Cork products, particularly those made with pure expanded cork, can have low embodied carbon and can be embodied carbon negative if biogenic carbon is counted. Pure expanded cork is also relatively simple to recycle and is currently recycled by some manufacturers. Agglomerated cork is more complex to recycle due to its composite material character and energy recovery is a more common LCA module D scenario.

APPLICATIONS
(Typical UK uses)
- Thermal insulation
- Internal finishes
- External cladding

PROS
- Strong environmental sustainability performance, including low embodied carbon emissions
- Specification can support historic, biodiverse farming landscapes
- Tactile, sensorially rich and aesthetically pleasing
- Good thermal, moisture and acoustic performance
- Biodegradable

CONS
- Poor fire performance
- Pure expanded cork is relatively friable as cladding and linings
- Relatively high cost as a thermal insulation board
- Limited number of suppliers for some product types

GLOSSARY

Agglomerated cork
Cork granules bound with polyurethane or another binder which accounts for around 10% of the product. Granules are ground from cork by-products, often from cork wine stoppers. Commonly formed using a compression mould.

Autoclave Machine used for industrial processes that require high temperatures and moist heat.

Dehesa (Spain), **Montado** (Portugal) Cork oak landscapes with a rich, biodiverse agro-sylvo-pastoral farming system where trees, crops and livestock are integrated.

Insulation Cork Board, ICB An expanded cork board that is used for insulation.

Pure expanded cork 100% pure cork, also known as pure expanded agglomerated cork. Made in billets by heating cork granules in an autoclave using superheated steam. The granules expand and darken in colour. The suberin in the cork melts and then resets, acting as a natural binder.

Quercus suber Commonly known as cork oak. An evergreen oak tree of medium size whose outer bark is stripped for cork.

Suberin A polyester biopolymer found in plants.

INTRODUCTION

Cork is the outer bark of the cork oak tree, *Quercus suber*. It is typically harvested once every nine years in a careful and skilled process of stripping that does not harm the tree. Cork is a natural closed-cell foam, under 20% solid by volume and with an average density of 150kg/m³. It is typically composed of 43% suberin, a waxy biopolymer, 22% lignin, a biopolymer that gives structural strength, and other compounds including cellulose.[5] These physical and chemical characteristics give cork a unique combination of properties, making it suitable for a broad range of construction uses. These include being thermally insulative, relatively water impermeable and vapour permeable, surprisingly resistant to surface wear, elastic and with some moderate load-bearing capacity.[6] Like other plant-based materials, **cork stores atmospheric carbon for the lifespan of the material**, and it is also relatively combustible. Cork is biodegradable, perhaps less so than most timber, and it generally does not readily biodegrade when allowed to dry out and is kept out of contact with soil microbes and other relevant biota.

Cork is harvested from cork oak landscapes around the Mediterranean basin totalling around two million hectares in area, large areas of which are historic, biodiverse mixed agroforestry landscapes. The cork processing and manufacturing industry is focused in Portugal and other parts of western Europe. The main revenue generator is wine bottle stopper production, and construction products make good use of by-product and lower-grade cork from branches and forestry management. The largest market for cork construction products is currently Germany.[7]

EXTRACTION AND MANUFACTURE

The bark of a cork oak is typically harvested for the first time when the tree is 25 years old, and then once every nine years after that for an average of 150 years. The bark is carefully stripped in planks in the late spring and summer months and then stacked to dry for some months before being processed. Cork has been used in construction for millennia; it was used as a roofing material in Roman times and today it is turned into a broad range of materials and products.[8]

Figure 7.1 Geographic distribution of cork production.[9]

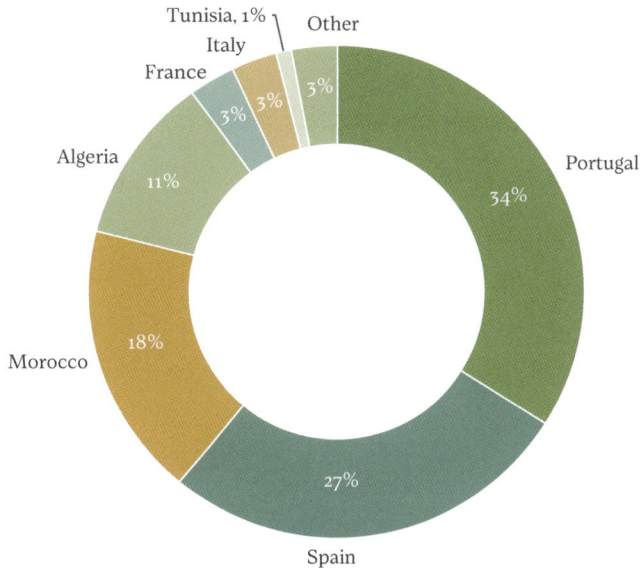

Cork bark in its natural form is occasionally used in planks and tiles as internal wall linings or external cladding. This is often done using first-strip bark, which has a rough, evocative texture, compared with later strips that are smoother. Agglomerated cork composites encompass a range of materials typically made with cork granules bound by polyurethane or another binder using pressure moulds, and sometimes including other materials. These are highly engineered materials and products, made in factories, some of which also produce parts for the motor and aerospace industries. Products include floor tiles and composite floor planks, wall tiles, carpet underlay and specialist antivibration mounts and pads. Declared cradle-to-factory-gate (LCA modules A1–A3) carbon for cork floor tiles produced by Amorim is 1.70 kgCO$_2$e per square metre.[10] The addition of the binder extends the performance and range of applications for these products and makes use of waste cork from the bottle stopper industry. It also adds to the embodied carbon of the resultant products and adds life-cycle complexity. The addition of the binder can yield products that degrade more quickly over time under UV light than pure cork products. The products are also more complex to recycle at the end of their service life, as they cannot simply be granulated and returned to the ground as pure cork can.

Pure expanded cork, also called pure expanded agglomerated cork, is a 100% plant-based material, invented by accident by John Smith and patented in New York in 1891.[11] Today, it is made using granules of lower-grade cork from upper branches, including that from routine forestry management. These are placed in an autoclave and cooked with superheated steam at a temperature of over 300°C, causing them to expand and darken, with the suberin in the cork melting and then rebonding to form a rigid billet

Figure 7.2 Freshly cut expanded cork insulation boards at Amorim Vendas Novas, Portugal, 2022.

of expanded cork. A common billet size is 1,000mm by 500mm with a thickness of up to 300mm after trimming. This method can be tailored to produce a range of billet densities, normally 100 to 160kg/m³. The main use for this type of cork is as a thermal insulation board; cork is the original foam insulation board, and most of the billets are sliced up and sold in a range of thicknesses, from 20mm to 100mm. Large Portuguese cork producer Amorim manufactures expanded insulation corkboard (also known as expanded insulation corkboard (ICB)) at 115kg/m³ and thermal conductivity 0.04W/mK,[12] with declared cradle-to-factory-gate (A1–A3) carbon of 82.8kgCO₂e/m³ of material, excluding biogenic carbon.[13] This, and similar products from Secil and other manufacturers, are available in the UK via suppliers including Ty-Mawr Lime and Mike Wye. Benefits of this product include the positive contribution it can make to indoor air quality and its carbon-negative character, when atmospheric carbon stored in the cork is counted. This is partly due to a high proportion of the cooking heat being generated by waste biomass. More uniform grades of board are also available, with more consistent granule size and visual character, for example MD Façade by Amorim, designed for use as facing material.[14] This has a wide range of uses but is also limited in facing applications due to its Euroclass E fire rating. It has been used as a thermally insulative external cladding and for internal wall and ceiling linings since the early 2000s, when it was introduced.

APPLICATIONS

Cork construction products can have good in-use and life-cycle performance, both in newbuild and retrofit applications. Expanded insulation corkboard (ICB) has been widely used for around 100 years, and the indication is that it is durable, with material being recovered today from demolished mid-20th-century buildings in Europe that is still in good condition. It can contribute to high indoor air quality, in addition to acting as thermal and acoustic insulation. It is commonly used as cavity, roof and floor insulation, including as structural insulation in floating floors. Further applications include external wall insulation with a lime render finish, or left fair-faced in certain applications, and as an internal wall lining. In addition to newbuild applications, these approaches can be used as part of energy-efficient retrofitting of solid wall buildings, where cork can be well suited due to its relative vapour permeability. Ty-Mawr Lime supplies a rendered external wall insulation system suitable for retrofits and uses cork granules in some of its lime render mixes for use in historic buildings to add character and improve thermal insulation performance.[15] Agglomerated and composite cork floor and wall tiles provide resilient internal finishes and can give acoustic and other benefits. Cork tiling was used to add comfort at Frank Lloyd Wright's Falling Water house, in Pennsylvania, and composite flooring has recently been installed at Gaudí's Sagrada Família, in Barcelona.

CONSTRUCTION METHODS

ICB is easy to work with on site and can be cut with carpentry tools. It is typically held in place with simple mechanical fixings when used as wall cavity insulation, in a similar manner to other ridged insulation boards. When used as external wall insulation, lime mortar or adhesive may be used in addition to mechanical fixings. ICB is typically loose laid to floors and roofs. Agglomerated cork tiles are most commonly adhesive fixed. ICB has some modest load-bearing capacity, demonstrated by some of its historical applications, including as self-supporting internal walls for municipal cold stores. It's not commonly used in a structural way today, other than as structural floor insulation, and so any use like this would have to be carefully considered by the design team.

END OF LIFE

Loose-laid and mechanically fixed ICB is relatively easy and simple to recover at end of building life for reuse or recycling. Recycling can include granulating the boards to give loose-fill thermal insulation or for adding back into the manufacture of new boards. Amorim does this at its Vendas Novas factory in Portugal, typically adding around 5% of recycled content to its insulation corkboard. This recycling would not currently happen with UK projects unless it was agreed with a manufacturer directly. Granulated ICB also has other uses, such as being a soil improver for certain soil types. Agglomerated cork products using polyurethane or another binder are harder to reclaim, as they are typically adhesive fixed, and are more complex to recycle due to their composite nature, so often end with energy recovery.

Figure 7.3 Cork House, Eton, Berkshire, Matthew Barnett Howland with Dido Milne and Oliver Wilton, 2019. Developed in tandem with a related research project, Cork House pioneers a radical form of plant-based construction, using pure expanded cork as its primary structure and envelope to demonstrate how cork might be more widely used in future. Bespoke interlocking cork blocks, manufactured off site and assembled dry with no mortar or glue, enable simple disassembly for reuse at the end of the building's life. Life-cycle environmental impacts were carefully considered at every design stage. Cork House has a unique character and very low life-cycle carbon.

DESIGNING FOR LOWER IMPACTS

Using cork products close to their place of manufacture can significantly reduce embodied carbon as many products have low cradle-to-gate carbon, so transport can add a significant proportion. Options for this in the UK are limited by the fact that the products are mostly made in Portugal or elsewhere in western Europe. Using ICB without any fixing adhesives or coatings can contribute to low life-cycle carbon as it makes it easier to recover and recycle or reuse at the end of a building's life. Where it is used as an external or internal wall or ceiling finish with no render, paint or other applied coating, this can mean no routine maintenance is needed over its lifespan, which can significantly reduce embodied carbon, as well as maintenance costs.

WIDER SUSTAINABILITY IMPACTS

Using cork in construction can contribute to strong life-cycle environmental sustainability performance that goes beyond the typical metrics, including at the start and end of the life cycle. Significant areas of cork oak forests – Montado in Portugal and Dehesa in Spain – are biodiverse, culturally rich and provide valuable rural jobs.[16] The management of these landscapes also helps to prevent desertification and some areas of Dehesa have been subject to environmental protection laws for several hundred years. Specifying cork can help to support and sustain these unique landscapes. Products such as expanded cork insulation board are 100% plant-based and have a range of uses at end of building life. These include reuse of the boards, recycling into new boards, grinding into granules for use as loose-fill insulation or simply returning to the ground as a soil improver.

QUESTIONS
FOR PROJECT TEAMS
AND SUPPLIERS

☐ What is the intended use for this cork product, and does it comply with the relevant ISO and other standards?

☐ What is the product made of, and in what proportions; is it pure expanded cork, or agglomerated cork using a binder and other treatments?

☐ What is the performance of the product, including fire performance and relevant other characteristics, such as thermal performance?

☐ Is an Environmental Product Declaration available for the product?

☐ Is it possible to confirm where the cork was sourced; is it sourced from a sustainably managed forest, and is it covered by PEFC or FSC certification?

FUTURE TRENDS

Architects are increasingly recognising the sustainability characteristics of cork and its unique tactile quality. The phenomenological character of cork has been discussed by Steven Holl and used in projects including his 2007 remodelling of New York University Department of Philosophy, with its cave-like, cork-lined lecture theatre. Herzog & de Meuron have also taken an interest in the material and put it to good use as 'made ground' in their 2012 Serpentine Pavilion, London, with Ai Weiwei.[17]

The Stirling-shortlisted Cork House, in Berkshire, designed by Matthew Barnett Howland with Dido Milne (CSK Architects) and Oliver Wilton (2019), shows the potential for considerably expanding the future uses of cork. It uses pure expanded cork blocks as its primary structure and building envelope, in a radically simple new form of plant-based construction that was developed in a combined industry and academic research project.[18] It is designed for simple, dry assembly, with no glues or mortars, and for easy disassembly at the end of the building's life for component reuse.[19] Cork House demonstrates how cork could contribute to future forms of construction that respond to our current environmental sustainability priorities and deliver strong life-cycle performance.

KEY TAKEAWAYS

- Cork for construction comes in two main forms: pure expanded cork, a 100% cork product used as thermal insulation boards for walls, roofs and floors, and for external cladding and internal linings, and agglomerated cork that includes a binder and is commonly used for floor and wall tiles. Agglomerated cork is much more difficult to recycle at end of life.

- Specification of cork from well-managed landscapes can contribute to the maintenance of biodiverse mixed agroforestry landscapes.

- Use of pure cork contributes to good internal air quality and cork finishes give a rich, tactile character to architecture that links to its provenance as the bark of a tree.

EARTH

Rowland Keable

Load-bearing strength of clay-bound materials
Typically between

1 N/mm² and **3** N/mm²

Rate at which 12mm of clay plaster holds internal humidity

A steady 50%

Amount of CO_2e locked up by a 50:50 clay/straw mix

100–150 $kgCO_2e/m^3$

Embodied carbon of rammed earth

47.5 $kgCO_2e/m^3$ [1]

Earth used in construction is a composite material which relies on clay as a binder and often with fibre as reinforcement. Clay is a powerful material with a range of qualities beyond simply holding other materials together. Widely available, it is the original and undisputed circular material which has no end of life. It can also bind insulators, buffer heat and coolth with mass, and control humidity.

A long vernacular tradition of building with earth in the UK and around the world signals the vast potential of earth construction. However, in conventional construction, earth excavated from a site is normally disposed of as waste. With increased awareness of the climate emergency, many practitioners, designers, academics and organisations in the UK and abroad have started to re-evaluate earth construction.

Earthen construction comes in many forms, but three forms are most widely used: monolithic (including rammed earth and mass earth (cob)), masonry (including moulded and compressed blocks and mortars) and framed structures (including infill to timber frames, masonry units, mortars and plasters). Choosing which form is most appropriate for a particular site is informed by the available local materials, the type of building and the height and load to be imposed.

APPLICATIONS

- Walls
- Flooring
- Wall finishes

PROS

- A widely abundant resource that can be sourced at or close to many sites
- Offers thermal mass, humidity control, fire and soundproofing
- Has no VOCs (volatile organic compounds)
- Entirely circular at end of life
- Well-designed earth buildings can last hundreds of years

CONS

- Requires protection from rain and water penetration
- Can be time- and labour-intensive, hence costly

GLOSSARY

Earth mortar A highly adaptable material, often faced with lime pointing. Can be used with stone, burned bricks, MEBs and CEBs.

Framed earth Typically, earth and fibre mix applied to a framework. Also called wattle and daub, mud and stud, stick and rice.

Light earth Clay and fibre or bio-aggregate are mixed in even proportions by mass. This mix produces a density of material around 0.3 tonnes per m³.

Mass earth A monolithic technique widely employed across the UK that uses a fine mix with natural fibres and a moisture content between plastic and viscous. Also known as cob, mudwall, clom and clay dabbins.

Moulded earth blocks (MEB), compressed earth blocks (CEB) Preferably used with clay mortars. Also known as clay lump, mud block, adobe.

Rammed earth A monolithic technique that uses a high percentage of aggregates (coarse, medium and fine) with a low clay content and is compacted in a moist condition, producing some of the densest and therefore strongest earth structures. Also known as *pise de terre*.

Turf A variety of cut and shaped blocks of varying consistency, from high bio content with low earth percentage to high earth percentage with low bio content, depending on the ground conditions from which it is cut. Highly durable in extreme climates (Scotland, Iceland, etc.)

EXTRACTION AND MANUFACTURE

The first step in earthen construction is testing. Typically, a site investigation is undertaken to assess a site's bearing potential. At the same time, samples can be taken for testing for suitability to build. This should guide the design process, enabling a site-specific response to environmental factors rather than an abstract idea of an earth building type. Often rammed earth has become a shorthand for all earth building, but it is in fact a particular form or technique which requires particular soil types or mixes. While these are commonplace in the mountains of eastern France, they aren't appropriate everywhere.

Soils used from the extraction of foundations should be a first consideration for use, with the soil type driving the design form, an option that is often overlooked in conventional practice. For instance, a heavy clay soil is not suitable for rammed earth, while a stony, sandy soil will be hard to mould, free-hand lay or make into blocks.

For earthen construction, the site material must be initially analysed before the designer selects the earth building method. Other materials can be brought to site to improve the mix but this should be done with great care to best suit the existing material. It is worth bearing in mind that in a pine forest it is not recommended to go searching for oak to use; likewise with soil, understanding the materials available at or close to site is the first step in integrating them into a design.

Once a process is aligned to the materials available, a specification needs to be developed. This may involve screening elements out or adding to the base material to improve its characteristics in some way. This might include increased strength (more clay), reduced shrinkage (more sand or gravel) or better particle size distribution. Additions should be low-processed materials or other waste sources, and not necessarily the clean sands and aggregates used elsewhere. Other typical additions include agricultural by-products such as fibres and bio-aggregates. Local production should be considered ahead of performance, unless very strict structural criteria are in place, to minimise transport impacts. Mass earth, for instance, is local to many areas. It could be substituted for rammed earth for a marginal increase in strength, but the emissions to bring a suitable mix to site are typically not justified for the increase in performance.

Figure 8.1 WISE Building, Centre for Alternative Technology, Machynlleth, Wales, Pat Borer and David Lea, 2010. The rammed earth wall enclosing the Sheppard Lecture Theatre is an example of load-bearing rammed earth in a public building with high mass and excellent thermal characteristics.

Another approach is that materials can be taken away from site and processed into products, such as blocks, and brought back. This is being trialled in the Cycle Terre project as part of the Paris Olympics 2024, as well as at the Tribeca: The Apex site in London's Kings Cross. The opportunity cost should compare the travel to and from site with the travel otherwise expected (e.g. waste to landfill and new materials brought to site) so that substitution is at least like for like.

Colour is perhaps clay's most noticeable feature, with a wide variety available, from pure white (Cornish kaolin, for instance), yellows, oranges, reds, greens, blues to grey and almost black. While it is the smallest particle in most mixes, it is the clay colour which will most strongly determine the colour of the final product.

Apart from colour, many of the qualities of soil are identified through site investigations which describe soil types, measure granular structure and often give vivid descriptions of the material qualities. These investigations tend to be led by the engineer, but architects can begin to recognise the types of soil and their corresponding qualities with a little guidance.

MATERIAL SOURCING

Manufacture may be on site or at a designated facility, with materials travelling from site to the factory, then brought back as products. Products include unfired bricks and blocks, plasters in tubs or bags, panels with pre-applied plasters, and panels from natural materials for plastering. Purchased materials such as moulded blocks are typically delivered dry to site and therefore can be used immediately, while site-made blocks must first be made, then dried, which requires space. These space and programme issues often favour factory-made products.

Plasters can be extracted and made at site, though specialist help is often required to get the mixes right. On-site production can bring many benefits, including low transport emissions and a unique product.

Unfired blocks may be made from site material which has been transported to a production facility or from material from off site. The producer may also make and supply mortar, either as a dry powder or wet, and the blocks can be assembled by bricklayers in a more-or-less standard fashion, though clay mortars offer the advantage that walls can be built higher than with cement in each session. When the earth used is from site, the mortar can be made from the same material so that the colour and material match the blocks. **Clay mortars generate no waste because they don't set chemically and can be rewetted.**

APPLICATION

Earth has two main applications, either as load-bearing structures or as finishes. However, other uses, such as indoor climate attenuation, fire and soundproofing, may also be thought of as increasingly legitimate applications.

The potential for using earthen construction as infill in timber frame could be huge across the UK. This includes low-carbon alternative products and processes to replace plasterboard, PUR (polyurethane rigid foam), PIR (polyisocyanurate), mineral wool and a range of other high-emitting, carbon-intensive solutions, including foamed concrete.

Wrapping earthen structures inside an insulating layer, with a body of dense mass to manage thermal and humidity fluctuations, reduces the risk of having earth materials externally exposed to the weather and helps reduce the risk of overheating. Any earth building type can be wrapped in a thermal envelope, ensuring it is kept dry and without thermal bridges.

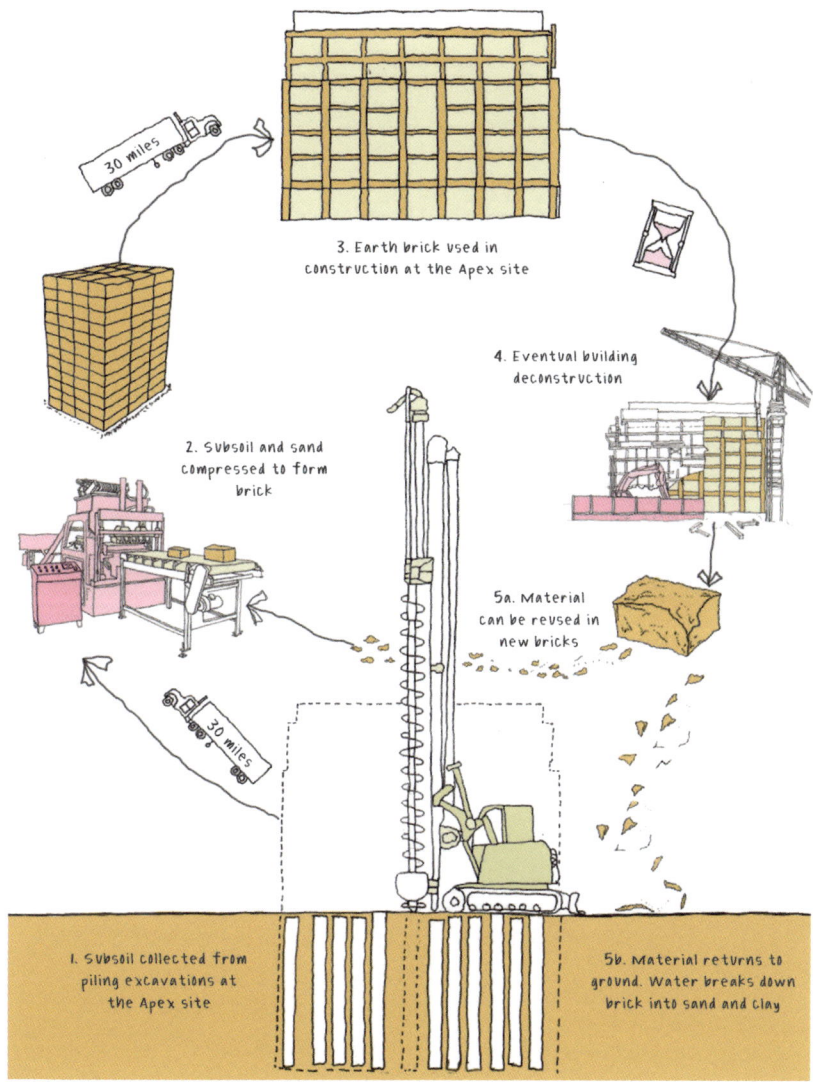

Figure 8.2 Bennetts Associates, diagram explaining circular use of excavated subsoil to make earth blocks for the Apex Building at King's Cross, London.[2]

3. Earth brick used in construction at the Apex site

4. Eventual building deconstruction

2. Subsoil and sand compressed to form brick

5a. Material can be reused in new bricks

30 miles

1. Subsoil collected from piling excavations at the Apex site

5b. Material returns to ground. Water breaks down brick into sand and clay

HOW STRONG IS CLAY?

For physical, load-bearing strength of clay-bound materials, typical values range between 1Nmm² and 3Nmm², so not high by the standard of fired brick and concrete but easily enough to build to five or six storeys. Different techniques produce different strengths, and engineers typically apply safety factors based on a number of metrics, not just the block strength but also the presence of mortar and other design details. While cement is often added as a stabiliser, this should be avoided because cement and clay are incompatible chemically and it fundamentally alters the potential circularity of earth construction.

Figure 8.3 Rammed Earth House, Wiltshire, Jonathan Tuckey Design, due 2024. Cylinders of excavated subsoil, used to test different soil types, colours and aggregate sizes, reveal the wide variety of clay colour within the 63-acre site. The architects worked with Austrian rammed earth expert Martin Rauch and the author to develop the soil mix and the detail design for the project.

Physical strength is just one parameter of clay and earth structures and finishes. Binding insulators such as hemp and flax shiv materials and other agricultural by-products provide less physical strength characteristics but reduce thermal conductivity, while not affecting the circular reuse of the components.

REDUCTION OF WASTE

Waste is a funny word, really coming into its own in the 20th century with the era of high-energy, single-use materials. Ironically, earth suddenly became waste as well, dug up and discarded although perfectly usable.

One of the key points to make regarding the use of earth as a building material is that during the extraction phase, materials should be kept as discrete as possible. Each layer of the soil excavated should be kept separate. This allows a high degree of fine-tuning in the production/building process. For instance, if clay overlays sand, the clay should be extracted and stockpiled and then the sand can be added in measured portions. This maximises the usefulness of the materials and the design options. It requires a change in building culture, but not a particularly complicated one.

In conventional practice, excavated site materials are frequently sent to landfill. By reusing site materials in situ, earth construction reduces both waste and landfill taxes.

LONGEVITY

Earth materials have existed for thousands or millions of years without undergoing any great chemical or mineral change. As building elements, they remain essentially the same unless exposed to water over a long period. **Correct design, maintenance and repair are essential to allow earth to remain functional in a built context over centuries.** The skills needed for maintenance are as important as those required for building and are different to those required for concrete or fired brick.

Design has an enormous part to play in the longevity of earth in construction. Modernist tropes of flat roofs without overhangs do not translate well to territories where wind and rain are year-round realities, such as the UK. Meeting building regulations typically means wrapping buildings with insulation, putting high mass earth walls inside, where they can either be left exposed or hidden behind earth plaster to suit the needs of the users.

When earth is used externally, the 'look' should be very carefully considered. Clear finishes, such as varnishes, do not work well in wet conditions because they break down over time, but this change cannot be seen in the way changes to a painted surface can and it can be worse than weathering with no finish at all.

The disaster for existing earth structures has been the application of cement renders. Cement produces strong bonds which are not compatible with clay strengths. This sets up a series of problems, with failures between surfaces leading to moisture ingress and damage that is hidden by the cement skin. Vapour permeability is also affected, with condensation risks inside the building at junctions with the render, leading to failure and mould.

Clay-bound materials are typically long lasting, if well detailed and maintained. A high percentage of earth-based buildings in the UK are more than a hundred years old. Clay and aggregates do not burn or rot and, as components of subsoil, have little or no organic content. Added fibres and bio-aggregates held in a clay matrix will not rot, and the clay prevents fire spread and holds humidity at a stable level, preventing degradation.

END OF LIFE

Earth is the pre-eminent circular material, reusable as components or recyclable as a material. Because no heat or chemical changes occur, it remains essentially virgin, requiring only extraction, possible screening and rewetting to be reformed.

Figure 8.4 Tribeca: The Apex Building, King's Cross, London, Bennetts Associates, due 2024. Earth blocks manufactured from excavated site subsoil have been used in non-structural basement walls, with a concrete block plinth of two courses to elevate the earth blocks in case of flooding.

Dismantling earth is typically both safe and easy. Provided cement and admixtures such as short plastic fibres have been avoided, then all the components are reusable. They can be screened or added to other fibres, bio-aggregates, aggregates or structural frameworks. The raw materials are unchanged through use and can be safely disposed of or stockpiled for later reuse without risk to people or planet.

Earth blocks are often the simplest thing to dismantle and reuse without too much processing, typically cleaning off the old mortar and plaster. Other wall types, such as daubs from framed structures, and mass and plaster mixes, just need rewetting before reuse, while rammed mixes require more breaking up before wetting for reuse. This is due to the relatively low moisture content of rammed earth compared to the other types.

DESIGN FOR LOWER IMPACTS

Designing with earth for lower impacts involves a number of considerations.

- **Sourcing materials on or close to site** This means site investigation, soil testing and interpretation. Expertise can be greatly enhanced by drawing on the knowledge of practitioners, local or otherwise, heritage builders and, where available, factories and other processing sites.
- **Design of the building in relation to the available materials** Selection of the most appropriate building type or technology for a particular site is crucial. This includes consulting with local practitioners where they exist, and undertaking test structures where possible and relevant.

- **Specification and processing of the mix** Again, local practitioners should be consulted and their time used in the development of materials in relation to the site and specific design.
- **Possible additions of pre-existing waste from site or nearby** Brownfield sites can yield all sorts of useful materials. Sands and gravels mix well with clay and, unlike cement and lime, are not adversely affected by the presence of clay in a mix.

WIDER SUSTAINABILITY IMPACTS

Earth is an extremely flexible medium, which can be combined with many different admixtures, fibres, aggregates and quantities of water. For instance, fibres can bring enzyme reactions, which improve the workability and durability of building and finishing materials. When mixed with straw, a light earth mix will lock away 100 to 150kgCO_2e per cubic metre.

Figure 8.5 Aria, Cornebarrieu, France, Philippe Madec, 2017. A community and media centre which pioneered the use of load-bearing compressed earth blocks.

QUESTIONS
FOR PROJECT TEAMS AND SUPPLIERS

☐ Has a site investigation been undertaken?

☐ Has the material been assessed for the building technique envisaged?

☐ Is there a closer source of supply for any materials not sourced from site?

☐ Has the material been tested for suitability to the building type?

☐ Have test results (BS, Eurocode etc) been discussed between the engineer and practitioner?

☐ Is the design adequate for the environmental exposure?

☐ Is the environmental exposure appropriate for the product and the intended design life?

☐ Are the materials at risk to achieve a particular aesthetic?

☐ Will the other insulations used for roofs and floor match the sustainability criteria of the earth walls and be both natural and breathable?

☐ Has a complete cost comparison been undertaken between the cost of using excavated site materials versus alternative types of construction (including dumping of excavated material and any applicable landfill tax)?

By concentrating on natural and low-carbon solutions, many agricultural 'wastes' can be considered; there is much scope for experimentation. Locality and availability should be considered.

A major potential sustainability impact with earth building is its combination with cement. Cement as a finish or as an admixture is unsuited to clay-bound materials and should be avoided. Cement produces a hard chemical set while clay never does, continuing to minutely shrink and expand through time. The chemistry of clay means that it attacks cement over time. Crushed construction waste can be used as an admixture where cement-bound materials have already undergone a chemical reaction, and can be screened out at deconstruction.

Clay also manages humidity in a unique fashion, adsorbing and desorbing humidity from the air faster and more effectively than any other material. Just 12mm of clay plaster can hold internal humidity at a steady 50% in common scenarios. Humidity control is inherent in the material itself, as is its ability to sequester volatile organic compounds, including formaldehyde.[3]

FUTURE TRENDS

Building with earth offers many benefits and many as yet unexplored opportunities. Its durable and ubiquitous heritage was largely forgotten in construction during the 20th century. Yet earth and clay-bound materials have much to offer, which can take design into new realms.

Earth building has so many different techniques and applications in both structural and surface finishing that predicting future trends is not easy. Mass earth, a technique common in the UK and France, is being researched to align with current energy regulations, combining with light earth in a layered system build.[3] Similar approaches with other building techniques could be imagined.

Retrofit is also well suited to earthen construction. Here earth can play an enormous role in developing a language which is yet to be fully articulated in modifying existing buildings with sound- and fireproof materials, while improving air quality.

The use of light earth to retrofit existing mass earth stock is a viable approach. Greater use of smart finishes using earth plasters to manage indoor environments should be a growth sector for both newbuild and retrofit. This is foreseeable due to easy wins in materials management and skills development.

There is also a push towards prefabrication, in blocks, rammed earth and structural insulation panels (SIPs) of all kinds. While these are promising, care is required to avoid expending enormous energy to carry heavy mass elements long distances at great cost.

New approaches include wrapping earthen structures inside an insulating layer to meet building regulations with a body of mass to manage thermal and humidity fluctuations. Any earth building type can be put inside a thermal envelope, ensuring it is kept dry and without thermal bridges. Many building materials are shielded from the elements by the external envelope, and earth is no different. The possibilities of form-making are endless.

The combination of carbon store materials, fibres, aggregates and timber with earth is a massive area for research, application and growth. Using low-emission soils together with materials that lock up carbon is a double win.

Going forward, we need a combination of technical and material solutions, based on local circumstances. Design based on local climate, local materials and local culture is the way forward.

KEY TAKEAWAYS

- Matching the construction technique to the available soil on a site is a crucial first step in building with earth.

- Different types of excavated site materials should be stockpiled separately so they can be remixed in specific quantities.

- The addition of cement should be avoided because it fundamentally alters the circularity of earth construction.

GLASS

Graham Coult and Stéphany Le Rhun

The amount of float glass produced in the UK annually

Around **750,000** tonnes[1]

Energy consumption of the glass industry in the UK

0.5% of total UK energy consumption[3]

Global float glass output annually

Over **76 million** tonnes[2]

Embodied carbon

3,600 $kgCO_2e/m^3$ [4]

Savings from recycling one tonne of glass

300 $kgCO_2e$

1.2 tonnes of raw materials, including **700kg** of sand, and **30%** of energy[5]

Glass is one of the few solid materials that is resistant, durable, transparent and colour-neutral, capable of providing daylight and views to the outside. Raw glass is usually modified through heat treatments, coatings, laminating and decorative inks and used in composite products. The permutations are almost endless. Glass can be used in a multitude of applications, from transparent or opaque envelopes to partitions, guardrails, doors or furniture. In recent decades, the opulence of energy and increasingly high-performance façade technologies have made it possible to provide ideal comfort conditions in fully glazed buildings. Buildings across the globe have adopted the 'all-glass' model as synonymous with modernity and wealth. Today, in the face of climate crisis, depletion of natural resources and energy sobriety, the proliferation of glass boxes must be questioned and curtailed. When a new building is deemed necessary, the challenge now is to put 'the right material in the right place' in a rational quantity. Two key design approaches – greater opacity and diversity of low-carbon materials in building façades – can greatly increase performance. The challenge is also to reduce the embodied carbon of the glass itself, through clever design and more virtuous production.

APPLICATIONS

- External windows
- Internal partitions
- Roof lights
- Finishes

PROS

- Transparency: allows daylight in and views out
- Energy: can provide high thermal, light and solar performance
- Aesthetics: offers versatility of treatments and customisation
- Safety: can provide high safety protection (guardrail, fire and anti-burglary protection)
- Size: available in custom format, up to very large panels

CONS

- Carbon footprint: high due to high temperatures in production
- Energy consumption: excessive glazing in façades can lead to overheating and high cooling loads
- Service life: limited for composite glazing (e.g. insulated glass units) due to reduced lifespan of secondary components
- End of life: disassembly difficult for composite glazing
- Recycling: difficult due to high contamination risk

Glass has a rich history going back to 2500 BC with the creation of amulets and beads. In buildings, Roman glass has been dated back to the first and second century AD. The oldest glass in Britain is considered to be Canterbury Cathedral's 12th-century stained-glass windows. The modern method of the 'float' process, developed in the 1950s, is the predominant process for flat glass production around the world today. It derives its name from the molten glass that flows or 'floats' over a molten tin bath to create two smooth and flat surfaces.

Glass is an amorphous solid – the atoms and molecules form a random network. It is inert and hardwearing, so the longevity of glazing is usually limited by secondary materials, such as spacer bars in insulating glass units (IGU) or coating degradation. **It is infinitely recyclable if it can be readily returned to the supply chain.**

EMBODIED CARBON

According to the Inventory of Carbon and Energy database (version 3.0), the global warming potential (GWP) of 'Glass, general' is 1.44kgCO_2e/kg (i.e. 3,600kgCO_2e/m³) for life-cycle assessment modules A1 to A3.[9] Its GWP increases according to its processing and manufacturing. For example, 'Glass, toughened' has a GWP of 1.67kgCO_2e/kg.

The glass industry in the UK accounted for 0.5% of total UK energy consumption and 0.4 to 0.6% of total UK greenhouse gas (GHG) emissions in 2019.[10]

The UK has three float glass plants producing around 750,000 tonnes annually. The EU produces around 10 million tonnes,[11] compared with a total global output of over 76 million tonnes.[12] The main markets are: 80% buildings, 15% transport, 5% other.

EXTRACTION AND MANUFACTURE

Float glass is produced in a carbon-intensive process that is generally responsible for about 90 to 95% of a manufacturer's direct CO_2e emissions.[13] Flat glass is produced in a float line, where a mixture of raw materials is melted in a furnace and refined before floating on a liquid tin bath, where it is given its correct thickness and flatness. It is then cooled, controlled, cut into standard sheets and stacked for storage. Float production is a continuous process

Glazing units An insulating glass unit is two or more panes of glass spaced apart and sealed in a factory with dry air in the unit cavity. The air may then be flushed out and replaced with a range of other gases to improve thermal or acoustic performance. The most common gas is argon.

IGU An acronym for 'insulated glass unit', which refers to windows with multiple panes of glass.

Post-consumer cullet This is post-consumer recycled (PCR) glass, retrieved from waste collection services.

Pre-consumer cullet Pre-consumer waste flat glass is made of cut-offs, losses during laminating, bending and other processing, including the manufacture of insulating glass units or automotive windscreens. This processing could be at the same facility as the production furnace but is separate to the flat glass manufacturing process.

Unitised curtain wall Unitised systems comprise narrow, storey-height units of steel or aluminium framework, glazing and panels preassembled under controlled, factory conditions. Mechanical handling is required to position, align and fix units onto pre-positioned brackets attached to the concrete floor slab or the structural frame.

Table 9.1 Whole-life overview: Distribution of CO2e in LCA stages – current snapshot for a typical double-glazed unit over its service life of 30 years.[14]

that can't be stopped because the materials (glass and tin) need to remain liquid during the first part of the process.

Globally, the combustion in the furnaces relies primarily on fossil fuels – with temperatures varying from 1,300 and 1,500°C.[15]

The raw materials are silica sand (SiO_2), soda ash (Na_2CO_3), calcium carbonate ($CaCO_3$) and dolomite ($MgCO_3$) and should be sourced as close to the factory as possible, preferably within the country. The raw materials are heated in the batch, decarbonation occurs and CO_2 is released in the air, a process responsible for up to a quarter of the total CO_2e emissions of the float process.[16] These emissions can be avoided by replacing the raw materials with cullet, recycled glass produced from production-line waste or from existing buildings, which has already undergone decarbonation.

Cullet also melts faster than raw materials, resulting in up to a 30% saving in energy consumption and related emissions.[17] It is therefore urgent to develop the cullet stream at an industrial and national scale. As designers, we can support this by ensuring the collection and recycling of glass from existing buildings, particularly during deconstruction before refurbishment.

Manufacturers can also reduce their emissions by using cleaner energy – with electrification or hydrogen.

LCA stage over service life of a product (30 years)		Typical proportion of whole-life CO₂e
A1 to A3	Cradle to factory gate	91%
A4 to A5	Transport and construction	8%
B	In use	Negligible
C	End of life	1%
D	Subsequent benefits and loads	Varies

Figure 9.1 Zayed Centre for Research into Rare Disease in Children, London, Stanton Williams, 2019. The building uses passive measures such as massing, varying the extent of glazing on different elevations and shading to maximise daylight penetration and provide a high-quality internal work environment while controlling solar gain. At street level, a fully glazed façade provides daylight to the double-height underground laboratory. Above, glazing is positioned where needed in framed openings balanced with opaque brick walls. Elsewhere, continuous curtain walls are protected by deep passive terracotta fins.

PRODUCTION PROCESSES

Glass is produced in a continuous ribbon, with adjustments to the process increasing or decreasing the final thickness. The ribbon is cut into lengths and then further processed within the same plant or delivered to glass processors.

GLASS COMPOSITION
Standard or low-iron

Glass as a raw material can be broadly distinguished between 'standard' glass and 'low-iron' glass. The iron content in the silica raw material produces a green tint in glass, and this has gradually decreased over recent decades as the market has sought clearer glass. Low-iron glass is mainly used for its aesthetic neutrality. Flat glass manufacturers (FGM) select particular silica deposits with a low-iron content, which produces a product with reduced levels of colour.

Body-tinted glass

Body-tinted glass is mainly used for decoration purposes today. The addition of metal oxides to the mix produces glass with a variety of colours. This does not affect the basic properties apart from light/energy transmission. This has been a way of controlling heat gain, but has waned with the increased application of coatings. Coloured interlayers are used to a similar purpose.

Figure 9.2 Process of producing float glass from raw material in a float line factory (above). Additional process to toughen glass through tempering (below).

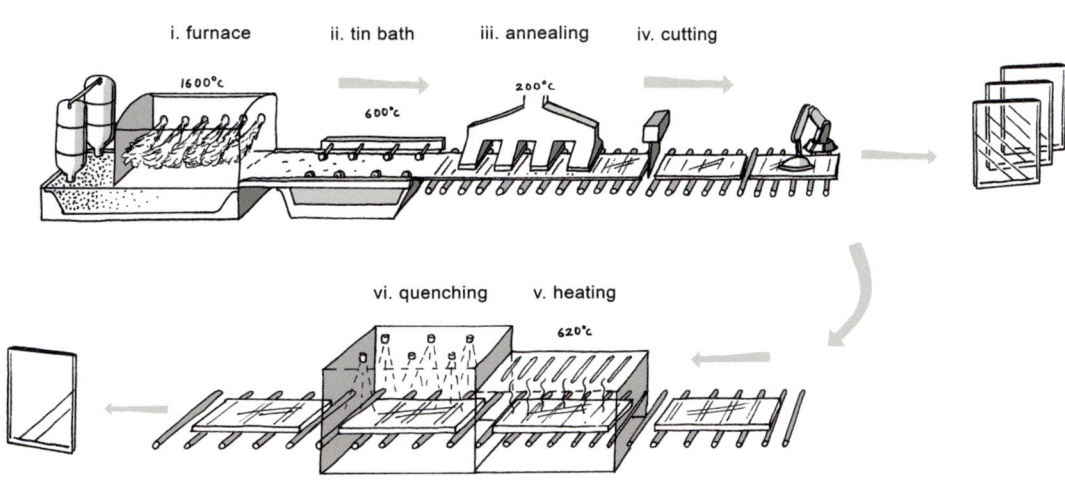

FLOAT GLASS

i. furnace   ii. tin bath   iii. annealing   iv. cutting

1600°c   600°c   200°c

vi. quenching   v. heating

620°c

TEMPERED GLASS

Figure 9.3 Applications of different types of safety glass and their respective breakage patterns when damaged.
FLS – fracture limit state
PFLS – post-fracture limit state
SLS – serviceability limit state
ULS – ultimate limit state

TEMPERING

Tempering is used for safety glass, and also when thermal shock breakage is a risk. It is a process of creating surface compression in glass by heating it in a tempering furnace at around 650°C and then rapidly cooling (quenching) it by jets of air. The glass surface is pulled into compression while the inner glass cools and contracts. It can be cooled quickly to achieve toughened glass (also called fully tempered) or cooled more slowly to achieve heat-strengthened glass.

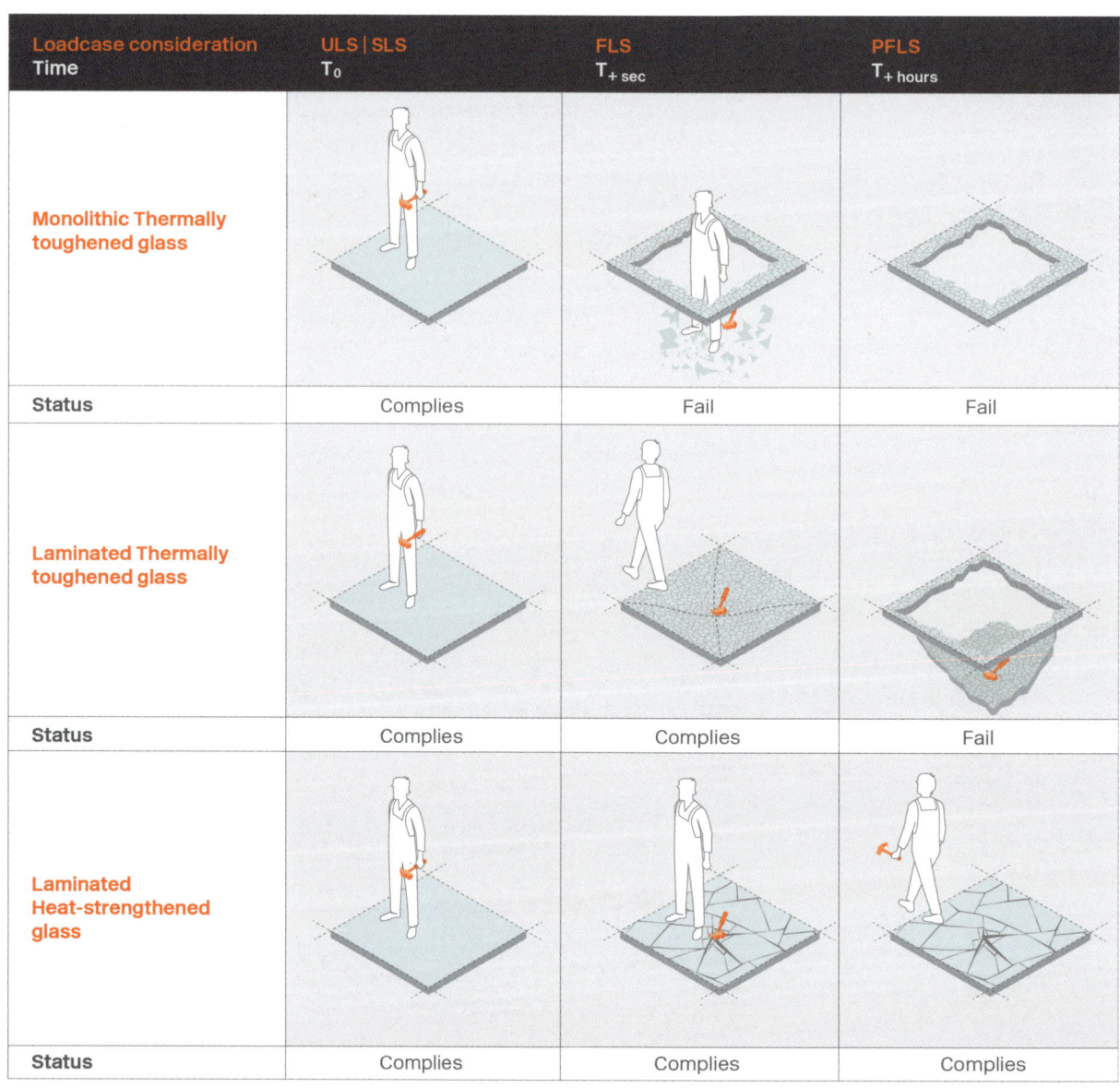

Loadcase consideration Time	ULS | SLS T_0	FLS $T_{+ sec}$	PFLS $T_{+ hours}$
Monolithic Thermally toughened glass			
Status	Complies	Fail	Fail
Laminated Thermally toughened glass			
Status	Complies	Complies	Fail
Laminated Heat-strengthened glass			
Status	Complies	Complies	Complies

Heat-strengthened glass is about twice as strong as typical annealed glass and toughened glass is about twice as strong as heat-strengthened. Heat-strengthened glass breaks into large pieces, while toughened glass breaks into much smaller, blunt dice.

Glass can be processed to achieve different shapes. Flat glass can be heated to droop over or into a form to take on a complex shape. Tempered and heat-strengthened glass can be formed into simple curved shapes while it is still hot from the furnace.

LAMINATING

Two plies of glass can be bonded together for strength and increased safety. The most common laminating material is polyvinyl butyl (PVB), but other sheet materials have been developed for increased strength and stiffness. Resins can also be poured between the individual plies to modify the performance.

At the end of its life, laminated glass is difficult to recycle because of its composite nature. In recent years, several techniques have been developed to allow delamination, for example by leaving the glass exposed to climatic conditions that allow moisture to penetrate, breaking down the bonding between the plies.

SURFACE TREATMENTS

The surface of the glass can be treated to change its performance, including energy transmission. Coatings for the management of energy transmission come in two main categories: hard and soft.

- Hard coatings (pyrolytic) are applied while glass is hot and soft so there is a strong bond at a molecular level. These coatings are tough.
- Soft coatings are applied to the glass in a magnetron. They are easily scratched and so are typically used on the cavity-facing surface of an IGU but can also be laminated. Their adaptability to toughening and curving varies.

Glass can also be fritted, where a ceramic 'ink' is applied and heated to provide permanence. The ink is composed of ultra-fine crushed glass and pigments. For simple patterns a silk screen is used, while for full coverage a roller is used. Digital printing is used for high-fidelity and non-repetitive decoration.

Polymer films can also be applied to glass to provide decorative effects, adjust energy or light transmission and to provide improved post-breakage performance. These films are easily retrofitted to existing glazing. In addition, the surface can be etched or ground using either chemicals or mechanically, creating a permanent surface.

INSULATED GLASS UNITS (IGU)

Insulated glass units (IGU) use two or more panes of glass with a gas cavity to provide thermal insulation. The most common IGUs are double-glazed units (DGUs) and triple-glazed units (TGUs). Coatings further increase thermal performance, most typically low-emissivity coatings.

An edge spacer seals the edge of the glass unit and includes a desiccant to remove moisture from the cavity. They can be aluminium, stainless steel or a thermopolymer. The simplest option for gas in the cavity is dehumidified air, but argon or krypton can be used for increased thermal performance.

Air and argon U-values (thermal transmittance) for a typical double-glazed unit with low-e coating are as follows:
Air cavity = 1.4 W/m²K; Argon/Krypton cavity = 1.1 W/m²K

Vacuum glazing is relatively rare and reduces the overall mass of gas in the cavity so that very high insulative values can be achieved. This is gaining momentum as it allows much thinner glass units to be fabricated, leading to possibilities for retrofitting historic glass without replacing the frame. The thermal performance is close to a triple-glazing unit, with an embodied carbon closer to a double-glazing unit. At the moment, it is still considered a 'special product' (AGC and Fineo manufacture these for existing/historic glazing frames).

LOGISTICS ASSOCIATED WITH TREATMENTS

The production of construction glass requires extensive logistics. From the float line factory, it is usually transported to another factory for coating, cutting and laminating. Often, it is transported again to a factory where it is assembled into an IGU, with further transport commonly required for any additional processing (special coating, frit, bending, etc).

The designer should understand the supply chain to estimate related emissions, inform project specification and streamline complex processing/manufacturing. Selecting a simple glass with standard – but efficient – coatings and treatments means simple procurement, and often lower emissions.

OPTIMISE MANUFACTURING PROCESS

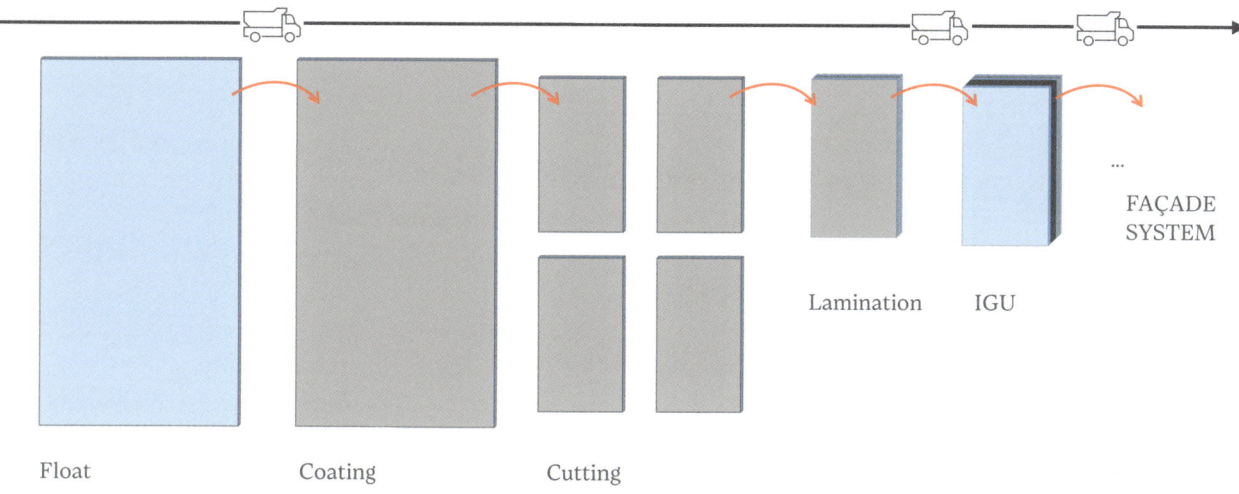

Float   Coating   Cutting   Lamination   IGU

... FAÇADE SYSTEM

Figure 9.4 Glazing production involves several treatments that generally require glass transportation. Multiplication of treatments generally means multiplication of related transportation.

FAÇADE SYSTEMS

Façade glass must provide thermal, solar, acoustic and safety performance. Glazed panels are generally inserted in a metal, PVC or timber frame as part of the façade system, such as a glazed curtain wall or window. Aluminium framing has dominated the office building sector due to its quality and durability. However, metal-framed façades have a high carbon impact compared to their bio-based counterparts, which tend to provide better thermal performance and are growing rapidly in the sector.

Glass can also be used for internal partitions and interior finishes, where its function relates primarily to a desire for transparency, aesthetics, acoustics and fire safety.

Reduction of waste in façade systems

Installation of glazing panels in a façade or partition system must be facilitated to reduce breakage and waste. Prefabrication of façade panels (unitised curtain wall, timber walls) can reduce waste, save time and improve logistics and risks associated with the construction site. Bonded and structural glass has been strongly favoured by architects in recent decades, but it is now recognised that this does not allow for easy circular design. **Façade systems should be detailed with (mechanical) fixings that allow an easy**

assembly and disassembly according to the project's Design for Manufacture and Assembly (DfMA) strategy. Prefabrication should be considered when possible.

LONGEVITY OF MATERIALS (DURABILITY, MAINTENANCE AND REPAIR)

EPDs typically consider a 30-year lifetime for IGUs, but the warranties are typically only for 10 years. Shorter periods of lifetime service can dramatically impact the overall whole-life carbon of a façade if the glass is not easily demountable.

The lifespan of primary glass is theoretically infinite. As seen above, glazing is generally a composite of glass, interlayers, a gas cavity sealed with edge spacers, and a coating. These elements are bonded to the glass for structural composite action, air and water tightness and durability. This composite nature explains why glazing assemblies have a shorter life than glass only.

For IGUs, the lifespan is dictated by when the spacer, the interlayer and/or the coating start to deteriorate and lose performance, leading to air leakage and water infiltration. As glazing units are difficult to disassemble, they generally must be dismantled and replaced entirely once they are damaged or degraded.

Despite the limited lifespan of insulating glazing, it is possible to increase durability by following best practices, such as ensuring edge protection (with the framing) against impact, water and sunlight.

The façade system in which the glazing unit is installed generally has a longer lifespan (e.g. 50 years for an aluminium curtain wall). In order to facilitate predictable replacement of glazing without damaging the integrity of the frame, the glazing should be easily removable. Mechanical fittings such as a capped system with flexible seals allow easy replacement of the glass panels. Silicone bonding makes removal more difficult, time consuming and expensive.

Interior partitions of monolithic glass should be designed to be dismantled and theoretically have an infinite lifespan, if properly cleaned and maintained. This way, they have a high potential for reuse, supporting a circular economy.

OPERATIONAL LIFE: ENERGY

Façades have a great impact on a building's operational energy use, and directly influence the sizing of mechanical systems. Glazing should be specified according to energy calculations provided by the mechanical engineers to ensure that it provides sufficient long-term performance and avoids oversizing of mechanical systems. Glazing should satisfy thermal transmittance (U-value), solar factor, light transmittance and light reflectivity requirements. Solar protection can be achieved with coatings but also with internal or external blinds, louvres and shading devices such as external fins. The balance between embodied and operational carbon should be carefully studied to provide the optimal approach. For example, in offices, triple glazing can emit more embodied carbon than operational savings.

END OF LIFE DISMANTLING, REUSE AND RECYCLING

Glass is 100% and infinitely recyclable. This quality has been largely underexploited and we now need to deploy recycling on a global scale. To achieve circularity, glass should be reused or at a minimum recycled into a material of the same quality (flat glass).

In recent best practice, only 6% of flat glass has been effectively recycled.[18] Most of this glass is downcycled (i.e. turned into lower-quality products such as glass bottles and glass wool insulation).

When flat glass is recycled back into flat glass of the same quality, it is crushed into small pieces, forming the cullet, which returns to the float line, where it is melted down with other glass-making materials. Most cullet used in today's flat glass comes from pre-consumer waste from manufacturing or processing prior to its use on site. In 2018 for Saint-Gobain, for example, less than 1% of the cullet came from post-consumer waste from renovation or demolition sites.[19] With incentives for increased circularity, this is starting to change.

Closed-loop recycling of flat glass presents the following challenges:
- New logistical and recycling streams need to be set up locally and nationally to collect glass on site.
- Non-glass components must be removed before cullet production to avoid contamination by other material, which can cause issues for float lines. Disassembly can be complicated, labour intensive and time-consuming.
- It is sometimes necessary to transport the glass intact to the off-site cullet production facility. This involves careful dismantling, which must be anticipated in terms of logistics, planning and budget at the beginning of a refurbishment project.

Closed-loop recycling has huge benefits. Recycling one tonne of glass saves up to 300kgCO$_2$e, 1.2 tonnes of raw materials, including 700kg of sand, and 30% of energy.[20]

To date, reuse of IGUs has not been widely adopted due to their rather short lifespan, rapid obsolescence, warranty requirements and lack of standardisation of sizes. Reuse of monolithic glass is more developed, especially for internal use.

For both circular approaches (reuse and recycling), the complexity lies in the fact that the façade glazing is a composite unit that is very labour intensive to dismantle. Sometimes, the glass is also bonded to other parts of the façade (window and curtain wall frames), which increases the obstacles to separate the elements and integrate them into a circular process. **To not repeat past errors, new façades should be designed as kits of parts, with demountable systems to facilitate reuse and recycling of glass.**

DESIGN FOR LOWER IMPACTS

To reduce the environmental impact of glazing in a building, its quantity needs to be reduced and its carbon intensity optimised. This can be done at different scales, as follows.

Reduce the surface of glazing at building scale

An efficient and compact building reduces the façade embodied carbon greatly. The shape of a building has a large influence on the total façade area. The more compact a building is (i.e. the smaller the form factor), the less façade area there is and therefore the less embodied carbon.

Reduce the volume of glass at system and panel scale

The embodied carbon of glass is directly proportional to its thickness, which should be kept to the minimum. The façade system supporting the glass has a large influence on the glass thickness. For example, a structural glazing with no vertical frame to support can be up to three times thicker than glass supported on all four sides with transoms and mullions.

Optimise the carbon intensity of the glass

Glass specification can greatly affect the carbon intensity:
- Glass with high recycling content has a lower carbon intensity.
- Low-iron glass has a higher carbon intensity than mid-iron glass because it requires higher temperatures during the float process.
- Thermally treated glass has a higher carbon intensity than annealed glass as it requires an extra heating of the glass.
- Cold bending has a lower carbon intensity than traditional hot bending as its forming doesn't require combustible energy.

As a general rule, any additional process to float glass requires extra energy and, occasionally, extra logistics and transportation.

Optimise panel size to reduce production waste

Offcuts and manufacture waste can be reduced if panel sizes are adapted to standard float glass sheet sizes. Glazing panels are cut from standard 3.21m by 6m float glass sheets. Designers can consider the glazed panels' dimensions to minimise losses and waste during the cutting process, see figure 9.5.

Dismantle and reuse existing components

A window is a building element that lends itself to reuse. Careful dismantling, retrofitting and reuse, either in the same building – as in Arup's One Triton Square, London, 2021 – or in another building, is a promising way forward.

WIDER SUSTAINABILITY IMPACTS

Glazing in buildings is an important contributor to health and wellbeing because it provides occupants with natural light, while also being an inert, non-toxic material.

Figure 9.5 Working with standard glass sheet sizes within the design of glazing can reduce the wastage during manufacturing.

OPTIMISE THE GLASS DIMENSIONS

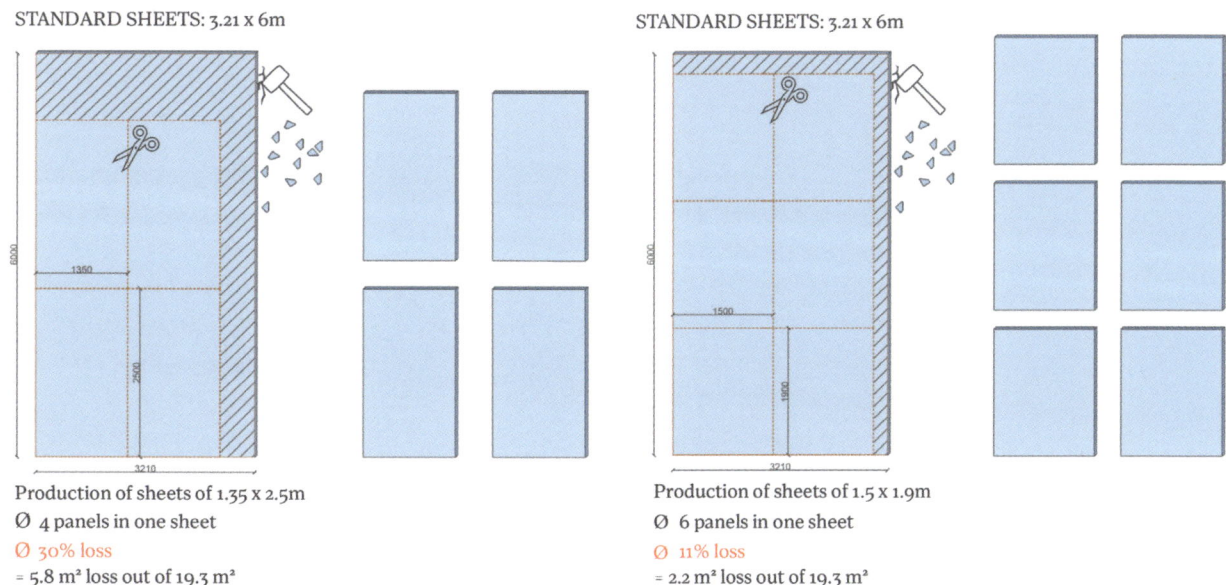

STANDARD SHEETS: 3.21 x 6m

Production of sheets of 1.35 x 2.5m
Ø 4 panels in one sheet
Ø 30% loss
= 5.8 m² loss out of 19.3 m²

STANDARD SHEETS: 3.21 x 6m

Production of sheets of 1.5 x 1.9m
Ø 6 panels in one sheet
Ø 11% loss
= 2.2 m² loss out of 19.3 m²

The extraction of the raw materials causes sustainability impacts comparable to other mineral extraction processes, including reduction in biodiversity and habitat loss. For sustainability and administrative reasons, it is currently difficult for industries to open new quarries in Europe and the UK to extract the sand required for glass manufacture. Manufacturers often use low-quality sand they have to wash, using excess energy and water, over higher-quality sand, although some are now investing in higher-quality sand to avoid this extra use of resources.

Regarding biodiversity when using glazing, it is advisable to limit the reflectivity of glass (generally to less than 15%) in façades to reduce the risk of bird collisions. Specific frit/PVB/coatings (if recyclable) can be specified to prevent this, which is even more important today with the increasing demand for vegetation which attracts birds.[21]

FUTURE TRENDS AND INNOVATIONS
Industry trends

To reduce carbon emissions from the float process, flat glass producers are investing in new, more efficient furnaces and greener energy sources, including the use of electricity instead of fossil fuels. They are also increasing the efficiency of the heating process by using electrodes that reduce glass melting temperatures.

The use of hydrogen is also being piloted. NSG Pilkingtons has carried out trials on the use of both biofuels and hydrogen with success. However, as for most heavy industries, the supply of clean hydrogen is in development, and it will be an expensive fuel, high in demand.

Greater use of cullet in glass production is also being explored. Saint-Gobain trialled its first production of 100% cullet float glass in 2022.[22]

Design and products trends

With regards to glazing products themselves, solutions such as vacuum glass or unsealed double glazing are now being developed to facilitate full recyclability of glass for reuse.

The demountability and reuse of existing glass is being studied and tested by many glass processors, who are developing new machines and techniques to separate existing glass panes from the insulating glass units and to assess their quality. Important aspects, such as qualifying the strength of the reused glass, are being studied at various universities.

Carbon reduction potential	Measure	Example	What to do now to enable it
Higher potential	Optimise the form of the building.	Optimise the compactness of the building (reduce surface of the façade/surface of the floor). Reduce the floor height.	Challenge the brief and the architectural and urban design. Optimise the coordination with MEP/structure and challenge the brief requirements.
	Reduce the window-to-wall ratio.	Allow for higher opaque ratio. Use opaque materials with low-carbon intensity (bio-based materials).	Challenge the design and choose a low-carbon cladding system (preferably using fast-growing bio-based materials).
	Reduce glass thickness.	Optimise the façade grid. Carry out detailed sizing analysis.	Dedicate time for detailed engineering analysis.
	Reduce carbon intensity of glass. Specify simple and standard glass.	Specify glass with high recycled content.	Determine resource availability and integrate in the specification.
		Select mid-iron glass over low-iron glass.	Accept a less neutral glass.
		Avoid fritting and aesthetical-only treatments.	Accept architectural impact.
		Avoid thermal toughening when possible.	Carry out engineering studies for feasibility.
		Prefer cold bending over hot bending.	Conduct engineering studies to study the feasibility.
Lower potential	Reduce glass waste.	Optimise panel size to reduce production waste.	Design with panels compatible with the standard float glass sheets.

Table 9.2 Carbon reduction potential.

User trends

Over the last few decades, designers and users have been demanding ever higher quality and transparency from glass and the industry has responded to these demands. Designers may not realise that these demands mean increased energy use.

Moves to incorporate more reused or recycled glass will impact glass quality (in terms of colour neutrality and presence of aesthetic defects). **Designers and users may need to accept a move away from 'perfect glass' in order to reduce its environmental impacts.** Acceptance of defects such as bubbles – which were the norm a few decades ago – could also lead to a reduction in glass furnace temperatures and associated emissions.

Figure 9.6 UNESCO V, Paris, Zehrfuss and Prouvé, 1970; refurbishment by Patriarche Architectes, ongoing. Existing façade before refurbishment. UNESCO V is a pilot circular refurbishment project led by the architects Patriarche, with Eckersley O'Callaghan as façade consultants. It involves reuse of the cladding and recycling of the flat glass. The glass was carefully dismantled from the existing façade and transported intact to the cullet producer Ares, where the cullet was inspected and sent to the Saint-Gobain float line to become part of its low-carbon production. The flat glass was recycled and the cladding was reused in a method audited and approved by the glass manufacturer. Thanks to glass recycling, up to 30 tonnes of CO_2e were saved.

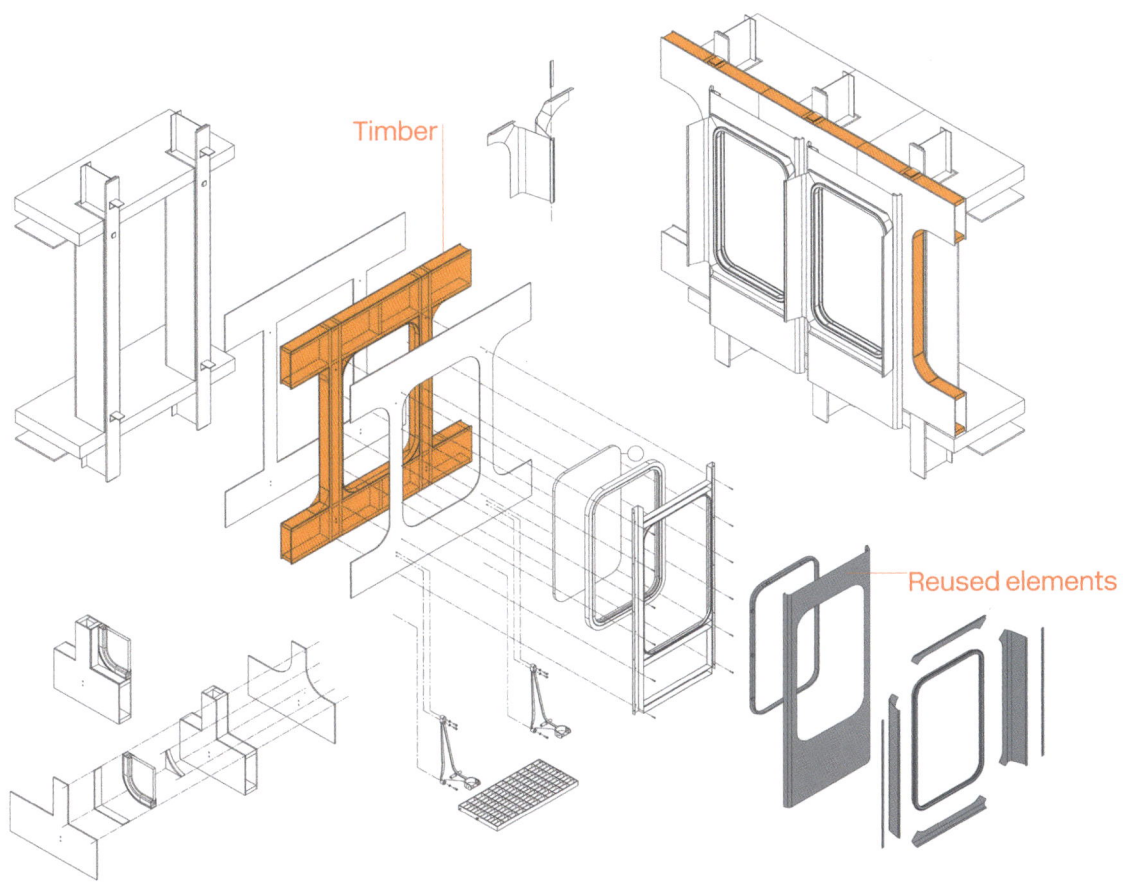

Timber

Reused elements

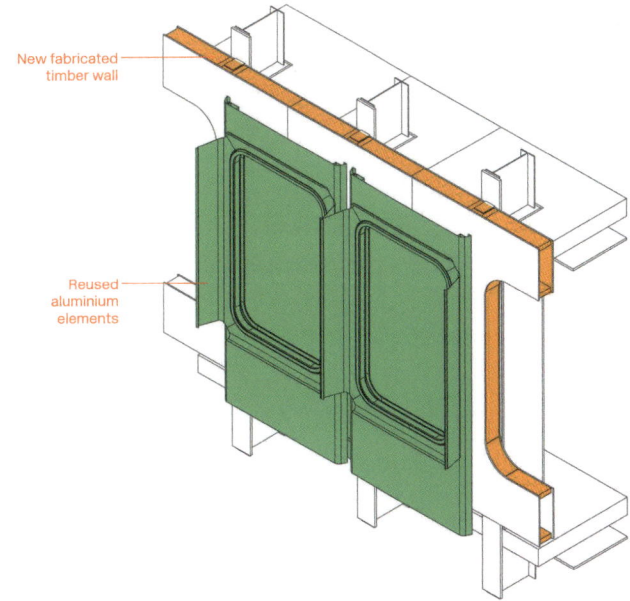

New fabricated
timber wall

Reused
aluminium
elements

Figures 9.7a and b At UNESCO V,
circular deconstruction has been
made possible thanks to the original
'kit of parts' façade, made with
components of 'human' dimensions
that can be simply handled. It has
been part of a larger low-carbon
design of the project, where the
reused materials have been installed
on top of a bio-based and high-
performing inner skin, with high-
performing glazed façades.

QUESTIONS
FOR PROJECT TEAMS
AND SUPPLIERS

☐ Is it possible to reuse the glass of the existing building we are about to deconstruct/refurbish? If not, is it possible to recycle it?

☐ Is the new glass circular (i.e. appropriate for reuse and recycling)? Is there a material passport to provide product information for the future users?

☐ Where does the new glass come from? What is its 'journey' to undergo different treatments? Is it possible to reduce its embodied carbon by changing/simplifying the specification?

☐ Is the ratio of opaque/glazed walls optimised for energy performance and embodied carbon? Are the opaque walls using low-carbon (fast-growing) bio-based materials?

☐ Are all the connections of the façade reversible to allow for easy installation, adaptation and deconstruction?

☐ Is the size of the glazing panels optimised to reduce glass thickness and reduce production waste?

In addition, designers have tried to combine all the functions of the façade within single glazing products: integrated blinds, photovoltaics, dynamic glass, etc. Although these elements can be very beneficial for the energy consumption of the building, they lead to increased embodied carbon, and less demountability, reuse and recycling potential. Simplicity and low tech are often the right choice for glass.

ENDING NOTES

We are all aware today of the considerable impact of building construction on global warming, and on environmental and biodiversity damage, and resource depletion. However, we do not yet commonly design our buildings and façades to take into account the impact they have on the environment. Yet, façades, and particularly their glazed portions, can play a considerable role in limiting the environmental impact and in adapting the building to the future climate.

As designers, our role is first to understand the impact of each of our decisions, and to try to limit the negative effects and move towards true carbon neutrality and regeneration.

For existing façades, we must first increase their lifespan by repairing and refurbishing them. When replacement becomes necessary, we must ensure that no existing resources are wasted by organising the reuse and recycling of materials and glass.

For new façades, we have to refuse the 'all-glass-box models' and to adopt a sensible glazing/opaque ratio, prioritising (fast-growing) bio-based materials for opaque walls.

We need to specify low-carbon glazing: reduced glass volume, efficient and simple glass processing and logistics, efficient sizing and handling and reduced waste.

We must also adopt a design that will allow for full longevity of the façade (timeless design, simple dismantling, best practice for durability).

We also need to anticipate the end of life of products to ensure that they are circular and the resources will be fully reused.

Finally, we have to contribute at scale to the research and development efforts to rethink glass products to ensure they become reusable components.

KEY TAKEAWAYS

- Promoting the collection of glass from existing buildings for recycling can greatly reduce the embodied carbon of new glass, as recycled glass (cullet) has already undergone decarbonation and can use lower temperatures for new glass production.

- Glass quality and aesthetics should be balanced with embodied carbon, recycling content and recyclability.

- Designers should understand the supply chain to estimate related emissions, inform project specification and streamline complex processing/manufacturing/logistics. Selecting a simple glass with standard – but efficient – coatings and treatments means simple procurement, and lower emissions.

- New façades should be designed as kits of parts, with demountable systems to facilitate reuse and recycling of glass. Façade components and connections should allow for easy installation, adaptation and disassembly by using accessible, reversible and mechanical connections and adequately scaled components.

- EPDs typically consider a 30-year lifetime for IGUs, and the glazing should be designed to maximise the lifespan, protecting the edges from damage, including knocks and high exposure.

- A great façade design using a good ratio of low-carbon opaque/glazed walls can directly influence the carbon payback time of the façade. It also has a great impact on a building's operational carbon and energy and directly influences the sizing of mechanical systems.

- Designers have tried to combine all the functions of the façade within single glazing products: integrated blinds, photovoltaics, dynamic glass, etc. Although these elements can be very beneficial for the energy consumption of the building, they can lead to increased embodied carbon, and less demountability, reuse and recycling potential. Simplicity and low tech often seem to be the right choice for glass.

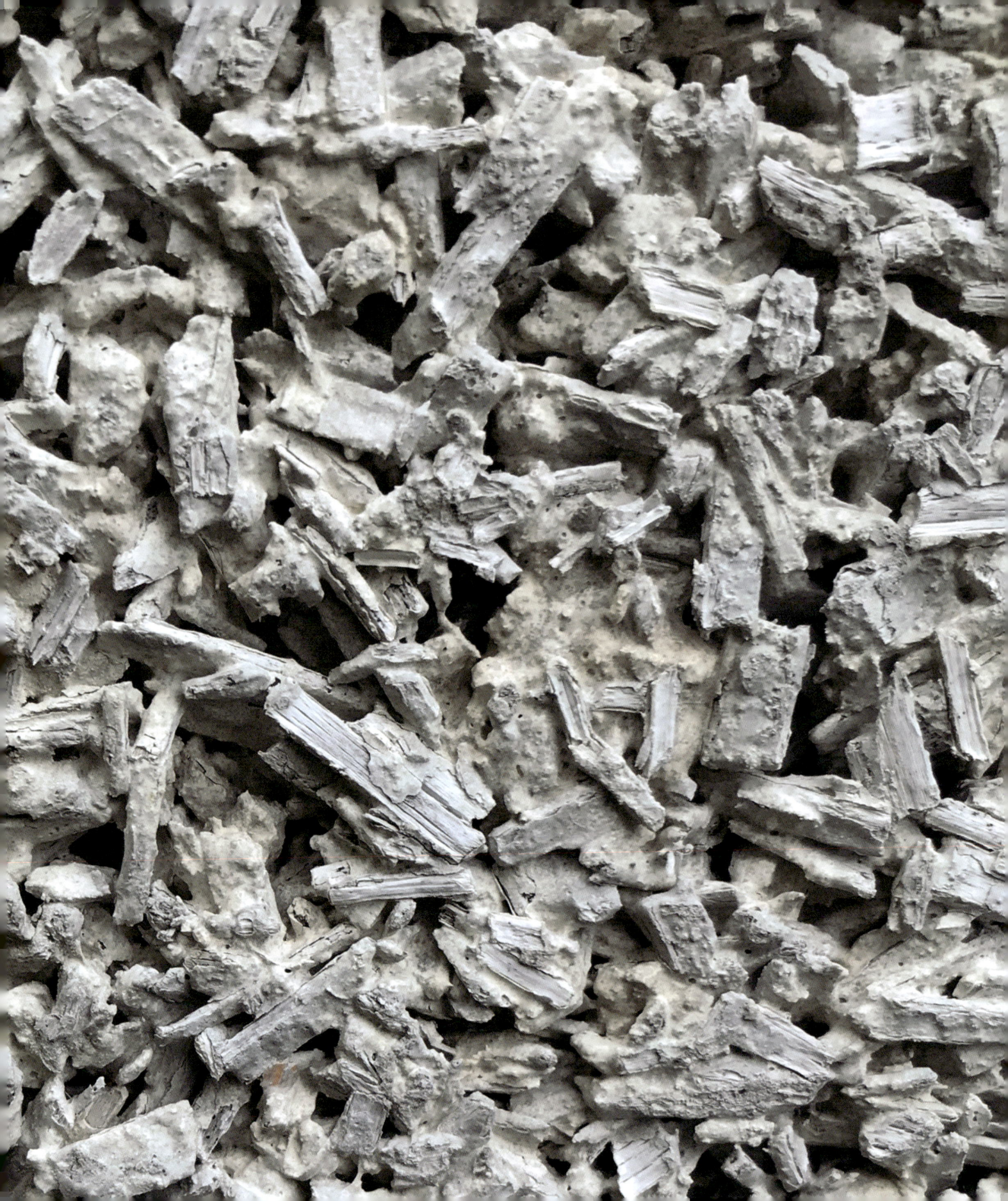

HEMPCRETE

Graham Durrant

Sequestered carbon for blocks

202 $kgCO_2e/m^3$ [1]

Density

300 kg/m^3

Embodied carbon for blocks

110–240 $kgCO_2e/m^3$ [2]

Heat capacity

1,500–1,700 J/kg

Thermal conductivity

$\lambda = 0.07$ W/mK

Hempcrete, or hemp-lime, is made from chopped-up stems (shiv) of the hemp plant bonded together with a lime-based binder. It came to the UK in 1999 having been developed some 15 years earlier in France. Used primarily in walls, hempcrete is an unusual material because it acts as the insulation, the main body of the wall and the plaster carrier all in one. It is non-structural, so in newbuild applications it is accompanied by a structural frame, usually of timber. It can also be used to insulate roofs and floors with the amount of binder adjusted to specific loads. Hempcrete can be cast into shuttering, spray applied or used in block form or prefabricated panels. It is usually plastered or rendered with clay or lime plaster, but alternatively can be covered with cladding. In retrofit, hempcrete can be applied onto the interior or the exterior of an existing wall to form a breathable, insulative layer.

Highly valued for its thermal mass, breathability, fire-resistance and acoustic properties, hempcrete has found early adopters among self-build housing enthusiasts. With the impetus of climate emergency, hempcrete is increasingly being deployed on low-rise residential buildings, as well as other building types. Hempcrete's strong eco-credentials mean that it is currently the subject of research and innovation.

APPLICATIONS

- Walls
- Insulation on existing walls
- Floor insulation
- Roof insulation

PROS

- Insulative
- Breathable (hygroscopic)
- Locks up carbon
- Good thermal inertia
- Non-toxic
- Fire-resistant
- Has ease of construction
- Good acoustic properties

CONS

- Non-load-bearing
- Not waterproof
- Requires greater wall thickness
- Requires adequate drying times
- Requires suitable temperatures during installation

GLOSSARY

GLOSSARY

Aggregate A broad category of coarse to medium-grained particulate materials used in construction, usually requiring a binder to hold the particles of varying sizes together.

Binder Lime-based binder acts as a glue to bind/hold the individual hemp shiv particles together to form hempcrete.

Shiv (or hurds) The broken-up stems of the hemp plant, usually once the fibre has been removed.

INTRODUCTION

In 2021, just 800 hectares were used for hemp cultivation in the UK, primarily in Yorkshire, compared to 17,000 hectares in France, Europe's largest hemp producer. The University of York's HEMP-30 project is working with farmers and other interested parties to increase the area of hemp cultivation in the UK by 2030.[3]

Industrial hemp has a rapid growth rate of up to 4m in 100 days, with a significantly high carbon sequestration level, locking up 1.5–2.1kgCO_2e/kg. It needs moderate levels of water, little fertiliser and virtually no pesticides or herbicides, and can grow in a wide variety of climates. Cultivation requires low energy use, with carbon emissions of 0.09–0.19 kgCO_2e/kg.

The firing of lime, using fossil fuels, emits carbon at 0.42–0.80kgCO_2e/kg, and another 0.70kgCO_2e/kg is released during manufacture by the chemical process of calcination. This carbon footprint is partly offset by carbonation, as lime reabsorbs CO_2 when it cures in the building, with about 55% of the CO_2 emitted during calcination being absorbed.

Due to the sequestration of carbon by the hemp plant and the carbonation of the lime binder, **the calculated embodied carbon value for hempcrete is negative**, between -0.30 and -1.00 kgCO_2e/kg, making it one of the few carbon-negative building materials.[4]

MANUFACTURE

Hempcrete for hand casting or spraying is manufactured on site by the installers, using hemp shiv and binder from different suppliers mixed with water in a drum mixer, pan mixer or spray machine. It is comparable to the manufacture of mortar for brickwork where aggregate, binder and water are mixed on site and installed without delay. A typical hempcrete mix uses a cubic metre of hemp, weighing approximately 110kg, plus 180–200kg of lime binder. This will produce a cubic metre of hempcrete, weighing approximately 310kg once dry.

Hempcrete blocks and panels are manufactured off site in a more standard manufacturing process, using largely similar ratios of hemp and lime. Hempcrete blocks tend to be approximately 500mm long by 200mm high and are available in a variety of thicknesses, ranging from 80 to 300mm.

Figure 10.1 A visualisation of Velindre Cancer Centre, Cardiff, White Arkitekter, planning approval gained in 2023. The 36,000m² clinic will have a hybrid concrete, glulam and CLT structure with 400mm hempcrete blocks behind a rain screen cladding. Through early engagement with the supply chain, the project aims to develop local Welsh supply chains for hemp without compromising local food crops.

MATERIAL SOURCING

The hemp shiv used in hempcrete is a by-product from the processing of industrial hemp for seed and fibre. Woody hemp stems are dried and chopped into varied lengths up to 20mm long, any fibre and excess dust are removed, and it is packaged into bales.

Global hemp cultivation is increasing rapidly, as markets expand for hemp products. Seeds are used to produce oil, foods, animal feed, biofuel and cosmetics. Hemp textiles provide a durable, more sustainable alternative to cotton and synthetic fabrics. In the UK, hemp is grown and processed for fibre to make natural mattresses and insulation batts for building.

Industrial hemp is a variety of *Cannabis sativa* with extremely low levels of tetrahydrocannabinol, the psychoactive compound found in cannabis. A consequence of the association with cannabis is that the cultivation of industrial hemp is strictly regulated. Licences are required to grow it, adding expense and extra complication for farmers. An additional barrier to the development of hemp production is the considerable investment required to establish processing facilities.

Figure 10.2 Flat House, Margent Farm, Cambridgeshire, Practice Architecture, 2020. Flat House is comprised of bespoke 4ft by 8ft structural panels, made from UK-grown and manufactured I-joists infilled with hempcrete. Clad on the exterior with bespoke hempcrete bio-resin panels, the hempcrete is left unplastered internally and finished with a light coat of clay paint.

The formulated binders used for hempcrete are made with either hydrated or hydraulic lime, with additives to improve performance, often including cement. They are supplied in bags, ready to mix on site. Lime and the other ingredients in these binders are in adequate supply globally, and lime binders for hempcrete are now being produced in the UK, using British lime.

APPLICATIONS

Hempcrete is an incredibly versatile, breathable, low-energy walling material in newbuild applications, and acts as an insulation layer and plaster carrier in renovations. For newbuild projects, hempcrete can be used in its simplest form, encasing a structural timber frame, lime-plastered on the inside, and lime-rendered on the outside.

Hempcrete can be applied over a wide range of substrates in renovation, insulating stone or brick walls, or replacing wattle and daub in timber-framed buildings. It is particularly useful for renovation when sprayed, as it can cover a wide variety of existing building materials and transform a building back into a coherent structure without losing the original character. It is quite a sculptural material, so curves and soft undulations are easily achieved, whether sprayed or hand-cast into formwork.

CONSTRUCTION METHODS

On site, hand-cast hempcrete is placed into formwork and the shutters are removed after a few hours. Sprayed hempcrete is applied onto a temporary backing board for newbuild or onto an existing wall in renovations.

From a designer's perspective, the timber-framing requirements of a hempcrete building are fairly standard. The frame sits in the middle of the wall when a rendered finish is desired. When cladding is required, the timber frame is set on the outside to hold the cladding and a ventilated cavity is left between the cladding and the hempcrete. Racking strength is necessary for both wall designs, and diagonal bracing should be used in preference to sheet materials. It is important to avoid the use of sheet materials which would introduce a barrier to vapour permeability.

Hempcrete should ideally start above the splashline, 250mm from exterior finished ground level, protecting it from rainwater splashing back up onto it, and avoiding the more humid microclimate of ground-level vegetation. Hempcrete can get wet as long as it completely dries out again but because hemp is carbon-based, permanent moisture would cause the hemp to break down.

Hempcrete drying times need to be carefully considered, but can be dealt with by employing strategies such as installing dry hempcrete blocks on the inside of the wall, with sprayed or hand-cast hempcrete on the outside. The advantage of incorporating the blocks is that the internal skin of the wall is installed dry, reducing humidity in the building and cutting overall drying times. The wet-cast or sprayed hempcrete is easily installed around the frame from the outside, and dries to the exterior, creating a monolithic, solid, gap-free wall.

Figure 10.3 Hempcrete being sprayed onto a flint-and-brick wall during a renovation project.

FIRE AND WATER DAMAGE

The greatest threat to most buildings is water damage, but hempcrete is a durable, problem-free building material that does not trap water for long enough to do harm if it is correctly detailed and installed. If damage does occur, hempcrete is easily repaired by simply cutting out the damaged area and installing new hempcrete.

Hempcrete has excellent fire resistance and will only char a little on the surface, even when a blow torch is held directly onto it. It has a

Figure 10.4 Paris Habitat, Butte Montmartre, Paris, Barrault Pressacco Architectes, 2020. The façade of this social housing project is comprised of a timber framework infilled with sprayed hempcrete and finished with an exterior coat of lime render.

fire rating of 1 hour (BS EN 1365-1:1999) although it often performs way beyond this rating. There have been two fires in hempcrete buildings in the UK and in both cases the hempcrete remained intact while other building elements were destroyed.

In timber-framed buildings, the hempcrete is more absorbent than the timber, so it draws moisture away from the frame and expels it, avoiding damage to both the frame and the hempcrete. This eliminates expensive and intrusive repairs and pushes end of life for the building much further into the future.

END OF LIFE

The subject of the longevity of hempcrete is often raised but with good design and detailing there is no reason why hempcrete should not last as long as the timber frame it encases. Hemp and lime were used in the Ellora Caves in India 1,500 years ago, and original samples look identical under an electron microscope to modern-day hemp and lime.

Hempcrete is not recyclable to any great degree, but it can be removed from a building at the end of its useful life and left exposed to the elements to compost.

It is very relevant to ask what the end of life for building materials would look like for less well-built buildings, containing synthetic or toxic building products. However, for bio-based, energy-efficient buildings built to high standards, it is less clear to see a rationale for their demolition. If we are seriously considering embodied carbon, we should ensure that new buildings are valued and able to be adapted rather than demolished, so that the demolition-and-rebuild cycle becomes a much more limited option in future.

WIDER SUSTAINABILITY IMPACTS

If tasked with inventing a building material to meet our needs in the face of climate change, we would almost certainly develop one we could grow. Bio-based materials incorporated into our buildings act as carbon sinks for the lifetime of the building, locking up carbon rather than emitting it. **Hempcrete locks up to 200kg of carbon per cubic metre in the building.** Conventional insulation materials emit carbon now, to save energy in the future, but bio-based materials can sequester carbon now and save energy throughout their lifetime. With the current climate crisis, the

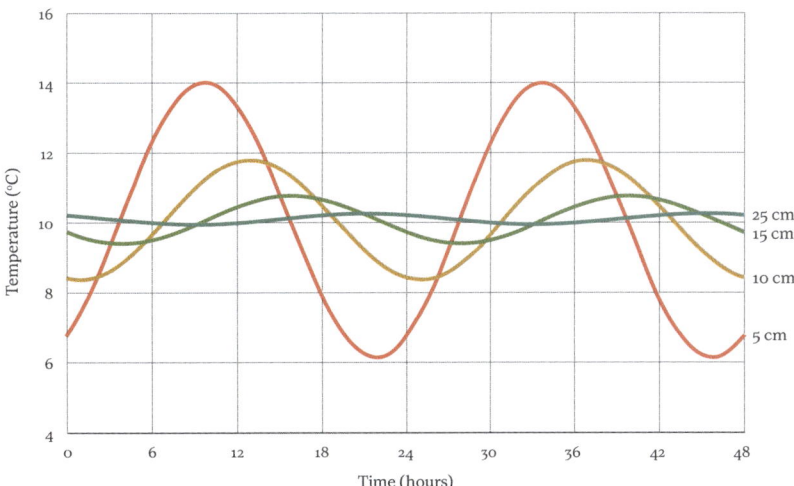

Figure 10.5 Graph showing how increasing depths of hempcrete reduce daily temperature fluctuations.[5]

QUESTIONS
FOR PROJECT TEAMS
AND SUPPLIERS

☐ Does the design allow the breathable hempcrete to breathe on at least one side?

☐ Has an experienced hempcrete contractor been consulted about the wall build-up?

☐ For a renovation project, will the hempcrete stick to the existing substrate, or will there need to be additional mechanical fixings?

☐ Will the hempcrete be 250mm above finished ground level on the outside?

☐ Are the expected drying times realistic and what strategies can be employed to make sequencing run smoothly?

☐ Has the lime binder that is being used been thoroughly tested?

☐ Is the plaster to be used adequately breathable?

☐ How experienced is the hempcrete builder and is it possible to talk to past clients?

Intergovernmental Panel on Climate Change (IPCC) suggests that the carbon we emit over the next decade is much more important than the carbon we might emit 30 years from now, so the choice of building materials has never been so important.[6]

USING HEMPCRETE TO REDUCE OPERATIONAL ENERGY

In most conventional buildings, insulation is placed near internal surfaces of external walls with the primary function of reducing heat flow through the wall. The problem with this strategy is that the fabric of the building envelope cannot warm up, so the only things in the building that are warm are the air and wall lining. This lack of thermal mass leaves buildings vulnerable to large daily temperature swings, exacerbated by air leakage and ventilation needs.

In hempcrete buildings, the hempcrete *is* the fabric of the building. Alongside a lambda value (a measure of thermal conductivity used for insulation) of 0.07 W/mK, it has the property of thermal inertia, holding onto heat in winter and keeping cool in summer. This allows buildings to be adequately ventilated without losing large quantities of heat. Hempcrete buildings are monolithic in nature with no voids or breaks in the insulation layer so are less likely to suffer from draughts.

Breathable materials like hempcrete help maintain lower relative humidity in a room by absorbing and desorbing moisture, while allowing some vapour to pass through the walls to the outside. As the human body is very sensitive to humidity, we easily feel cold in damp conditions, but will feel comfortable at reduced temperatures with lower levels of relative humidity. This comfort at lower temperatures saves money and energy on heating, resulting in lower carbon emissions.

FUTURE TRENDS

The UK is considered to be at the forefront of hempcrete building in the English-speaking world, while France probably has the most developed industry. Australia has some established contractors now, and in America the hempcrete industry is expanding at speed. Notable newbuild exemplars in the UK include Mikhail Riches' Clay Field (2009) in Suffolk, Glenn Howells Architects' The Triangle (2011) in Swindon for Kevin McCloud's HAB Housing, as well as Aukett Fitzroy Robinson's M&S Cheshire Oaks (2013). At the time of writing, it is estimated that between 1,500 and 2,500 buildings have been built or renovated with hempcrete in the UK, though it is difficult to give precise figures because, unlike France, the UK currently has no hempcrete trade body. France has had a code of practice (*règles professionnelles*) for hemp construction in place for over a decade.

Controlling the hempcrete installation process is one key to its successful expansion. The current shortage of competent specifiers and contractors means there is limited access to the material. This problem should be relatively easily solved as increased demand instigates investment in training.

There appears to be a requirement for new materials to be perfect, without drawbacks, even if they are less damaging than the materials they replace. For instance, working in low temperatures can cause difficulties when using lime-based materials so certain trades are limited by colder weather. The construction industry values the convenience of being able to work in low temperatures above the need to build with materials which help combat climate change. If the construction industry continues with its current set of priorities, there will be no need to work in cold conditions as the planet will be hot enough for lime work all year round!

The proliferation of hempcrete buildings with comfortable interior environments and low energy use is proof of concept, but the expansion of hempcrete use will depend on industry willingness to innovate in the face of the climate emergency.

KEY TAKEAWAYS

- Hempcrete locks up to 200kg of carbon per cubic metre in the building.

- Hempcrete is suitable for newbuild or renovation over a variety of substrates.

- Hempcrete is breathable and moderates relative humidity in buildings, keeping buildings dry.

- Hempcrete is versatile and easy to work with; a site operative can learn to install it in a couple of days.

- Hempcrete cannot trap moisture, which is the most prevalent cause of deterioration in buildings.

- Hempcrete enables the fabric of the building to hold onto heat, levelling out daily temperature fluctuations and saving energy.

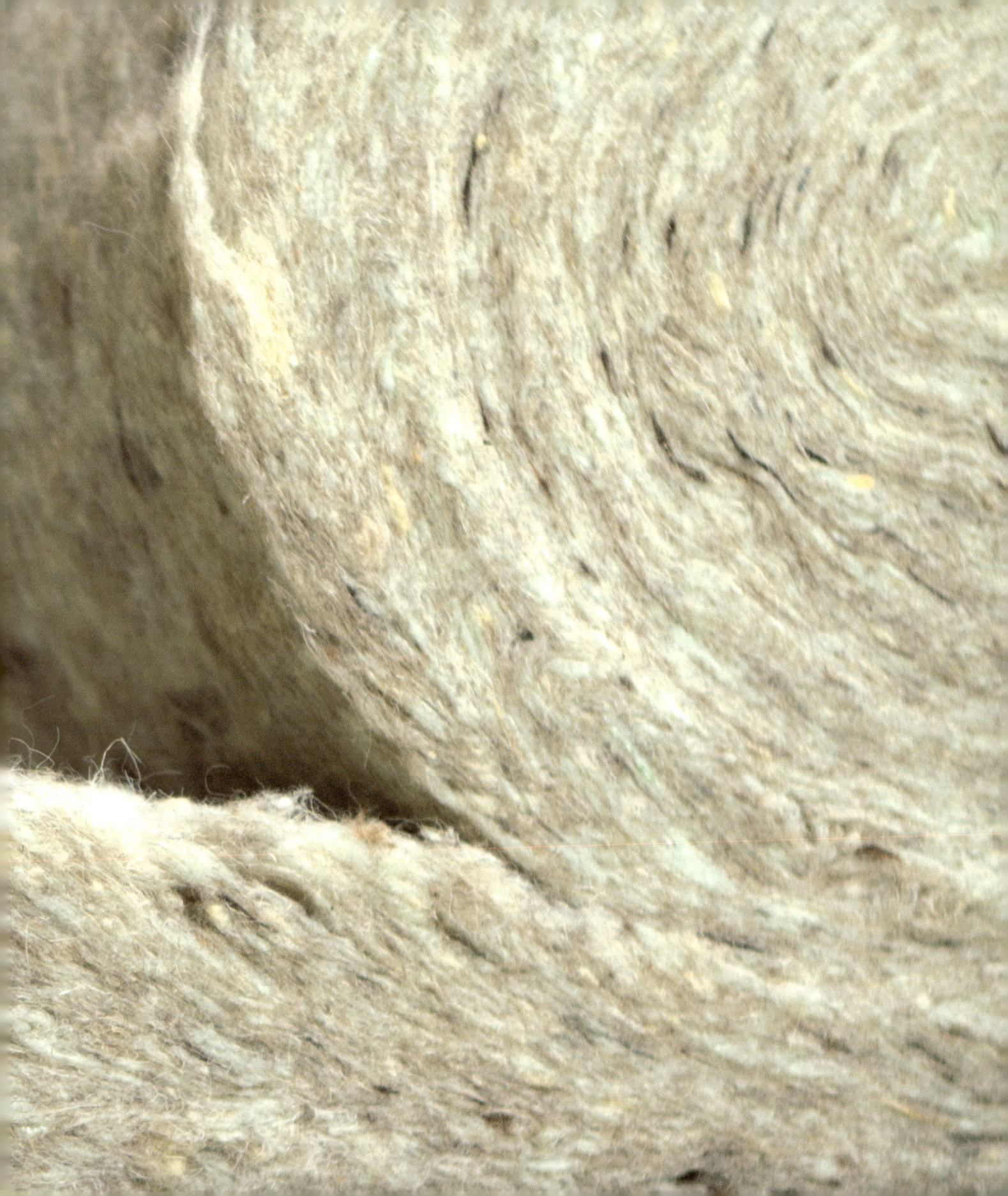

INSULATION

Mark Lynn

The amount that building emissions need to fall by 2035 (relative to 2022)

43%[1]

Natural fiber insulation market share Germany and France

> 10%

UK **< 1%**[2]

UK market share of insulation derived from fossil fuels

63%[3]

At first glance, insulation is a simple concept: we install materials that reduce heat loss, saving energy and money. Scratch beneath the surface, however, and there is much more to consider. Aside from heat loss, we need to understand other performance aspects and how these influence what type and how much insulation is most appropriate to the needs of the building.

The amount and type of insulation, and the way it is installed, impacts energy savings, heat gain, damp and humidity, fire risk and indoor air quality, and contributes to the embodied and operational environmental impacts of the building. It is important to know what types of insulation are available and to understand the relative merits of each.

Different types of insulation all have benefits depending on the context in which they are used. Insulation types include mineral fibre quilt, which is often considered when fire risk is high, petrochemical foams, which ensure lower heat loss with thinner fabric, and natural fibres, which help balance humidity and act as a carbon store. Reflective materials can improve the thermal performance of voids and insulated plasters can augment performance or help provide energy efficiency where other insulation measures are not possible. Lastly, the role of insulation in sound reduction should not be overlooked.

APPLICATIONS

- Internal wall insulation
- External wall insulation
- Roof and loft insulation
- Floor insulation
- Acoustic absorption

PROS

- Improved energy savings and reduced heat loss
- Reduced condensation risk and better indoor air quality
- Limited sound transmission and improved acoustics
- Lower summertime heat gain providing a more stable environment
- Material choice can reduce embodied impacts

CONS

- Bad design or installation can lead to damaging condensation and poor indoor air quality
- Certain materials can act as pollutant sources impacting indoor air quality
- Material choice can increase embodied impacts

Choosing the most appropriate insulation for a project is one of the most important and complex decisions designers face. Not only does insulation prevent energy loss and help save energy, it contributes to many other aspects of building performance and has a significant influence on the overall environmental impact of a building.

The tragic events of Grenfell Tower illustrate, in the starkest way, the need to understand the properties of insulation options and the importance of specifying appropriate materials. Priorities vary, so specification will vary depending on the needs, benefits and risks associated with each project. This requires a full understanding of what functions insulation performs.

Insulation prevents energy loss, minimises heat gain, helps maintain a healthy moisture balance, promotes good indoor air quality, ensures fire safety, and contributes to lowering operational and embodied carbon, improving the overall sustainability of a building.

The importance of insulation is refleced in growth expectations for the sector, predicted to be 9% in 2022.[4] Mineral fibre and phenolic foam boards dominate the UK market. Natural-fibre insulation (NFI) has arguably been the fast-growing segment in the UK market, averaging more than 20% growth since 2018.[5] However, NFI remains a small part of the UK market (<1%) compared to our near European neighbours, including France and Germany, where market share is greater than 10%.

The need to improve the UK's existing buildings is a key driver for growth because insulation plays an essential role in the upgrading of existing properties. The UK's 30 million homes, the least energy efficient housing stock in Europe, account for 21% of the country's total carbon emissions.[6]

Insulation comes in many types and forms, including man-made mineral fibre, open- and closed-cell petrochemical foams, natural-fibre insulation, multi-foils, and other mineral materials such as foam-glass and aerogel. With a diverse range of insulation types available and many performance aspects to consider, it is no surprise that guidance and advice can often seem confusing and contradictory.

Sorptive The ability to undergo sorption.

Thermal bridging Occurs when materials with a higher thermal conductivity (such as solid timber) run within the building envelope and bypass the insulation line, conducting heat to the outside. This has the effect of increasing the U-value so a greater depth of insulation may be required to compensate or additional non-bridged layers installed.

Thermal conductivity Also known as the lambda (λ) or K-value, expresses the degree to which heat can pass through a standard thickness of insulation (1m) at a steady state. A low thermal conductivity denotes a material that is a poor conductor of heat and therefore a better steady state insulator, whereas a high thermal conductivity indicates the reverse.

Thermal diffusivity The rate at which heat moves through a material relative to its bulk. Heat moves rapidly through materials with a high thermal diffusivity compared to those with a low thermal diffusivity.

Thermal effusivity A measure of the ability of a material to exchange heat with its immediate surroundings at a surface. Materials with a low thermal effusivity feel warmer to the touch compared to materials with higher thermal effusivity.

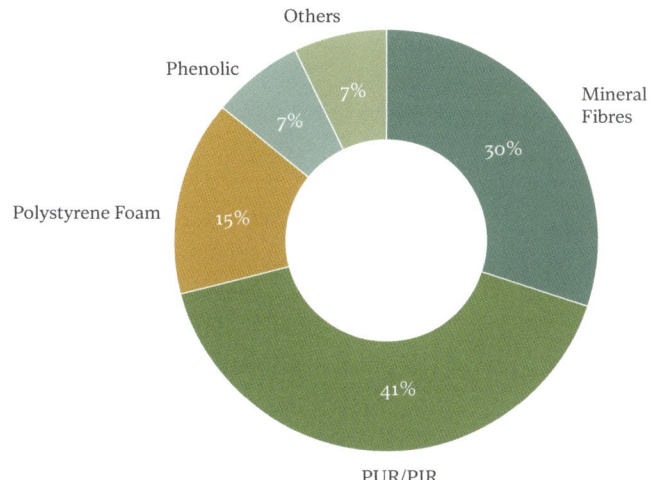

Figure 11.1 Share of UK market by insulation type, based on market value, from 2018. 'Others' include natural-fibre insulation, multifoils, insulated fibres and other speciality products. PUR/PIR (polyurethane/polyisocyanurate) remains the largest segment by value, but the lower unit cost of mineral fibre makes mineral fibre the largest segment by volume. Source: AMA Research Ltd.[7]

It is important to consider in more depth the main roles that insulation plays in order to help set priorities at the early design stage and enable the most appropriate materials to be specified. It is usually at the early design stage when lesser-known technologies are overlooked, often on the basis of long-held misconceptions.

COMMON MISCONCEPTIONS

It is helpful to address some frequent misunderstandings about insulation which often lead to ineffective or even damaging design decisions.

It's best to have the biggest insulation depth possible. Not always. If we look at energy savings for upgraded elements, then the majority of energy savings come from the first 100mm of insulation depth. There are still savings to be made by increasing insulation depth but it's not always necessary or desirable. Optimum insulation thickness should be assessed in context through U-value calculations, condensation risk analysis and hygrothermal modelling in the higher-risk situations.

Hygroscopic (water-absorbing) materials accumulate moisture and cause building damage. Hygroscopic materials such as natural-fibre insulation are often beneficial because they contribute

Thermal mass A measure of how hot a material gets as it absorbs energy. Materials with a high thermal mass can absorb a lot of energy without their temperature rising significantly. Materials with a low thermal mass will warm up quickly as they absorb energy. This means that insulation with a higher thermal mass will absorb energy more effectively, which limits the rate heat can penetrate the property during the hotter months.

U-value The amount of energy lost from a building element (such as a wall or roof) is expressed as the U-value and measured in W/m²K. The U-value tells us how much energy (heat) is lost across a specific area (1m²) for every degree of temperature difference either side of the element at a steady state. The lower the U-value, the lower the energy loss and the more effective the insulation. The U-value is the reciprocal of the R-value, i.e. 1/R.

Vapour open A property of materials with a high vapour permeability.

Vapour permeability The degree to which a material allows the passage of gases such as water vapour through them. Materials with a low vapour permeability are more resistant to movement of water vapour.

Vapour resistance The degree to which a material resists the passage of water vapour. The higher the vapour resistance, the less water vapour can pass through a material.

to breathability when installed in an appropriate setting. Some extremely hygroscopic materials can create harmful effects, including corrosion, but this is not the case with commonly used materials such as natural fibres, timber or lime.

Hydrophobic (water-repellent) and non-breathable insulation damages building fabric. It depends where the insulation is installed. There are situations where water repellence is an advantage, such as the void in a brick and block cavity or when installed below ground.

ENERGY LOSS, ENERGY SAVINGS AND HEAT GAIN
The primary role of insulation is to reduce heat loss, save energy and minimise external heat gain, so it is important we understand all the fundamentals behind each of these. The U-value (measured in W/m²K) tells us the rate of heat transfer from a building element and from that we can determine energy savings. U-values are steady-state measurements, which means they don't indicate how much energy the insulation itself can store or release and how it minimises external heat gain. For this we need to look to thermal mass.

The lower the U-value, the less energy is lost. The depth of insulation required to meet a given U-value depends on the thermal conductivity of the individual components sometimes indicated by the lamda [λ], and measured in W/mK) and the degree of thermal bridging. Lower thermal conductivities require a thinner fabric to achieve the same U-value or a lower U-value for the same insulation depth. That's why materials such as PIR, which have very low thermal conductivity, have become widely used.

In the case of retrofit, U-values can prove useful in calculating energy savings. Energy savings are the difference between the U-values of the insulated and uninsulated element. Comparing U-values without reference to the uninsulated element greatly exaggerates differences, skewing the result in favour of the lower U-value. It is often more informative to compare energy savings rather than U-values.

It's important to give special consideration to the impact of thermal mass because it is overlooked in U-value calculations. Because different insulation materials can absorb different amounts of heat

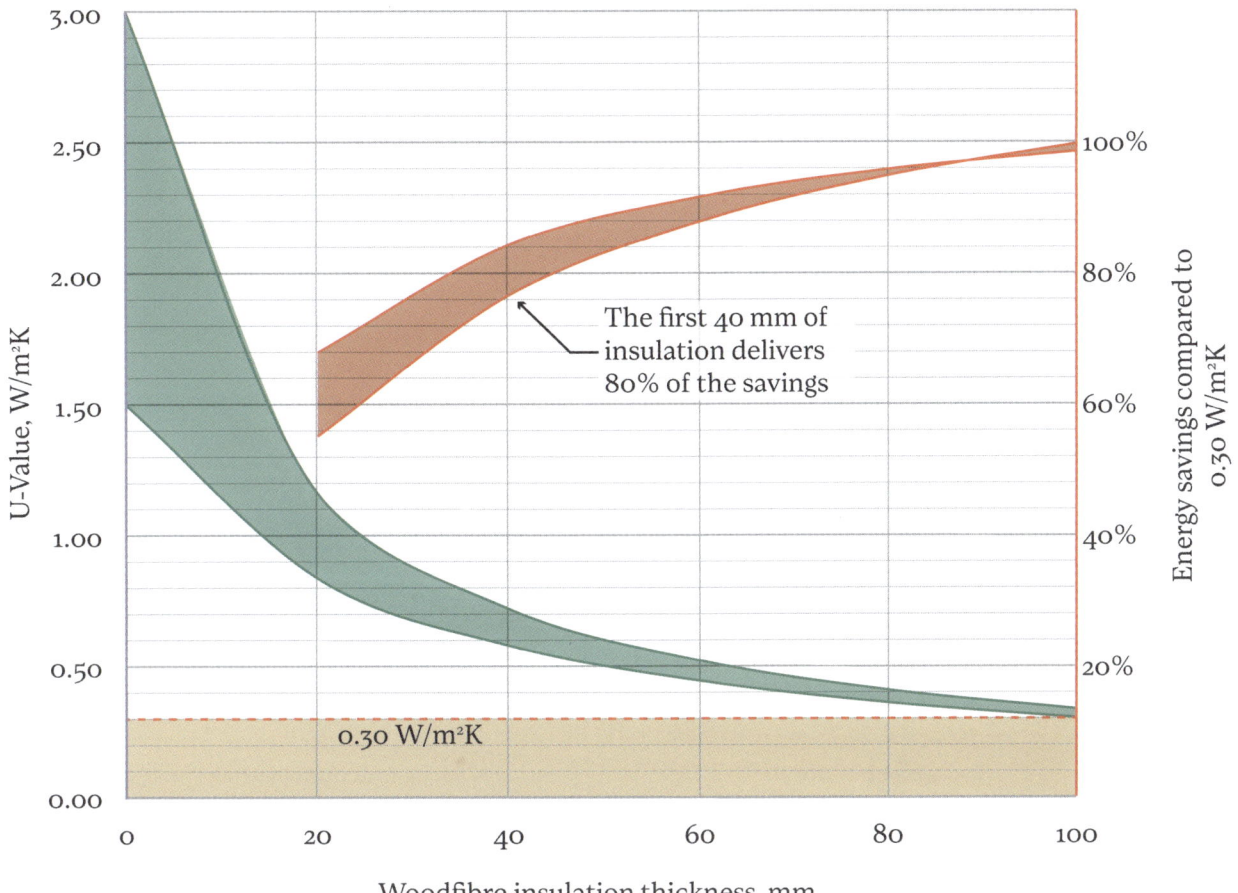

The first 40 mm of insulation delivers 80% of the savings

0.30 W/m²K

Woodfibre insulation thickness, mm

Figure 11.2 This graph shows the impact of increasing insulation depth on the energy savings from an uninsulated wall. The blue zone shows the changing U-value with increasing insulation depth, starting at different initial U-values. The red zone shows the energy savings compared to a recommended minimum U-value of 0.30W/m²K for the increasing insulation thickness. Because of the non-linear nature of the graph, nearly 80% of energy savings are achieved with the first 40mm of insulation. This shows us that it does not always follow that double the insulation provides double the savings.

at varying rates, choosing the right insulation can have a significant impact on the way the building absorbs and releases heat. This allows insulation to further help control summertime overheating within the building.

For example, wood-fibre insulation generally has a higher thermal mass than PIR insulation. A roof insulated with PIR will achieve a better U-value than wood-fibre insulation at the same depth, but the thermal mass of the wood-fibre insulation means it will absorb more energy and reduce overheating. The PIR may provide more energy savings during the colder months but is unlikely to offer the same protection against overheating in the warmer months compared to the wood-fibre insulation.

BREATHABILITY AND MOISTURE BALANCE

Persistently high humidity and moisture causes significant damage to the building structure and is the single biggest contributor to poor indoor air quality within dwellings. Using breathable insulation is often key to maintaining a healthy moisture balance.

Breathable insulation is both sorptive and vapour open. This allows humid air to diffuse into the insulation, allowing water vapour to attach and detach itself to the insulation fibres. This has the effect of helping decrease humidity as temperature falls and increase humidity as temperature rises. Sorptive insulations 'store' this moisture in a form that prevents damage. Because the movement of moisture is through diffusion, materials such as wood fibre boards can be resistant to air pressure (air-tight) while still being breathable.

There are three broad types of insulation:
- **non-breathable**, such as PIR, which has high vapour resistance and is non-sorptive;
- **vapour open**, such as mineral fibre, which has low vapour resistance and is non-sorptive;
- **breathable**, such as natural-fibre insulation, which is both vapour open and sorptive.

Breathability is not a substitute for ventilation. Most moisture within the building fabric should be removed through well-designed ventilation. Breathability is a means of helping create a safe moisture balance, dampening fluctuations in humidity and helping ensure moisture is held in a safe form within the fabric when humidity levels become temporarily elevated.

INSULATION AFFECTS INDOOR AIR QUALITY

The products we use within buildings can have a significant bearing on the quality of indoor air. Our choice of insulation not only affects moisture and humidity, but also the materials themselves can be a source of harmful indoor air pollutants. Materials can release harmful emissions as they age (off-gassing), most notably volatile organic compounds (VOCs). VOCs are a significant contributor to poor indoor air quality so choosing low- or zero-VOC materials is a key control measure.

FIRE

The behaviour of insulation in the development and propagation of fire within buildings is of paramount importance. The degree to which insulation needs to react to fire depends on the building element and context of the building and risks posed to occupants. Certain situations require the use of insulation that is classed as non-combustible, others require a level of fire performance that can be provided by other components, such as the internal lining board or timber cavity barriers.

Whereas it is essential to use non-combustible insulation where mandatory, using non-combustible insulation in areas where this is not necessary may be at the cost of other priorities, such as reduced breathability or embodied carbon. It may also result in lower indoor air quality if the fire retardants used in the insulation have the potential to generate harmful emissions.

Insulation materials are generally rated according to the Euroclass system, although British Standard fire ratings are still acceptable and referenced. Ratings range from A to F. A denotes non-combustibility (A1) or limited combustibility (A2) with F being the lowest class for materials that have either not been tested or that have failed all tests. Ratings B, C, D and E indicate different degrees of reaction to fire and the production of smoke and droplets.

GENERAL TYPES OF INSULATION

Insulation can also be categorised by the type of material or materials that make up the insulation. Categories include man-made mineral fibre comprising stone or glass wool; natural fibre comprising materials such as wood, sheep's wool, hemp or jute; petrochemical foams comprising open- and closed-cell foams made from materials such as polyurethane and polystyrene; reflective foils including low-emissivity membranes or multi-foils that combine foils and wadding; lastly other mineral insulation that include materials such as foam glass or insulated plaster.

Man-made mineral fibre (MMMF)

MMMF comprises stone or glass wool that is formed by spinning molten diabase rock, slag materials or glass into a fine fibre. The fibres can be bonded into a quilt of insulation that can be converted into insulation slabs or rolls. The fibres can also be cut to a short length and left unbonded to be blown into cavities within the building.

Both rock and mineral fibre are manufactured in the UK, where readily available raw materials are present. It is energy intensive, requiring melting of the substrate, glass or a particular type of volcanic rock called diabase for rock fibre, and sustaining temperatures of around 1,500 °C during manufacture. The raw materials for both types of mineral fibre are mined, although recycled glass can provide a feedstock for fibreglass.

Mineral-fibre insulation is ideally suited to applications where non-combustibility of the insulation is required or where water-repellent properties are required, for instance within a brick and block cavity.

The performance of mineral-fibre insulation relies in part on its density. Low-density products are commonly used in loft spaces, whereas denser mineral fibre is used extensively in acoustic absorption applications. Where lower thermal conductivity is required between rafters or in walls, then denser forms of mineral-fibre insulation with a lower thermal conductivity can be used.

The binders used in the products can often be formaldehyde based, although formaldehyde-free alternatives are available, so this should be considered if source control is a priority. Loose-fill MMMFs used for blown-in applications are free from binders.

Foam-based insulation

Foam insulation can be categorised as closed cell or open cell. Closed-cell foams are denser and have cells (bubbles) within the foam that are smaller and closer together, preventing gas escaping from within the foam and restricting moisture entering the foam. Closed-cell foam insulation is made from polyethylene (PE), polyurethanes (PUR, PIR) and extruded polystyrene (XPS).

Raw materials derived from crude oil are reacted and extruded in the presence of a blowing agent (gas) that creates the foam and remains in the foam cells once the foam has cured. XPS now tends to contain CO_2, whereas cells in PIR and PUR are filled with pentane gas rather than air. Pentane has a thermal conductivity lower than air (0.013W/mK versus 0.024W/mK) which enables the insulation to achieve a thermal conductivity lower than air. The relatively low global warming potential (GWP) of pentane (5kgCO_2e/kg) and CO_2 make them preferable as blowing agents compared to gases such as HFCs (hydrofluorocarbons) that can have GWPs in the thousands. High GWP blowing agents are now very uncommon.

Open-cell foams such as expanded polystyrene (EPS) are less dense and have bigger cells (bubbles) within the foam. They are made by bonding beads of foam into a single block. Open-cell foams have a higher thermal conductivity than closed-cell foams. Open-cell foams also have a lower vapour resistance than closed-cell foams, which limits their use to situations where this is not critical, such as wall cavities.

Foil backing on PUR and PIR foams boards helps retain the blowing agent and create a low-emissivity cavity when used adjacent to an air space. They can take the form of rigid boards with a reflective foil backing on both sides or a filling between orientated strand board (OSB) to create structurally insulated panel systems (SIPS). PIR and PUR lend themselves to use in thinner building elements where their low thermal conductivity is an advantage. Large insulation sheets also reduce the degree of thermal bridging in the structure and, due to their relatively high compressive strength, they are also used under floors to create an insulated floating floor. Expanded polystyrene open-cell foam is commonly used to provide insulation in concrete systems within concrete walls and floors.

Foams offer less sound insulation and due to their relatively low thermal mass and thickness, they provide limited protection against overheating. Petrochemical foams are energy intensive and rely heavily on fossil sources that will impact the embodied carbon of the products.

Natural-fibre insulation

Natural-fibre insulation (NFI) uses natural fibres such as wood, wool, hemp, jute, etc and takes the form of flexible quilt, rigid board, loose fibre or straw bales. Regardless of the material used, natural-fibre insulation takes advantage of the unique way that natural fibres absorb and release moisture, as well as the regenerative nature of the main raw material.

Figure 11.3 A studio barn in Cumbria has been retrofitted using Thermafleece CosyWool, made from UK sheep's wool, to improve the thermal performance of the walls and roof.

Natural fibres are sorptive, allowing all natural fibres to create truly breathable insulation. Natural fibres are made of cellulose, hemicellulose, lignin or keratin. On a molecular level, these fibres have slightly charged areas that water molecules attach to like weak magnets, creating their characteristic sorptive behaviour.

Flexible insulation is typically made by blending the natural fibre with a polyester (PET) binder and thermally bonding the insulation. This prevents the fibres from settling and the insulation from slumping and allows for easy handling on site. Flexible insulation can be made without a binder but may be prone to slumping and may delaminate during installation. Flexible NFI can be used in most areas where conventional mineral wool is used, with a few exceptions.

Rigid NFI is made from either wood fibre or cork. Wood fibre boards are widely used and come in a variety of densities depending on the end use. They are used under or over timber frames, for flooring and lining solid walls, and can be used internally or externally. Because rigid NFI can span timbers or form a continuous layer on walls, thermal bridging can be minimised.

Loose cellulose or wood fibres can be blown into cavities and voids. The fibres are injected under pressure to make sure a minimum density is achieved throughout the insulation and to prevent settlement of fibres.

Straw bales can form part of the wall structure in straw-bale construction. The depth and density of the straw bales also creates protection from overheating and has excellent acoustic performance.

Natural-fibre insulation is an excellent choice for all-round performance and is particularly useful when embodied carbon and breathability are key. Natural-fibre insulation is not suitable in situations where the insulation could be exposed to prolonged wetting, for example in certain ground-floor applications where the insulation is in permanent ground contact, or where a non-combustible insulation is mandated.

Foils

Foil-based insulation relies on highly reflective surfaces that limit the radiation of heat. Heat radiation is similar to light in the way some surfaces reflect radiation and some absorb it. High-reflectivity surfaces are most commonly referred to as low-emissivity surfaces (emissivity being the opposite of reflectivity). Foils are generally made from pure aluminium that is coated to prevent oxidation.

A low-emissivity surface adjacent to a cavity can improve the insulation performance of the cavity because radiant heat transfer is reduced. These voids are referred to as low-emissivity cavities, for example the air space between a foil-backed foam sheet and plasterboard.

ASPECT	Performance measure	Mineral fibre	PIR/PUR	XPS	EPS	Natural fibre	Multi-foil	Insulated lime plaster	Foam glass
Thermal performance	Thermal conductivity	Beneficial with limitations	Beneficial	Beneficial with limitations	Beneficial with limitations	Beneficial with limitations	Beneficial	Beneficial with limitations	Beneficial with limitations
	Thermal mass	Beneficial with limitations	Beneficial with limitations	Not beneficial	Not beneficial	Beneficial	Not beneficial	Beneficial	Beneficial
	Thermal diffusivity	Not beneficial	Beneficial with limitations	Not beneficial	Not beneficial	Beneficial	Beneficial	Beneficial	Beneficial with limitations
	Thermal effusivity	Not beneficial	Beneficial	Beneficial	Beneficial	Beneficial with limitations	Beneficial	Beneficial with limitations	Beneficial with limitations
Moisture	Breathable	Beneficial with limitations	Not beneficial	Not beneficial	Not beneficial	Beneficial	Beneficial	Beneficial	Beneficial with limitations
	Vapour open	Beneficial	Not beneficial	Not beneficial	Not beneficial	Beneficial	Beneficial	Beneficial	Beneficial
	Non-breathable	Not beneficial	Beneficial	Beneficial	Beneficial	Not beneficial	Beneficial with limitations	Not beneficial	Not beneficial
	Hydrophobic	Beneficial	Beneficial	Beneficial	Beneficial	Not beneficial	Beneficial	Not beneficial	Beneficial
Fire	Non-combustible applications	Beneficial	Not beneficial	Not beneficial	Not beneficial	Not beneficial	Not beneficial	Beneficial	Beneficial
	Other applications	Beneficial	Beneficial	Beneficial	Beneficial	Beneficial	Beneficial	Beneficial	Beneficial
Sustainability	Embodied carbon	Beneficial with limitations	Not beneficial	Not beneficial	Not beneficial	Beneficial	Not beneficial	Beneficial with limitations	Not beneficial
	Renewable materials	Not beneficial	Not beneficial	Not beneficial	Not beneficial	Beneficial	Not beneficial	Beneficial	Beneficial with limitations
	End of life	Beneficial with limitations	Not beneficial	Not beneficial	Not beneficial	Beneficial	Not beneficial	Beneficial with limitations	Beneficial with limitations
Air quality	Formaldehyde	Beneficial with limitations	Beneficial with limitations	Beneficial with limitations	Beneficial with limitations	Beneficial	Not beneficial	Beneficial	Beneficial
	VOCs	Beneficial with limitations	Not beneficial	Not beneficial	Not beneficial	Beneficial	Not beneficial	Beneficial	Beneficial

Table 11.1 Each type of insulation has strengths and weaknesses depending on the end use and the specific performance requirements. Here, the main types of insulation have been rated against each key performance measure.

Category	Code
Beneficial	(green)
Beneficial with limitations	(yellow)
Not beneficial	(orange)

Foils can be mixed with wadding material in layers to create multi-foil blankets. These blankets span timbers to create an insulating layer and low-emissivity cavities. To ensure multi-foils work efficiently, it is important to install them in conjunction with other insulation and to maintain air spaces as directed.

Other mineral insulation

Foam glass made from recycled glass is lightweight aerated glass similar to pumice that is used in floors to provide insulation to the subfloor. It can be used under a conventional concrete slab or as an insulating layer in limecrete floor systems.

Aerogel is a specially formed silica material that has an extremely low thermal conductivity, making it an excellent insulation choice in areas where depth is extremely limited, such as window and door reveals.

Insulating plaster combines lime plaster with insulated particles such as short hemp fibres or cork. The insulating plaster provides a surface that captures less heat and is warmer to the touch. Insulated plasters are particularly useful in situations where installing other insulation measures may not be possible or practical, such as internal traditional solid walls.

COMPARING DIFFERENT INSULATION PROPERTIES

The importance of performance measures should be viewed in the context of the overall design requirements. It is vital to choose the materials that provide the most benefit and least risk.

COMPARING ENVIRONMENTAL IMPACTS OF DIFFERENT INSULATIONS

When comparing the environmental impact of one insulation over another, it is important to make a comparison on a level footing. As such, Environmental Product Declarations (EPDs) for insulation use a functional unit that achieves a U-value of 1. The thickness of insulation in mm that achieves a U-value of 1 can be calculated by multiplying the thermal conductivity of the material by a factor of 1,000.

DESIGNING FOR LOWER IMPACTS

During a climate crisis where every gain counts, we should be looking at choices that provide the most operational, in addition to the most embodied, impact gains. We also need to see impacts in a light that goes beyond climate change, impacts that create healthy, safe and sustainable built environments.

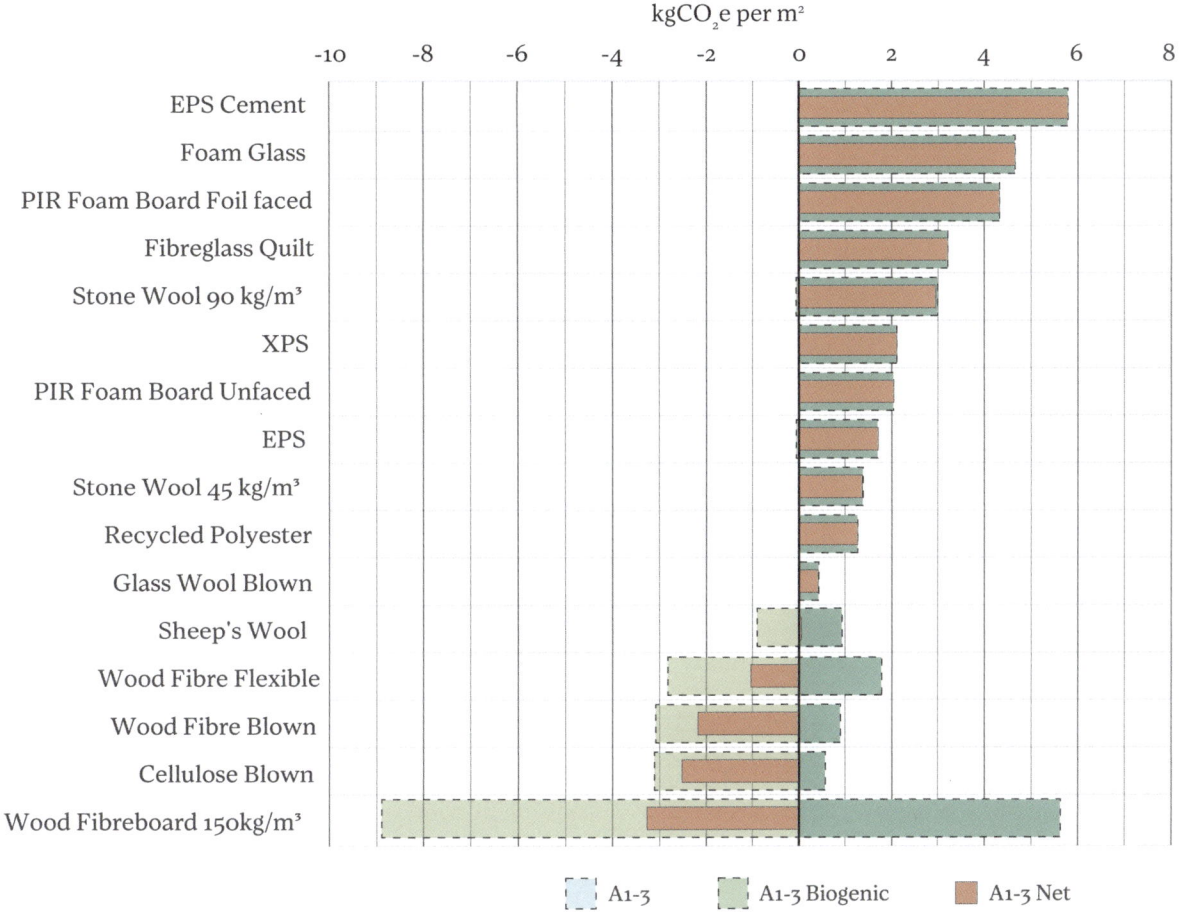

kgCO$_2$e per m^2

A1-3 emissions, kgCO$_2$e per m^2 to achieve a U-value of 1.0 W/m^{2}K

Figure 11.4 A chart showing the product emissions (A1 to A3) for different insulations that achieve a U-value of 1.0W/m²K, with both emitted and biogenic carbon shown.

In a world where all products need to become ever more sustainable, it is even more important to understand what each type of insulation does to ensure we have a well-balanced mix of product types available for ever-lower-impact design. This can only be achieved by the use of transparent and verifiable data at the product level, emphasising the importance of EPDs in particular.

FUTURE TRENDS
Biogenic carbon
Awareness is growing that biogenic carbon in natural materials has a measurable benefit to climate change mitigation. This is broadly recognised in Europe, where the use of natural-fibre insulation is accelerating as value is placed on the biogenic carbon.

**QUESTIONS
FOR PROJECT TEAMS
AND SUPPLIERS**

☐ Can the manufacturer provide a Declaration of Performance (DoP) for the product?

☐ Is the insulation UKCA mark compliant if applicable?

☐ Can the manufacturer provide evidence relating to the declared performance values in accordance with any norms (e.g. BBA etc)?

☐ Does the supplier have liability insurance and warranty position?

☐ Is system information such as BIM, CAD designs and drawings available?

☐ Are installation guides and documents/videos available?

☐ Does the insulation have a verified and current Environmental Product Declarations (EPD)?

☐ Are relevant case studies available?

Circularity

The move towards greater circularity has a significant influence across the insulation category. Insulation tends to be problematic at the end of life due to its bulkiness and contamination from dust and debris. Older insulation tends to degrade, making removal and reprocessing difficult. Using insulation that maintains its structural integrity is helpful for end-of-life options. Carbon can remain sequestered in natural materials when reused or recycled.

Innovation

Finding replacements for petrochemical materials, lowering energy intensity of production, making better use of underutilised materials and improving end-of-life options are all key to minimising the impact of the insulation. Equally important is a greater focus on resource efficiency, ensuring only the materials we need are put to use and reuse.

We can't lose sight of the fact that innovation goes beyond products and technology. One of the most important innovations is the way we use existing technologies in combination, for example combining different types of insulation to gain the benefits of each. By gaining a greater understanding of what insulation does, we can design elements that make use of each product's relative strengths.

On the horizon are novel natural materials such as mycelium that hold some promise as a replacement for conventional rigid foams. A greater use of natural fibres in the UK is almost inevitable, given what we see in other countries such as France or Germany. It's also likely that more sustainably derived plastics will gain a greater foothold, as will technologies that allow reprocessing of petrochemicals back to their constituent parts.

THE RETROFIT CHALLENGE

When it comes to retrofitting solid walls, over-insulating can cause moisture problems at the interface of the wall and the insulation if the wall temperature is dramatically reduced. In this instance, it is critical to not only conduct condensation risk modelling but also to appreciate how the different types of insulation react to humidity.

KEY TAKEAWAYS

- All insulation saves energy, which in turn saves money, so any type of insulation installed correctly pays for itself over time, so don't discount any insulation solution purely on upfront cost.

- Insulation does much more than prevent heat loss, so take other fabric requirements into account at the start of the design process.

- Choose a 'no regrets' approach to insulation. The costs and impacts of upgrading at a later date far outweigh the upfront cost of the optimal design first time.

- In retrofit, when considering different insulation systems, compare energy savings rather than U-values (energy loss). Considering energy savings provides a realistic comparison of the benefit from retrofitting, whereas energy loss doesn't.

- When we upgrade existing elements, a large proportion of the energy savings come from the initial 40–50mm of insulation depth, an important consideration when upgrading solid walls in particular.

- Insulation significantly impacts both operational and embodied carbon. A focus purely on operational carbon is a missed opportunity.

- Good workmanship is the key to good insulation performance.

MYCELIUM

Duncan Baker-Brown

Estimated number of fungi species
2.2–3.8 million[1]

Identified number of fungi species
120,000[2]

Fungi share of our DNA
50%[3]

Fungi are extremely plentiful: they are omnipresent in our lives, yet we don't often notice them unless we are consuming them or they are damaging our belongings. They are found literally everywhere on planet earth – in oceans, soils, the atmosphere – and, of course, they are on and inside us in the form of yeasts. Without fungi, trees and plants would not be able to sustain themselves: they would die. Fungi consume dead matter and create nutrient-rich soils. They can also consume rock and even synthetic human-made materials such as plastic. Humans are just at the point of getting to know more about fungi, exploring some of their potential, and over the last 30 years or so we have discovered that if we feed mycelium certain foodstuffs, we can create organic, non-toxic material alternatives to fossil-fuel based products. Companies around the world are currently investing multimillions of dollars developing mycelium-based projects in what most people think will be a business opportunity as large as that of the current petrochemical industry. Mycelium products are easy to manufacture and do not require land currently occupied by forests or farms. They can be produced in factories and even in urban environments as manufacturing processes do not emit pollutants. Current mycelium materials are creating organic, compostable and non-toxic alternatives to plastic packaging, acoustic panelling and, very soon, high-performing rigid insulation for the construction sector. In many cases, these alternatives are providing higher levels of performance than conventional synthetic materials. But we are only at the beginning of our exploration into the potential of this hugely diverse and beneficial world of fungi, one that could enable us to once again live in harmony with our host planet.

APPLICATIONS

- Consumer packaging
- Non-load-bearing bricks for internal partitions
- Internal finishes
- (Hopefully, very soon) building insulation

PROS

- Non-toxic, organic & biodegradable, i.e., compostable
- Doesn't require land to grow
- Feeds off organic and even synthetic waste streams
- Potential for highly varied uses
- Insulation rating similar to expanded plastic (PIR) insulation products
- Better fire rating than any petroleum-based insulation product

CONS

- Currently an emerging material with much research ongoing Therefore, the product range is currently limited
- Many product certifications are yet to be formally achieved
- Much development is being protected by patents, limiting future supply chains

GLOSSARY

Biotrophic Feeding on living cells of other organisms.

Chlamydospores Asexual spores formed by the breaking up of fungal hyphae.

Cortina A cobweb-like partial veil consisting of fine, silky fibres.

Endomycorrhiza Mycorrhiza in which fungal hyphae penetrate cell walls of the host plant, creating a form of symbiotic relationship that occurs between a fungal symbiont, or mycobiont, and the roots of various plant species.

Hyphae (pronounced Hy-fee) The long, fine tubular structures that branch out in many directions and fuse together, forming a 'tangled filigree' collectively known as mycelium. Hyphae are the main mode of vegetative growth for fungi.

Lactarius A genus of mushroom-producing, ectomycorrhizal fungi, containing several edible species.

Latex Milky fluid that oozes from cut surfaces of *Lactarius* species.

Mycelium The body of a fungus, most of which is underground or hidden within wood.

One area of research that is occupying start-ups, as well as more established multinational corporations around the world, is that of harnessing the many, almost supernatural, qualities of fungi, and in particular, the highly complex micro-thread-like root systems known as mycelium. Completely ubiquitous, we literally live and breathe fungi spores every day. These spores can be discharged at speeds '10,000 times faster than a Space Shuttle directly after launch, reaching speeds of up to a hundred miles per hour – some of the quickest movements achieved by any living organism'.[4]

Famously referred to as 'the wood-wide web',[5] vast networks of subterranean mycelium roots are crucial to sustaining trees and plants – which wouldn't exist without them. Merlin Sheldrake, author of *Entangled Life* (2020), describes mycelium as 'ecological connective tissue'.[6]

Without fungi and mycelium, there would be no life on earth, it's as simple as that.

Most fungi produce spores without growing mushrooms. They also form vast networks of cells known as 'hyphae' which form (mostly) subterranean networks of fine tubular structures that are constantly on the move, looking for food and nutrients. These networks of cells appear to have a collective and singular intelligence. For example, they are able to choose the most effective route around an obstacle, seemingly able to change direction at will until one opportunity presents itself as the most beneficial to the whole. This characteristic has led researchers to utilise slime mould to model the Tokyo train network. Researchers at the University of the West of England's 'Unconventional Computing Lab', together with researchers from the Democritus University of Thrace in Greece, even used slime mould to calculate the most effective fire-escape routes from buildings.[7]

Sheldrake also speculates that if 'one teased apart the mycelium found in a gram of soil – about a teaspoon – and laid it end to end, it could stretch anywhere from 100m to 10km'.[8]

Mycorrhiza The symbiotic association between a green plant and a fungus. Mycorrhizas are located in the roots of vascular plants and play a vital part in plant nutrition and soil health. They are the structure by which a fungus and a plant mutually exchange nutrients.

Necrotrophic Feeding by killing and consuming (part of) another organism.

Slime moulds A group of fungus-like organisms that use spores to reproduce.

Spore Reproductive structure of a fungus, usually a single cell.

Sporophore Fungal fruitbody.

Stipe Stem of a mushroom.

MANUFACTURING

Mycelium-based products are almost certainly going to be ubiquitous, replacing many of the plastic-based artefacts we rely on today. This fact is not lost on the major companies investing huge sums of money in the development of numerous mycelia mixes suitable for creating the products we need. However, at the moment, an equally huge amount of money is being invested to ensure that valuable intellectual property is protected by corporate copyrights for the small number of biotech start-ups and the larger multinationals who have submitted and gained patents. As a result, the actual specification of all the ingredients and the processes that produce the highest-quality acoustic and thermal insulation panels, load-bearing bricks and packaging is not public knowledge. To be fair to the manufacturers, these products are still in the research stage of their development, and therefore the manufacturing processes are still evolving and being refined.

However, for the purposes of this publication, we can be sure that the manufacturing process of all mycelium products starts with feeding the fungi different forms of commercial agricultural waste, such as hemp, grass or sweetcorn husks (plastic waste will come along soon). This mixture of fungus and fungus food is placed in a mould (in the shape of the product being manufactured) and then left for a few days to allow the mycelium roots to consume the foodstuff and grow rapidly, filling the mould that surrounds it. When the appropriate weight and density of mycelium mix has been achieved, it is dehydrated by heating. This kills the mycelium and stops it growing further. It is at this point that the manufacturing of the product is complete, save for quality checking.

APPLICATIONS OF MYCELIUM

Mycelium is an emerging product and there are many exciting ways that these incredible organisms could work with us to provide low-carbon, biodegradable, regenerative, nature-based products for the construction and other industries.

Many companies, together with academic institutions, have been developing products as varied as bricks, textiles, shoes, leather, insulation and packaging. Another company, Ecovative,[9] has a portfolio of organic materials that are marketed as 'safe, healthy and certified sustainable'.[10] Myco Board, manufactured by Ecovative,

Figure 12.1 The Hy-Fi, New York, by The Living with structural design by Arup, 2014. Over 10,000 mycelium bricks were used to create the 12m-high building.

is like a timber particle board, but instead of using glues to bind the fibres, mycelium is used. The same company produces Myco Foam, which could replace much of our plastic packaging, as well as thermal and acoustic insulation. Mycelium is grown to order, by mixing it with waste organic matter such as corn husks, waste timber, etc.

There is an ever-growing choice of interior finishes and fittings available made of 100% mycelium-based material. Lamps and tiles created by Biohm use food waste, and even spoil from construction sites, as the nutrient source for mycelium to consume while it grows into products.[11]

3D-printed partitions

New York-based David Benjamin, architect, academic and founder of design practice The Living, created an algorithm that emulated the way mycelium grows. He then applied this to create a 3D-printed prototype for a super-lightweight partition for Airbus airliners that was 45% lighter than a standard Airbus partition.[12]

Bricks

Mycelium bricks were used by The Living in 2014 to create the Hy-Fi, the Museum of Modern Art's (MoMA's) PS1 space in midtown Manhattan.[13] This is still one of the most ambitious projects embracing the potential of mycelium, because Benjamin used the bricks to create three interlocking towers that were 12m high, load-bearing and left outside for six months. Working with structural engineers Arup, Benjamin developed the compressive strength of these bricks at Columbia University to ensure that they were up to the job. The bricks were grown in moulds over a period of about five days by mixing corn husks (waste agricultural by-product) and mycelium. The resultant brick was solid, lightweight and durable to a point (though not as durable as a clay brick – yet).

It should be pointed out that mycelium bricks have a compressive strength of only around 30psi, which is a long way from the 1,000 to 1,500psi compressive strength of standard bricks. However, relative to its weight, a mycelium brick is stronger than brick, with a cubic metre of mycelium brick weighing just 43kg compared to a cubic metre of bricks weighing anything from 1,600 to 1,920kg.

Insulation

To grab any part of the construction sector's massive insulation market, mycelium products will need to perform as well, if not better, than the petroleum-based plastic insulation that, together with other high-carbon alternatives, currently has over 95% of the insulation market. The good news is that Biohm's mycelium insulation will have a thermal conductivity as low as 0.03W/mK, which compares very favourably with PIR plastic insulation at 0.022W/mK. In addition, mycelium performs very well as an acoustic insulation with absorption rates of at least 75% at 1,000Hz (the typical frequency of road traffic noise). And, best of all, initial fire tests demonstrate that mycelium has a far superior fire rating to its competition because it doesn't contain the synthetic, resin-based materials that cause the harmful toxic smoke and quick spread of flames during fire.[14]

FUTURE TRENDS

All of the above has to be understood within the context of mycelium insulation still being in the process of undergoing testing with BBA (British Board of Agrément). Testing is quite correctly thorough, and is taking longer than normal partly because

Figure 12.2 Growing Pavilion, Dutch Design Week 2019 and Floriade Expo 2022, by Pascal Leboucq, Krown Design and Biobased Creations. Demountable mycelium panels were fitted on a timber frame to form a circular pavilion to showcase the potential of mycelium as a construction material. Edible oyster mushrooms were harvested from the mycelium panels daily during the exhibitions.

Biohm's insulation is the first mycelium-based product being tested for the construction sector, and so there are no similar approved products to benchmark this one against. So, until mycelium-based products get full BBA certification, they are consigned to creating interior fixtures, finishes and fittings and intriguing demonstration pavilions such as 'The Growing Pavillion' at Dutch Design Week in 2019.[15] However, once the performance certifications are in place, they have the potential to create a huge positive impact for the construction sector. In the short term, they can provide an organic, non-toxic, genuinely sustainable and regenerative alternative to the currently ubiquitous toxic petroleum-based, plastic insulation products with no end-of-life strategy beyond floating in our oceans for millennia.

Biohm recently created a huge amount of interest because of its research into mycelium that grows by consuming plastic. The discovery was an unintentional result of testing different strains of mycelium in its laboratory. Biohm found that a sample of mycelium had broken down and assimilated the plastic sponge used to contain it, treating the plastic as a new food source.[16] The mycelium had actually evolved in a matter of weeks from eating organic matter such as leaves and tree bark, to eating plastic. This might appear to be a huge evolutionary 'jump' that would take humans millennia of natural selection to achieve, but for certain fungi, it is a lot more straightforward, and crucially, quick. The consequences of this laboratory accident are profound. For a start, they have intensified the worldwide research into this strain of mycelium. It has also raised the prospect that, one day soon, toxic waste plastic insulation could be the food supply for organic, non-toxic mycelium insulation that, obviously, has an end-of-life strategy providing compost for the biosphere as opposed to plastic that tends to end up in our oceans for millennia.

The worldwide potential for mycelium-based products supplying many sectors is massive. Mycelium can absorb carbon from the atmosphere, provide filtration of chemical pathogens in our water supply and, of course, sustain the cultivation of forests and gardens. In addition, it can absorb and clean toxic waste, oil even, while consuming plastic without any toxic by-products. Growing mycelium in moulds to create interior fittings, external bricks and insulation is potentially a big deal, but it may well be only the first step, the beginning of a huge development in the way we design and construct a built environment that leaves no negative impact on the natural world.

As magical as they are ubiquitous and **everyday, fungi are only just beginning to be understood by us**. At the moment, we are mainly asking this most intelligent of life forms to perform relatively straightforward tasks, emulating a material we are used to and that we understand more completely. Looking forward, by understanding the potential and characteristics of mycelium more fully, humans will learn how to work with it while it is still alive, to create organic, self-generating, self-healing, three-dimensional environments and products that can adapt to changes in climate and even changes in building use. In short, we will work in partnership with living communities of mycelium.

KEY TAKEAWAYS

- Mycelium provides a closed-loop system, being non-toxic, organic and biodegradable.

- Crucially, mycelium does not require land to be grown, even at a commercial level. Therefore, it is not competing with land for growing crops for food or for sustaining biodiverse environments.

- Moving forward, mycelium has the potential to change the way buildings, and the built environment, are designed, occupied, maintained, adapted and deconstructed.

PLASTICS

Carol Costello

Global production of plastic per year
380 million tonnes

UK construction industry use of plastic per year
900,000 tonnes

Embodied carbon of high-density polyethylene (HDPE) plastic
1,900 kgCO$_2$e/m^3

Embodied carbon of recycled high-density polyethylene (HDPE) plastic
665 kgCO$_2$e/m^3

Embodied carbon of polyethylene terephthalate (PET) plastic
4,830 kgCO$_2$e/m^3

Embodied carbon of recycled polyethylene terephthalate (PET) plastic
1,380 kgCO$_2$e/m^3 [1]

While there is considerable public awareness regarding plastic pollution from household waste, the issues surrounding plastics in the construction industry have only recently come to light.

The plastics industry is the fastest-growing source of industrial greenhouse gases in the world. The UN Environment Programme estimates that the greenhouse gas emissions from plastic production, use and disposal could account for 19% of the total global carbon budget by 2040.[2] Environmental campaigners are now calling for the elimination of the production of virgin plastic to combat its harmful impacts on the planet.[3]

An estimated 50,000 tonnes of plastic packaging is created every year in the construction industry, much of this is either sent for energy recovery or to landfill. Very little is recycled.[4] Construction has the highest quantity of plastic 'stock in service', defined as more than a year in use, at 20 million tonnes out of a total of 42 million tonnes in use per year in the UK,[5] and is a significant user of recycled material. Recycling, while imperfect, is currently a better option than the alternatives of landfill and incineration, which produces more carbon emissions than natural gas.[6]

APPLICATIONS

- Packaging
- Membranes
- Mouldings (such as window frames or doors)
- Flooring (such as vinyl or carpet)
- Paints
- Cables and pipes
- Insulation
- Lighting

PROS

- Versatile material that can be made into a myriad of products
- Packaging protects other materials
- Lightweight
- Low cost
- Waterproof and airtight membranes rely on plastics
- Some types can be recycled
- Construction industry capable of using plastic recyclate

CONS

- By-product of fossil fuel industry
- Relatively high embodied carbon (by weight) compared to other materials
- Less durable than other materials: may discolour, flake, leach overtime when exposed to elements
- Frequently enters wastestream rather than being recycled with detrimental environmental impacts

INTRODUCTION

Most of us were born into a 'plastic age'. [7] Before the 1950s, plastic use was rare compared to the enormous global prevalence of this material today. In the postwar era, plastic seemed to be the answer to so many problems: it's lightweight, malleable, strong, waterproof, colourful and cheap. But now there is undisputed scientific evidence that plastic production, use and waste is causing significant harm to the planet's animal and human population. Some nations are slowly introducing legislation to mitigate the damage. Meanwhile, responsible architects need to be proactive in educating themselves about plastic, advising clients on its usage and demanding change from both the supply chain and policymakers to stem its toxic consequences.

EXTRACTION AND MANUFACTURE

Plastics are made from materials such as cellulose, coal, natural gas, salt and crude oil through a polymerisation or polycondensation process.[8] Plastic materials can also be produced from renewable sources such as wood fibres or algae, but this industry currently represents only a small fraction of global plastic production (less than 1%).[9] Plastics are primarily a by-product of the fossil fuel industry. In 1950, global production was 1.5 million tonnes; in 2021, production was 391 million tonnes.[10] The World Economic Forum predicts that plastic production will double in the next 20 years as petrochemical companies increase plastic production to diversify in an economy turning away from fossil fuels in response to climate change.[11]

ENVIRONMENTAL AND CARBON IMPACT OF PLASTICS

Greenhouse gas emissions caused by the plastic life cycle are not only from the production of primary plastics, but also from the incineration of plastic waste. In 2019, the production and incineration of plastic produced more than 850 million tonnes of greenhouse gases – equal to the emissions from 189 500-megawatt coal power plants.[12]

According to the publication *The P-Word* by the Resource Efficiency Collective at Cambridge University, the use of plastics in the UK generates 26 million tonnes of CO_2e every year across the life cycle of the plastic product, with production accounting for 80% of these emissions.[13] Recycling plastics could greatly reduce emissions, but rates of recycling in the UK are pitifully low due to lack of recycling facilities and the difficulty of recycling certain plastic types, including composite materials and contaminated products.

Figure 13.1 Whole-life carbon emissions from UK plastic consumption. The use of plastics in the UK generates 26 million tonnes of CO₂e every year, across the life cycle of plastic products.[14]

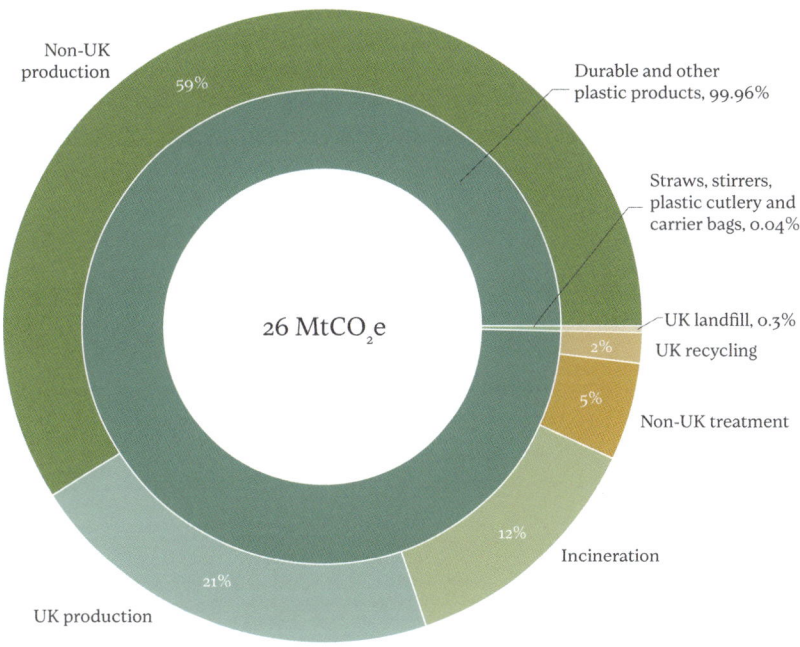

WHERE ARE PLASTICS USED IN CONSTRUCTION?

The construction sector accounts for 16% of all plastic use.[15] The construction sector has the highest quantity of plastic 'stock in service', defined as more than a year in use, at 20 million tonnes out of a total of 42 million tonnes in use per year in the UK.[16] Plastic stock in service has the advantage of not entering the waste flow. For example, a polythene damp-proof membrane installed under foundations remains in use for the life of the building. However, this figure does not include product packaging.

Construction packaging presents a significant problem because it is often single use; that is, items that are not reused or recycled due to poor management or cannot be recycled due to plastic type or contamination. Of course, any plastic building product can enter the waste stream. For example, vinyl flooring contaminated with adhesives would likely end up in landfill after an average 10-year lifespan.

Reducing production of plastics and recycling plastics offer ways to mitigate the negative impacts of greenhouse gas emissions and plastic pollution. For those working in the built environment, a first step is to understand the types of plastics that exist, how they are used in construction, and their ability to be recycled. Six types of plastic account for 80% of the plastic used by the construction sector (see Table 14.1).

Symbol	Abbreviation	Full name	Recyclable	Ease of recycling (in pure form)	Waste management in construction and examples	Typical construction uses (many construction products are made from a composite of several types of plastic)
⟲1	PETE (PET)	Polyethylene terephthalate	Yes	Easy	Recycling for catering/site facilities more common. Can be segregated for recycling but does not often happen. Otherwise sent for energy incineration or to landfill	Recycled PET carpet, packaging such as banding, water bottles
⟲3	PVC	Polyvinyl chloride	Yes	Difficult	Recycling of UPVC window frames and vinyl flooring via Recofloor and Recovinyl, also, some piping recycling; otherwise sent for energy incineration or to landfill	Single ply membranes, flooring, tubing, piping, ducting and guttering, door and window frames and other external profiling such as cladding, soffits and fascia boards, flooring and cabling, waterproofing and linings
⟲2	HDPE/ MDPE	High-density/ medium-density polyethylene	Yes	Easy	HDPE is a very common waste product on construction sites and worth focusing on reducing; it can be segregated for recycling but this does not often happen; there is some recycling of hard hats and waterproof membranes, but HDPE is otherwise sent for energy incineration or to landfill	Tubing, piping, ducting and guttering, waterproofing and linings, shrink wrap, hard hats
⟲4	PE-LD/ PE-LLD (LDPE/ LLDPE)	Low-density polyethylene	Yes	Manageable	LDPE packaging film is a very common waste product on construction sites and worth focusing on reducing; it can be segregated for recycling but this does not often happen; otherwise it is sent for energy incineration/ to landfill	Packaging such as bags, bandings, stretch wrap, shrink wrap, hoods
⟲5	PP	Polypropylene	Yes	Manageable	Depends on the product, but recycling is generally very limited; some limited recycling of carpets and PPE recycling, e.g. by Reconomy and Bryson; also temporary protection recycling; easier to recycle PP films	Carpets, tubing, piping, ducting and guttering, packaging such as shrink wrap, plastic buckets, woven PP bags

Table 13.1 Plastic types, use and recyclability.

Symbol	Abbreviation	Full name	Recyclable	Ease of recycling (in pure form)	Waste management in construction and examples	Typical construction uses (many construction products are made from a composite of several types of plastic)
♲6	PS	Polystyrene	Yes	Manageable/ not collected	Not usually recycled; sent for energy incineration or to landfill; difficult to transport economically due to its volume	Packaging
	EPS	Expanded polystyrene	Yes	Difficult/not collected	Not usually recycled; no take-back schemes for insulation; sent for energy incineration or to landfill; older EPS may contain persistent organic pollutants (POPs), which should be destroyed	Thermal and acoustic insulation, packaging
♲7	PA	Polyamides	Yes	Manageable	If used as paint and adhesives, it will be bonded to substrate and this will possibly affect the recycling of the product. If used in textiles (nylon), there may be some small-scale recycling	Paints (exterior treatments/ industrial), adhesives, textiles
	PC	Polycarbonate	Yes	Easy	Not usually recycled; sent for energy incineration or to landfill	Transparent roofing sheets, interior linings, light fittingsw
	PMMA	Polymethyl methacrylate	Yes	Difficult	Not usually recycled; sent for energy incineration or to landfill	Transparent sheet, windows, smart screens
	PUR	Polyurethane	Yes	Manageable	Not usually recycled; sent for energy incineration or to landfill. PUR insulation (pre-2004) may include ozone-depleting substances which should be recovered	Thermal and acoustic insulation
	PES	Unsaturated polyester	Yes	Easy	Not usually recycled; sent for energy incineration or to landfill	Fibre-reinforced plastics, sanitary-ware, tanks, pipes, gratings
	S	Silicone	Yes	Manageable	Not usually recycled; sent for energy incineration or to landfill; residue of sealant in tubes etc. can affect their recycling	
♲ABS	ABS	Acrylonitrile butadiene styrene	Yes	Easy	Not usually recycled; sent for energy incineration or to landfill	Light, rigid, moulded products such as pipes, enclosures and hard hats and helmets
♲	OTS	Other thermoset	No	Difficult	Not usually recycled; chemical recycling is required for most; sent for energy incineration or to landfill	Powder coatings

Figure 13.2 The National Automotive Innovation Centre, University of Warwick, Cullinan Studio, 2019. Multiple ETFE pillows span the 7m by 7m roof-light openings in the timber roof structure, providing a lighter-weight and lower-carbon solution than glass-and-steel roof lights.

CONSTRUCTION METHODS

Plastics may be specified as a key building element where their performance has advantages. An example would be ETFE pillows that can span large roof-light openings with less structure, reducing embodied carbon compared to a steel-and-glass assembly. With a 30-year lifespan, the ETFE pillows are a good example of plastic stock in service; however, their end of life needs careful consideration.

For primary systems, Environmental Product Declarations (EPDs) should be available to enable carbon analysis and advise on recyclability, to aid decision-making. However, on most building sites, significant quantities of secondary plastics are used

Figure 13.3 Mountain View, London, CAN, 2020. Recycled plastics are already widely used in interior finishes and furniture, as in this south London kitchen.

throughout the process for packaging and protection, often single use. The carbon data for packaging and protection is generally not included in older EPDs, and therefore not scrutinised.

END OF LIFE

Unlike natural materials such as stone, metal, fired clay and even timber, plastic products have less opportunity to be disassembled and reused, therefore recycling must be considered to avoid the last resorts of landfill or incineration. As Table 14.1 illustrates, the likelihood of recycling depends on the composition of the product, the technical capability and market viability.

DESIGN FOR LOWER IMPACTS – REDUCING PLASTICS IN CONSTRUCTION

There are an increasing number of organisations carrying out research on the negative effects of plastics. The Ellen MacArthur Foundation is a leader in this area, raising public awareness and offering advice on steps towards a circular economy for plastic.[17]

Architects selecting building components need to apply additional research and quantification at specification stage. For architects who have committed to the RIBA 2030 Climate Challenge, measuring the whole-life carbon of building systems and components will be part of the process to reduce material carbon emissions. However, to really tackle plastic pollution requires further interrogation. The Alliance of Sustainable Building Products' (ASBP) Reducing Plastics in Construction Group, which includes developers, designers, contractors and manufacturers, is a useful resource for further guidance.[18]

In 2022, housing association Green Square Accord completed the first plastic-free homes in the UK, in Redditch, as part of an EU-funded research project.[19] The homes are virtually plastic-free, but some plastic products could not be avoided due to warranty issues. These included the radon barrier under the ground floor slab and the vapour barrier, both made from recycled plastic. For all other products, non-plastic alternatives were researched and found technically possible.

- Windows are timber frame with mill-finish aluminium outer frames; polyester powder coatings were specifically avoided.
- A plastic-free sealant was applied in lieu of airtightness tapes around the windows.
- Natural, plastic-free finishes and paints were used on the interiors.
- Solid wood was used for kitchen units in lieu of MDF, which contains phenolic resins and polymers.
- Bathroom fixtures are all ceramic.
- Plumbing pipes are copper, which was selected for its durability compared to plastic.
- MVHR (mechanical ventilation with heat recovery) kit and ductwork, conventionally plastic, are galvanised metal.
- Electrical cables are mineral-insulated copper cables.

However, the housing association acknowledged that plastic and recycled plastic can have lower embodied carbon than some alternative metals and glass.[20]

Cullinan Studio is working with the Bankside Open Spaces Trust (BOST) to design a community building for the Marlborough Sports Garden in south London with circular economy principles which includes a rethink on plastics. The approach is threefold: to reduce operational use of plastic, to employ wise specification of permanent materials, and to create zero waste during construction.

Reduce operational use of plastic: The building's café will be designed for the preparation of fresh, healthy food without relying on plastic-packed fast food or single-use plastic dishes and cutlery. This requires adequate area for crockery and dishwashing facilities. Where plastic packaging is unavoidable, space will be allowed for recycling. Water refill stations will be installed to discourage single-use plastic bottles.

Wise specification of permanent materials: Natural finishes are specified where possible for the Marlborough Sports Garden, but where plastics cannot be avoided, investigation is being carried out

Figure 13.4 Green Square Accord's plastic-free homes for Redditch Borough Council, designed by the housing association's in-house architect, 2022. The project was part-funded by the EU's Interreg North-West Europe programme as part of the CHARM (Circular Housing Asset Renovation and Management) initiative involving social housing organisations from four different nations.

QUESTIONS
FOR PROJECT TEAMS
AND SUPPLIERS

For installed products made from plastics

☐ What is the polymer composition type, is it virgin or recycled plastic?

☐ What is its carbon footprint?

☐ Does the product leach chemicals or emit VOCs during its service life?

☐ Can it be reused or recycled at end of service life?

For any product which arrives on site in packaging or is used for protection

☐ Has the manufacturer taken steps to reduce packaging waste and its harmful impacts?

☐ Can the manufacturer send images of the product packaging layers, including the transport layer?

☐ Can a material description of all packaging elements be provided, including plastic type, percentage of recycled content, size, weight and carbon impact?

☐ What take-back schemes are offered?

☐ What means of recycling are available and how, if at all, does the manufacturer track the packaging journey?

on material options and end-of-life disposal. One example is the substrate layers for the biodiverse flat roof. Flat roof membranes are either polymer or bitumen-based – without any viable alternatives to achieve the performance and warranties required. Among the polymer options, TPO/FPO (thermoplastic or flexible polyolefin) single-ply membranes are deemed less damaging to the environment and health than EPDM (elastomeric membranes) or PVC (polyvinyl chloride) membranes. PVC, in particular, has been found to release harmful toxins in production and when incinerated. When researching all the component layers of a biodiverse roof, the specifier will find a variety of plastics, some with options for recycled content. The polymer roof is an example of plastic 'stock in service', and, if made of a significant amount of recycled material, could be a circular way to keep plastic out of waste streams while providing the benefits of a biodiverse roof installation.

Create zero waste during construction: The Marlborough Sports Garden is also a case study in the ZAP Toolkit published by the Alliance of Sustainable Building Products (ASBP) and funded by Ecosurety. Cullinan Studio and BOST collaborated with ASBP, Mace and Morgan Sindall to devise the toolkit. Included are model contract preliminaries that the Marlborough Sports Garden will use. No skips or rubble sacks will be allowed on site and excess material will be referred to as arisings to be sustainably dealt with, rather than waste to send to landfill. Designers and the supply chain will be required to reduce the amount of plastic waste on site through careful specification, and to create a plan for unavoidable plastics that involves supplier take-back, local recycling schemes or for the plastic waste to be used on site.

WIDER SUSTAINABILITY IMPACTS
There is wide public awareness of plastic pollution through people's own encounters with plastic litter and through media campaigns. Plastic can degrade over time when exposed to UV light, temperature variations and chemicals, causing leaching and flaking into the air and watercourses, and there is growing concern about microplastics being inhaled and ingested by humans. In addition, toxic gases are emitted from the burning of plastics in landfill, while some types of plastics emit volatile organic compounds (VOCs) which can be inhaled during use.

FUTURE TRENDS

While enlightened individuals, organisations and businesses strive to take action to prevent plastic pollution, it is worth highlighting changes in policy and legislation that could speed the pace of change.

- In 2021, a new British Standard for biodegradable plastic was introduced called PAS 9017:2020 The biodegradability of polyolefins (in open air, not water). Plastic claiming to be biodegradable will have to pass a test to prove it breaks down into a harmless wax which contains no microplastics or nanoplastics.
- In March 2022, at UNEA-5, the United Nations Environment Assembly endorsed a landmark agreement to end plastic pollution and forge an international, legally binding agreement by 2024. The agreement addresses the full life cycle of plastic.[21]
- The Plastic Packaging Tax (PPT) came into force in the UK in April 2022 and is charged at £200 per tonne on packaging that is predominantly plastic by weight and does not contain at least 30% recycled plastic content.

KEY TAKEAWAYS

- Work across the industry to support a move away from plastics.
- Select products that use less packaging and address packaging in specification.
- Use plastic from recycled sources.
- Screen plastics for leaching of chemicals or emitting VOCs.
- If plastic must be used, try to ensure a long service life.
- Avoid landfill to reduce likelihood of plastic entering the natural environment.

STEEL

Will Arnold, Ana Girao-Coelho and Michael Sansom

Embodied carbon (A1–A3)

2,614 - 19,232

kgCO$_2$e/m³ [1]

Global emissions from constructional steel

3-4% of total emissions

(approximately) [2]

UK steel production per year (all steel)

7.2 million tonnes

(0.4% of global production) [3]

Recyclability

100%

UK end-of-life destination from demolition

93% recycled 7% reused [4]

Steel is one of the strongest materials used in construction and is well known in the UK for its uses in long-span commercial and industrial buildings, stadia and bridges. In 2021, nearly 2 billion tonnes of steel was produced worldwide, of which approximately half went into construction.[5] This figure is expected to increase as economies continue to develop.

Two main methods of production are used to create the majority of the world's steel. These are basic oxygen furnace (BF-BOF) and electric arc furnace (EAF) steelmaking. These two methods, also known as primary and secondary steelmaking, respectively, have significantly different levels of upfront carbon emissions associated with them; however, the lower-carbon EAF production route is currently constrained by the amount of scrap steel available globally, and so cannot be specified for every project.

Steel is the most recycled industrial material in the world, with over 650 million tonnes recycled annually. It can be fully recycled without losing its inherent material properties. Globally, 85% of all scrap steel from all sectors is recycled through existing well-established supply chains. In the UK, 100% of steel recovered from demolition sites already re-enters the supply chain, and no specific steps need to be taken by the designer to ensure this.

The reuse of reclaimed steel offers carbon savings of up to 95% over the use of new primary or secondary steel.

APPLICATIONS

- Structural elements, such as beams
- Purlins and side rails
- Decking and cladding
- Reinforcement bars ('rebar') in concrete

PROS

- Strong
- Durable
- Quality-assured off-site manufacture
- Quick and easy to construct with
- Easily adaptable
- Completely recyclable without loss of properties
- Reusable

CONS

- Fire and corrosion protection often required
- Relatively high carbon emissions associated with manufacture
- Limited options today for significantly reducing emissions due to scrap supply constraints

BF-BOF A form of steelmaking in which molten iron is produced from iron ore, in a blast furnace (BF) using coke, and refined into steel using a basic oxygen furnace (BOF). BF-BOF steelmaking is often termed 'primary steelmaking' as the method is based on creating new steel, mainly from virgin materials.

Cellular beams Light, deep fabricated sections with a series of circular holes along their length, used to accommodate services distribution within the depth of the beam, or for aesthetic reasons.

DRI A form of ironmaking that can then be fed into an EAF. Direct reduced iron (DRI), also called sponge iron, is produced by reducing iron ore into iron, using a reducing gas or elemental carbon, typically produced from natural gas or coal.

EAF A form of steelmaking typically involving melting scrap steel (up to 100%) using an electric arc furnace (EAF). EAF production is often called 'secondary steelmaking' as this method is typically based on recycling scrap.

Slag A co-product of the ironmaking and steelmaking processes, which can be used in construction (typically as an additive to concrete, or for road construction).

KEY PROPERTIES OF STEEL

Steel is an alloy of iron, carbon and other elements. It is a versatile material, available in more than 3,500 variants or grades that differ by composition, by how the steel and final product are made and by their physical and metallurgical properties. Approximately half of all steel produced globally goes into construction.[6]

Most steel used in construction is mild or low-carbon steel. This term relates to the carbon content, typically 0.3 to 0.6%, not the carbon emitted during steelmaking.

Steel has many construction applications, the most common being:
- **structural steel**, also known as 'hot-rolled steel', which includes:
  - open sections such as I-shaped beams, L-shaped angles and C-shaped channels
  - hollow sections, including circular, square and rectangular tubes
  - flat plate used to make bespoke structural members and connections
- **light-gauge steel**, also known as 'cold-formed steel', which includes:
  - secondary structural elements such as purlins and side rails
  - steel decking and cladding
  - studwork used in light steel framing, modular construction, infill walls
- **reinforcement bars** (or 'rebar') used to reinforce concrete.

The carbon footprint of steel construction products is dominated by the impact of making the steel. Although, in practice, environmental impacts are product and project specific, as a rule of thumb, the typical breakdown of embodied carbon for structural steelwork in the UK is as shown in Figure 15.1.

The inherent recyclability and reusability of structural steel means that it has excellent circular economy credentials. This benefit, in terms of environmental impacts, is reflected in module D values for steelwork within whole-life carbon assessments.

This chapter focuses primarily on structural steel, commonly used in commercial and industrial buildings, stadia and bridges. Some high-level information is given below about other types of steel products that are used elsewhere within the built environment.

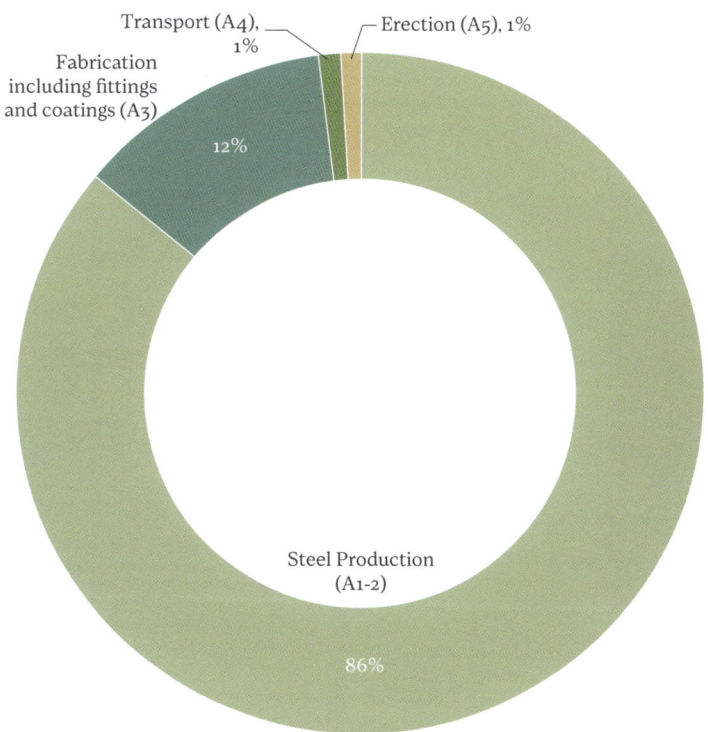

Figure 14.1 Emissions breakdown (LCA modules A1 to A5) of structural steel, based on average UK consumption for structural steel sections, from BCSA (British Constructional Steelwork Association) research.

Transport (A4), 1%

Erection (A5), 1%

Fabrication including fittings and coatings (A3) 12%

Steel Production (A1-2) 86%

Steel is one of the strongest structural materials used in construction, with tension and compression yield strengths typically between 275 and 460MPa (approximately 10 times that of concrete or timber). It also has high elastic modulus, toughness and ductility (i.e. low brittleness).

Durability of structural steel is typically achieved, where required, through painting or galvanising, which may require maintenance during a building's life. Other types of steel with built-in corrosion resistance include stainless steel (created by adding chromium and other elements during manufacture) and weathering steel, including Corten steel (commonly used in exposed, outdoor applications such as bridges and façades). Stainless and weathering steels both lower maintenance emissions during the structure's life and therefore can reduce whole-life carbon emissions.

Steel is non-combustible, although it does lose strength at high temperatures and therefore generally requires fire protection, typically in the form of intumescent paint, plasterboard or cementitious spray. Applied/sprayed protection will have implications for any planned future strengthening, alterations or deconstruction.

By-products of iron and steel production are often used to reduce emissions in concrete construction.

STEEL PRODUCTION

Global crude steel production was 1.95 billion tonnes in 2021 and continues to rise, mainly because of industrialisation in developing economies – and particularly the recent growth in China, which accounted for 53% of all steel production in 2021. UK steel production is currently around 7.2 million tonnes per annum; 0.4% of global production.[7]

Most steel is manufactured in one of two ways.
- **BF-BOF** Molten iron is produced from iron ore, in a blast furnace (BF) using coke (a fuel made by heating coal in the absence of air), and then refined into steel using a basic oxygen furnace (BOF) to which scrap steel is also added (up to 30%) as a coolant. BF-BOF steelmaking is often termed 'primary steelmaking' as the method is based on creating new steel, mainly from virgin materials.
- **EAF** Most EAF production is today a scrap-recycling route, which involves melting steel scrap (up to 100%), using an electric arc furnace (EAF). EAF production is often called 'secondary steelmaking' as this method is mostly based on recycling scrap. Other forms of EAF production are also available that create primary steel rather than secondary steel. These methods use iron ore pellets, produced mainly using natural gas, often in combination with scrap. Primary EAF steelmaking is less common in Europe and the UK, but is important going forward, because using green hydrogen (in place of gas) to reduce the iron ore enables EAF steelmaking with nearly zero carbon emissions.

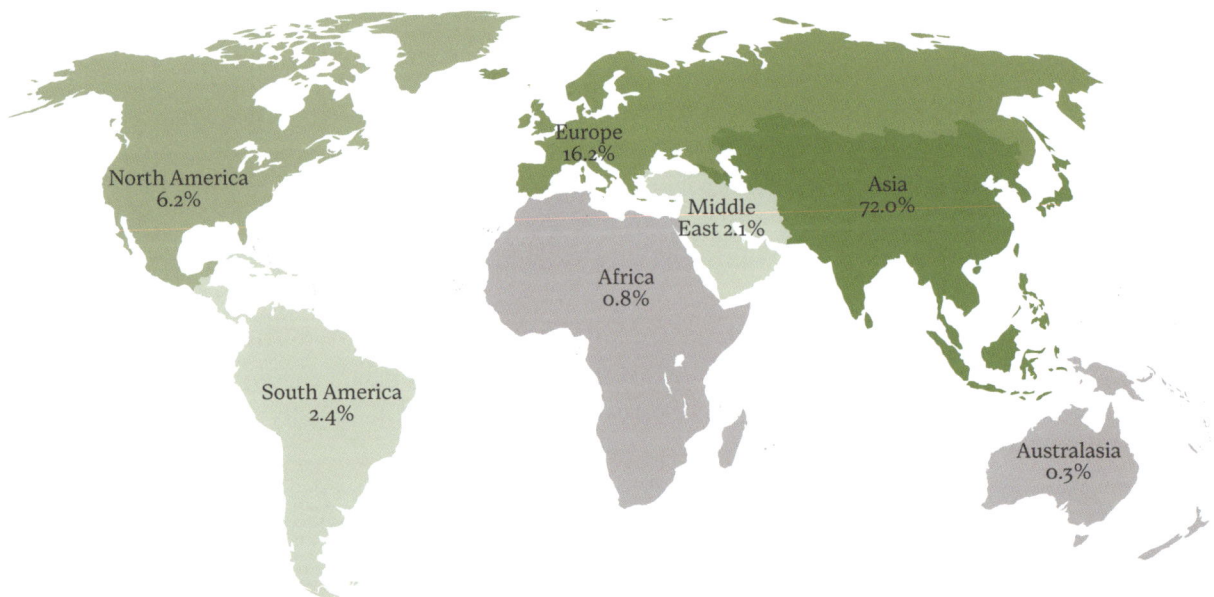

Figure 14.2 Map of global steel production, by region.[8]

In 2021, BF-BOF primary steel accounted for 71% of global crude steel production, and EAF production 29%.[9] Both routes produce the same structural properties in steel.

Importantly, **100% scrap-based EAF steel has a carbon footprint of around 20% that of primary steel**. However, this does not simply mean that the designer can specify EAF steel over BF-BOF steel to reduce emissions, as the scrap steel used to supply EAF manufacturers is globally constrained. This means that asking for more EAF (and thus, more scrap) on one project will simply result in other projects receiving more BF-BOF (and thus, less scrap) on theirs – resulting in no net-benefit to global emission levels.

Finished and semi-finished products

Before molten steel can be rolled or formed into finished products, it is formed into standard, semi-finished products. Today, most steel is produced by continuous casting, in which molten steel is squeezed or extruded through moulds at a controlled rate, to produce basic shapes called billets, blooms or slabs, which are cut to length.

These semi-finished products are then reheated and rolled using different arrangements of rollers to produce two main classes of product – flat products such as plates, sheets or strips of uniform thickness, and long products which are lengths of a particular cross section, ranging from rectangular bars to double-flange I and H sections. Hollow sections (tubes) are typically made by bending and welding flat sheets. The other most common products are reinforcement (rebar), for use in concrete, and hot-dip galvanised coil, used for floor decking, cladding and secondary structural elements, including purlins and side rails and light-gauge steel framing.

Structural sections are available in a range of standardised strengths or grades, cross sections and sizes, and connections are also standardised, with comprehensive design guidance readily available (including Blue Book SCI-P363,[10] Green Books SCI-P358[11] and SCI-P398[12]).

All steel products can be produced using either BF-BOF or EAF routes; however, some products are typically produced by one or other of these routes. Today, only long products (rolled open sections, and reinforcement) and stainless steel are typically produced from scrap in electric arc furnaces, as shown in Table 15.1. This is because it is more difficult to make thin, flat steel products (including plate and hollow sections) from scrap steel due to the variable nature of the input material.

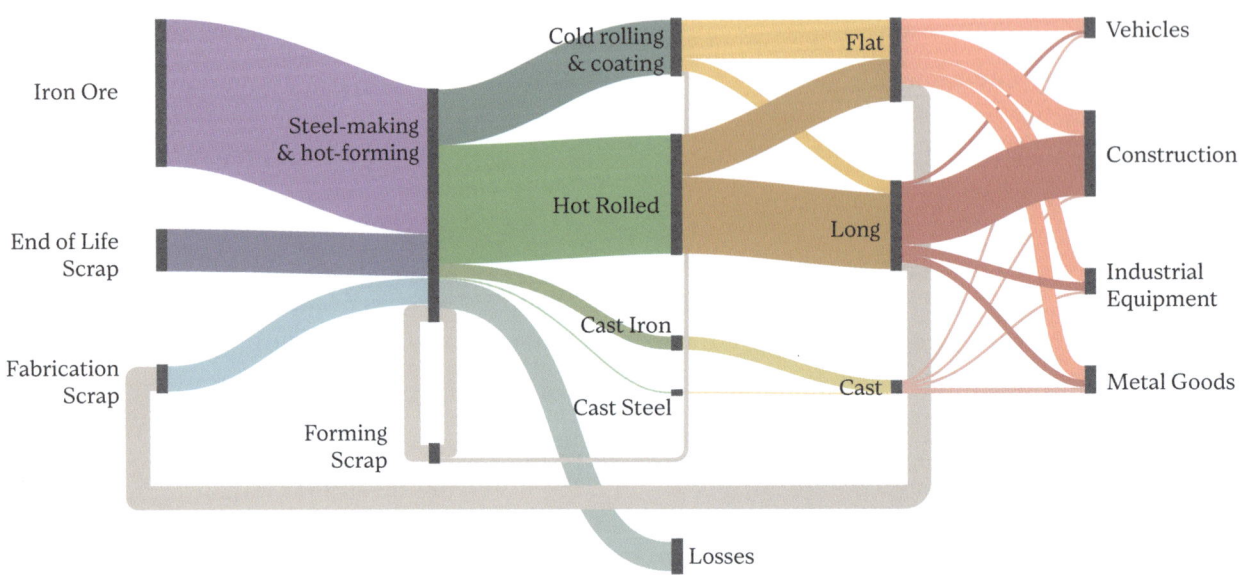

Figure 14.3 Steel production by type.

Iron and steelmaking by-products

The production of iron and steel generates by-products, the most notable of which is slag. On average, the production of one tonne of steel results in around 400kg of co-products in BF-BOF steelmaking and 200kg of co-products in EAF steelmaking. Globally, around 98% of all slag produced is recovered and used. Different slags are produced in the blast furnace and the steel furnace; each has different properties and uses. Most slag comes from blast furnaces, and around 85 to 90% of this is processed to produce ground granulated blast furnace slag (GGBS), used by the concrete industry as a cement replacement. Remaining slags are generally used as aggregates, typically for road-base construction.[13]

Almost all GGBS produced globally is used by the cement industry today, with no surplus or 'stockpiles'. As such, GGBS use in concrete is the same as scrap in steelmaking – it's a constrained material that's already being near-fully exploited to bring down global greenhouse gas emissions and specifying its use to achieve low-carbon building targets does not reduce global carbon emissions.

Gases from iron- and steelmaking processes are routinely cleaned and reused, reducing energy demand. Coke oven gas contains about 55% hydrogen which is fully reused within the steelmaking plant, and can provide up to 40% of the plant's power.

Other by-products include dusts and sludge, which are collected in filters and recycled back into the steelmaking process.

OTHER STEELS, PRODUCTS AND USES
Reinforcement bars (rebar)

Steel and concrete are often used together, as concrete has negligible tensile strength. In reinforced concrete, steel rebar is cast within concrete to enable the steel to work in tension and the concrete in compression. Rebar can also be used in masonry walls to add tension capacity (for example for earthquake resistance).

In the UK, rebar is typically produced using the EAF process described above and, as such, shows a low embodied carbon factor. Appropriate detailing is essential to ensure that the concrete protects the rebar from rusting to meet the design life of the structure.

Composite steel-concrete

The term 'composite' in reference to steel structures typically refers to the action between a concrete floor slab and a steel beam in which the concrete slab resists compression and the steel section is largely in tension. Composite action increases the bending resistance and stiffness of the beam and leads to a reduction in steel weight of 30 to 50%.

Composite action is generally achieved using shear studs welded to the top flanges of beams and cast into concrete floor elements which are supported on light-gauge steel decking. Demountable, bolted shear connectors are also available, enabling the beams to be uncoupled from floor slabs so they can be more easily deconstructed and reused. The Steel Construction Institute provides design guidance on these (SCI P428 [14]).

It is also possible to use steel beams compositely with other flooring systems, including cross-laminated timber (CLT) or precast concrete.

Light-gauge steel

Light-gauge steel construction products and systems generally have the same sustainability credentials as hot-rolled steelwork. In addition, their relative lightness can lead to further carbon savings.

Product category	Typical manufacturing route	Products
Flat products	Typically BF-BOF	Hollow sections, plate (and fabricated sections), hot-dip galvanised coil
Long products	BF-BOF and EAF	Open sections, reinforcement, stainless steel

Table 14.1 Typical manufacturing routes of different steel products.

Light-gauge steel is thin steel, typically 0.4 to 3.0mm thick (compared to the 4 to 140mm range of structural steel). It is used in a range of construction applications, including purlins and side rails, floor decking, cladding and studwork used in infill walling and modular construction.

Light-gauge steel comes in large coils which, in the final production process, are uncoiled, slit into appropriate widths and then cold roll-formed into the final product.

Light-gauge steel is available in a number of grades or strengths, ranging from 220 to 550N/mm2. Most light-gauge steel used in construction is galvanised prior to forming into the final product. The type and weight of metallic coating applied depends on the product and climatic conditions it is to be used in. The coatings, usually zinc based, do not prevent recycling and are recovered during EAF steel production.

In addition to metallic coatings, some light-gauge steel construction products, particularly cladding, have organic coatings applied to improve durability and the aesthetic properties of the product.

Stainless steel

Stainless steel is a family of corrosion- and heat-resistant steels containing a minimum of 10.5% chromium and other alloying elements, including nickel, manganese and molybdenum. The chromium in stainless steel reacts with oxygen in the atmosphere to produce a very thin, self-healing chromium-rich oxide film on the surface of the steel that provides corrosion resistance. There are more than 200 stainless steels, differing mainly by chemical composition and varying levels of corrosion resistance, strength and weldability.

Stainless-steel construction products and systems have the same sustainability credentials as hot-rolled steelwork; however, note that the production emissions are typically higher than normal structural steel, due to the additional alloying elements. Stainless steel is produced using the EAF process and the actual production emissions will depend on the proportion of scrap used.

Stainless steel is used in many construction applications, due to its corrosion resistance, strength, toughness and appearance. Some applications are highly visible, such as architectural cladding and roofing. Others are practical, safety-related and sometimes hidden, like masonry and stone anchors and rebar, handrails, bollards and safety railings and drainage and water systems and fixings. As their corrosion resistance provides longevity and low maintenance benefits, the upfront carbon must be weighed against the whole-life benefits to arrive at the most sustainable overall solution.

Weathering steel (Corten)

Weathering steel, often referred to as Corten steel (which is a trade name), is a family of high-strength steels that contain additional alloying elements that increase the steel's strength and corrosion resistance.

The carbon emissions associated with the production of weathering steel are very similar to those of normal structural steel. However, weathering steel requires no maintenance and therefore generally has a lower whole-life carbon impact compared to structural steel, which requires coating maintenance.

With weathering steel, the rusting process is initiated in the same way as conventional structural steel, but the alloying elements in the steel produce a stable rust layer that adheres to the base metal and is much less porous. This 'patina' acts as a barrier that protects the steel from further corrosion, meaning that corrosion protection is not required.

Weathering steel is commonly used in outdoor applications, where it provides an attractive, very low-maintenance solution.

FABRICATION

Typically, structural steel components are accurately manufactured to quality-assured standards in the controlled environment of a fabrication factory. This is more efficient, faster and safer than site construction – and yields high-quality, accurate products with fewer defects. Waste material produced during the fabrication phase (e.g. offcuts or shavings) is recycled and used again in the steelmaking process.

Fabrication of the steel components ready for installation in buildings typically involves the following processes:
- shotblasting, to clean and prepare the surface of the steel for welding and painting
- cutting, using saws or flame cutting
- drilling and punching holes
- section and plate bending
- welding
- corrosion protection, using paint or hot-dip galvanising.

The carbon impact of steelwork fabrication and erection typically represents 10 to 15% of the overall impact of structural steelwork, although the actual impact is project specific, depending on the complexity of the steelwork, corrosion and fire protection requirements and location and type of building.

Web openings and cellular beams

Cellular beams are light, deep fabricated sections with a series of circular holes along their length, used to accommodate services distribution within the depth of the beam, or for aesthetic reasons. These sections are typically produced by cutting hot-rolled steel sections and rewelding them to create deeper beams with a series of circular holes, or by fabricating a section from steel plates with holes cut along the web element.

Cellular beams are lighter weight and so reduce material use and thus emissions. The additional processing impacts to fabricate cellular beams are more than outweighed by material reduction savings. They are the most common way of designing for long spans – typically 13 to 18m – because the web openings allow the structural and service zones to be combined, reducing floor-to-floor heights.

DELIVERY AND CONSTRUCTION

Manufactured structural steel components are brought to site for final assembly either by bolting or welding elements together. Structural steel products are delivered to site with minimal packaging using timber pallets and bearers and strapping.

Bolting is more common in the UK as it is quicker than welding, but in some countries the preference is to weld. Bolting results in quick erection times, little waste, small construction teams and minimised local disruption and impacts compared to site-based construction systems. The predictability and accuracy of the components means that the structure can be erected quickly.

Steelwork erection is generally undertaken using mobile or fixed (tower) cranes and mobile elevating work platforms (MEWPs) either on the ground or, on high-rise structures, attached to the partly erected structure. Such machinery is transitioning towards electrification in order to reduce A5 emissions.

IN-USE LONGEVITY

Buildings like the National Liberal Club in London (1887) and structures like the Forth Rail Bridge (1890) demonstrate the potential longevity of steel structures. When properly designed and protected where required, steel structures provide long-term durability.

In many instances, steelwork is located in warm, dry interiors where it will not corrode, and its structural stability will not be threatened during its design life. Inside dry buildings with benign atmospheres, such as offices, shops or airport terminals, no corrosion protection is required.

Figure 14.4 Ancora 40, Madrid, Buckley Gray Yeoman, 2021. An example of using lightweight composite steel and cross-laminated timber structural frame to connect two existing buildings.

However, when steelwork is exposed to moisture, corrosion will occur at a rate depending on the severity of the environment and an appropriate coating system is required.

Standard paint systems, which may require one, two or three layers of different paint types, are available to protect steel in various environmental conditions. The carbon emissions relating to paint use will typically be small compared to that of the steelwork. Specifying the correct protection for steelwork is important and can require specialist advise from steelwork contractors and coating manufacturers.

The detailing of connections, particularly externally, is also important to minimise moisture build-up, for example detailing holes in connections to avoid rainwater collecting in crevices.

Stainless and weathering steels do not require additional protection against corrosion, though discolouration can occur.

ALTERATIONS AND REPAIR

Structural steel, because of its strength and relative lightness, can also be used to extend the life of concrete and masonry buildings, for example by façade retention and/or by adding additional storeys. The embodied carbon of additional steel must be considered in evaluating refurbishment options.

Steel buildings can easily be strengthened (by welding on additional steel plate) and/or reconfigured to accommodate change of use. This can be useful when considering future possible loading conditions, vertical extensions to buildings or changes in use. Commonly, steel structures are used in retrofit to open up the internal structure of a building to ensure its commercial viability.

Lightweight steel structures can be used to extend steel or concrete-framed buildings vertically to provide additional floorspace, without the need for strengthening the existing structure.

Steel structures are intrinsically resilient, and they perform well during and following extreme events, such as earthquakes. Elements can be repaired or cut out and replaced if repair is not possible.

DECONSTRUCTION AND REUSE AT END OF LIFE

Steel structures can be deconstructed at end of life and directly reused in new buildings, with the potential to save as much as 95% of the embodied carbon compared to using new steel. Reuse is on the increase in the UK but it is far from mainstream. For more information, see 'Use of reclaimed materials in new designs', on p 203.

Recycling

Where refurbishment is not viable and a steel structure cannot be deconstructed and reused, steel structures are demolished and recycled into new steel products. Because of the economic value of scrap steel, it is efficiently recovered and recycled through a well-established and efficient supply chain.

Steel is 100% recyclable without loss of its inherent material properties and is the most recycled industrial material in the world, with over 650 million tonnes recycled annually – an estimated 75% of steel products ever made are still in use today. **In theory, all new steel could be made from recycled steel using EAF production with a much lower carbon impact; however, this is currently not possible because global demand for steel exceeds the supply of scrap by a factor of approximately 3. Hence, it is crucial to use less steel and to decarbonise primary steelmaking.**

Module D benefits of recycled and reused steel

Steel products produced today all contain a proportion of recycled steel from one or more previous incarnations. Therefore, as long as recycling continues, the life of a steel product (or, more accurately, the steel within it) is, in effect, infinite and individual incarnations or uses of a steel product are merely parts of the larger life cycle of the material.

Whole-life cycle assessments, using international standards, on the other hand, generally limit the scope of assessment to the building only. This incompatibility between the life of the building and the life of steel leads to methodological challenges in accounting for the environmental impacts and benefits of products and materials outside the scope of assessment.

Sustainability assessment standards account for this by means of module D (see Primer, Figure 1.1) which applies to all environmental impacts, not just greenhouse gas emissions (GHG), and is an important metric for quantifying the reuse and recycling potential of products and materials. This aspect of whole-life assessment is important for steel because of its inherent closed-loop recyclability and therefore it should be quantified and considered.

Looking only at modules A to C (whole-life carbon) fails to account for these future benefits, and while the future recyclability of a product does not reduce today's emissions, it is important to minimise future emissions where possible. Reporting module D helps understand the future impacts/benefits of decisions taken today.

DESIGNING FOR LOWER CARBON IMPACTS
Using material efficiently

As with all materials, **the most effective way to reduce emissions today is to minimise the amount of material used.** Steel structures should minimise complex load paths and transfers, as well as optimise spans and grids for efficiency.

Ensuring that structural zones are deep enough to facilitate efficient beam depths is key. Alongside efficient layout of the structure on plan, the vertical zoning of the floor-to-floor section is particularly important for steel. While shallow versions of typical steel beams are available, these are heavier (and thus higher carbon) than deeper beam profiles because extra steel is added to the top and bottom flanges of the beam to compensate for the lack of depth. Cellular beams are structurally efficient, long-span structural solutions which allow services to be integrated within the structural zone, reducing the building height and envelope area.

Seek a structural engineer's input early

Input by the structural engineer at the earliest possible design stage can ensure that an efficient layout and zoning is incorporated from the start. Where long spans are required or justified in terms of building form and functionality, the size of steel beams is often dictated by the need to limit floor deflections or vibrations. The decision to use more structural materials (and therefore more upfront carbon) to provide longer, clear spans needs to be considered in the context of the broader sustainability impacts of the building, including its utilisation, flexibility and commercial viability.

To reduce the weight of individual members, consider shape optimisation, such as asymmetric built-up sections or cellular beams, and specify higher-strength steel grades for columns and transfer trusses where they are most effective and economically feasible.

Although higher-strength steels differ in terms of their manufacture, their embodied carbon impact is around the same as normal-strength steels. Any additional impact is more than offset by the material savings where high-strength steels are used in the appropriate application, for example columns.

Consult a steelwork contractor early

It is important to consult a steelwork contractor early, in order to discuss the art of the possible. Final refinements of the structural design and details can be made during RIBA Stage 4 if sufficient time and design fees allow. This enables the designer to optimise member sizes, by avoiding over-rationalisation (e.g. using the same-sized beam across a whole floor) and improving detailing (e.g. pre-cambering beams by bending them to counteract expected deflection due to adding the floor slab on site) to find final weight reductions. The practical and cost implications on fabrication and erection of optimising members also need to be considered.

Assign time in the process for the designer and contractor to work together

Practical aspects of construction should be considered during design to avoid later increases in member sizes by the contractor, for example, where the flanges of floor beams are too narrow to install shear studs. This enables the best balance of optimisation by reducing sizes to respond to structural requirements and rationalisation to aid fabrication and constructability.

Review temporary works requirements

Where required, temporary steelwork should be reviewed, minimised and, where possible, designed to be reused in other temporary or permanent works.

Use of reclaimed materials in new designs

Steel structures can be deconstructed at end of life and reused in new buildings, which can save around 95% of the embodied carbon compared to using new steel. This typically takes place by cutting beams out from a frame. Even when the frame was bolted, it is generally easier and safer to cut than unbolt. Entire structures can also be carefully unbolted and re-erected in the same configuration at a new location – this is often the case with single-storey industrial and agricultural buildings.

Table 15.2 shows the hierarchy of how steel-framed buildings can contribute to a circular economy, reducing carbon impacts today.[15]

Reuse of structural steel does not pose many technical barriers, but the chemical and physical properties of the reclaimed steel, along with its condition (section loss, damage, straightness, etc.) must be understood prior to reuse. Currently, the barriers to reuse relate primarily to practical and logistical issues such as availability of reclaimed sections and safe deconstruction of structures.

Identify a source for reclaimed materials as early as possible

If reclaimed structural steel elements are being considered, flexibility in concept design is essential to allow design iterations to reflect the availability of reclaimed materials.

In the UK, reuse of steelwork produced after 1970 is easier because it meets the material properties assumed in British product and design standards. The Steel Construction Institute has published guidance on reusing steel produced since 1970 (SCI P427)[16] and produced between 1932 and 1970 (SCI-P440).[17]

Note that in terms of the climate crisis, the benefit that is obtained through the reuse of reclaimed steel sections comes from avoiding the energy needed to produce new steel (alongside the other benefits to nature that come from reducing our reliance on extraction of materials). Every tonne of steel reused is a tonne not recycled. As a rule of thumb, if reusing reclaimed sections inefficiently increases the steel weight by more than 20 to 30%, compared to a solution using new steel, then, in terms of carbon emissions, recycling is likely to be the better option.[18]

Design for adaptability and deconstruction

As steel structures are easily strengthened and adapted, it is good practice to review design options to consider future flexibility, deconstruction and reuse. Some approaches to increasing flexibility, such as larger column grids, come with an upfront carbon cost which must be assessed to see whether it can be justified in terms of whole-life carbon.

Seek solutions which minimise upfront emissions while also increasing adaptability and deconstruction

Simple measures to minimise upfront emissions while also increasing adaptability and deconstruction include the following.

- Use bolted connections rather than welded joints.
- Use standard connection details, including bolt sizes and spacing of holes.
- Ensure easy and permanent access to connections.
- Where feasible, ensure that steel is free from coatings or coverings that will prevent visual assessment of its condition.
- Minimise the use of fixings to structural steel elements that require welding, drilling holes or fixing with nails; use clamped fittings instead, where possible.
- Identify the origin and properties of components by bar-coding, e-tagging or stamping.
- Maintain an inventory of products.
- Save detailed, as-built BIM models so that steel structures can be safely and efficiently adapted, deconstructed and reused.

Records of as-built structures, along with any adaption, deconstruction and reuse strategies, should be passed on to the client to facilitate future changes. Permanent labelling of components (e.g. via stamping, RFID (radio frequency identification devices) tagging, etc.) with key properties such as member sizes and steel grades will facilitate future reuse by providing traceability of steel properties.

Responsible sourcing

All structural steel should be procured based on the principles of responsible sourcing, as defined in BES 6001 (or equivalent), which requires an Environmental Management System compliant with BS EN ISO 14001 (or equivalent).

Primarily used by the structural engineer, the UK National Structural Steelwork Specification for Building Construction (NSSS)[19] specifies requirements for the steelwork, such as workmanship, welding and erection. Annex J specifically addresses the specification of sustainable structural steelwork. The NSSS states that, where possible, steel should be sourced from producers who have defined and are implementing a strategy to reduce greenhouse gas emissions, and who have made a public commitment to decarbonise in line with national and/or international targets. This includes:

- an emissions reduction pathway compatible with the goals of the 2015 Paris Agreement;
- a validated science-based target, for example a target approved by the Science Based Target Initiative (SBTi);
- ResponsibleSteel certified steel or steel meeting an equivalent international standard.[20]

Measure	Explanation
1. In situ reuse	Extends the life of buildings by retaining and adapting the structure. Steel buildings can be strengthened and/or reconfigured. Steel structures can also be used to extend the life of concrete buildings by adding additional floors. The reuse of a structure eliminates all carbon emissions associated with creating a new structure.
2. Deconstruction and reuse	Entire steel frames and structures can be deconstructed and re-erected on a new site, with the individual members made available for reuse. The emissions due to reusing steel frames are up to 95% lower than creating a new structure.
3. Demolition and recycling	Where reuse is not viable, a steel building is demolished and all the steelwork is recovered and recycled into new steel products. The carbon impact of scrap-based EAF steel is around 70 to 80% lower than that of primary BF-BOF steel.

Table 14.2 Hierarchy of circular economy opportunities with steel.

Specifying the ResponsibleSteel Standard, created by not-for-profit organisation ResponsibleSteel, supports responsible sourcing and production by tackling the economic, social and environmental issues associated with steel production and procurement. The Standard includes requirements around corporate leadership, health and safety, labour rights, human rights, communities, greenhouse gases and biodiversity.

WIDER SUSTAINABILITY IMPACTS
Routes to decarbonisation require step changes in current steelmaking technologies. Such technologies are already being piloted by the steel industry but will take time and significant finance to commercialise at scale.

Corporate commitments and implementation plans are one way a supply chain can demonstrate that it is working to reduce emissions. Many steel producers have published credible, long-term emissions reduction pathways that are compatible with the Paris Agreement and include medium- and long-term, science-based GHG emissions targets.

Many steel producers are also becoming ResponsibleSteel-certified. Organisations using steel can sign up to global initiative SteelZero to demonstrate their commitment to procuring, specifying or stocking 50% low-carbon steel by 2030, and 100% net-zero steel by 2050.[21] For definitions of 'low carbon steel' and 'net zero steel', SteelZero refers to the ResponsibleSteel Standard (see 'Responsible sourcing', on pp 204-05).

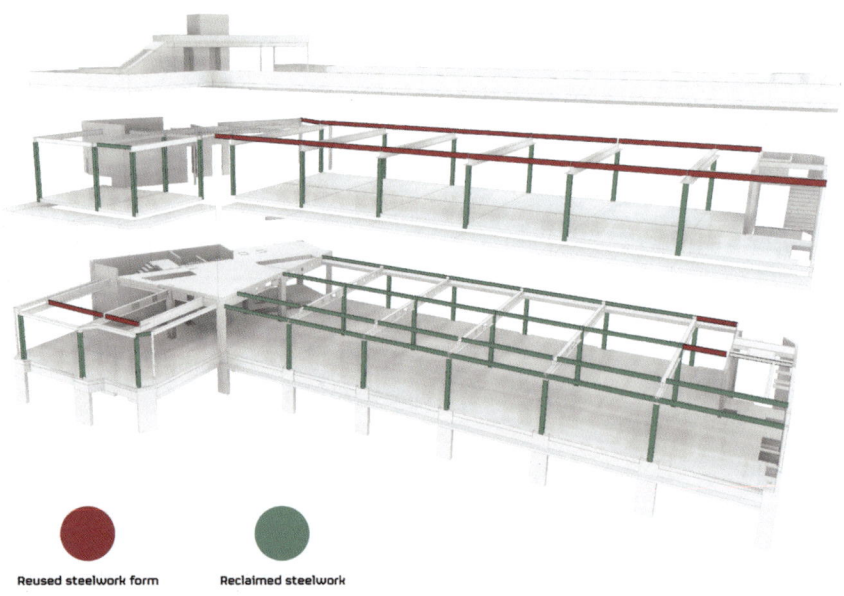

Reused steelwork form   Reclaimed steelwork

Figures 14.5a and b Holbein
Gardens, Belgravia, London, Barr
Gazetas, 2023. The project utilised
25 tonnes of reclaimed steel sections,
taken from other projects within the
Grosvenor Estate and from a general
stockist, with Figure 15.5b illustrating
the sourcing of the steel.

British Constructional Steelwork Association (BCSA)
Sustainability Charter and Roadmap
Launched in 2005, the BCSA Sustainability Charter was updated
in 2021 in response to the climate emergency.[22] Under the Charter,
steelwork contractors formally commit to sustainability principles
and are assessed and monitored against a range of environmental,
social and economic criteria.

Published in 2021, the BCSA's UK Structural Steelwork: 2050 Decarbonisation Roadmap sets out how the sector will decarbonise to meet UK net-zero 2050 targets (see figure 15.6).[23] Decarbonisation requires action across the supply chain. **Designers, steelwork contractors, steel producers, stockholders and demolition contractors must collaborate on both demand-side and supply-side reduction measures.**

To address the largest carbon impact within the supply chain, the biggest reduction measures must come from steelmaking: so-called supply-side measures, levers 3, 5 and 6. **Designers have an important demand-side role to play in reducing the amount of steel used**, through specification.

Figure 14.6 The BCSA's UK Structural Steelwork: 2050 Decarbonisation Roadmap is based on six decarbonisation strategies or 'levers', each of which is either already available or being piloted. Transition to net zero will involve all these different approaches, and the precise mix is likely to change as technologies and national and international policies and support evolve.

Biodiversity

As with all materials, the indirect environmental, ecological and social impacts associated with the production and processing of raw materials are significant – over two billion tonnes of iron has been mined each year since 2011.[24] As steel supply chains are global and complex, it is difficult to control the provenance and impact of the raw materials used in steelmaking. The use of sustainable sourcing standards such as those outlined above is a step in the right direction.

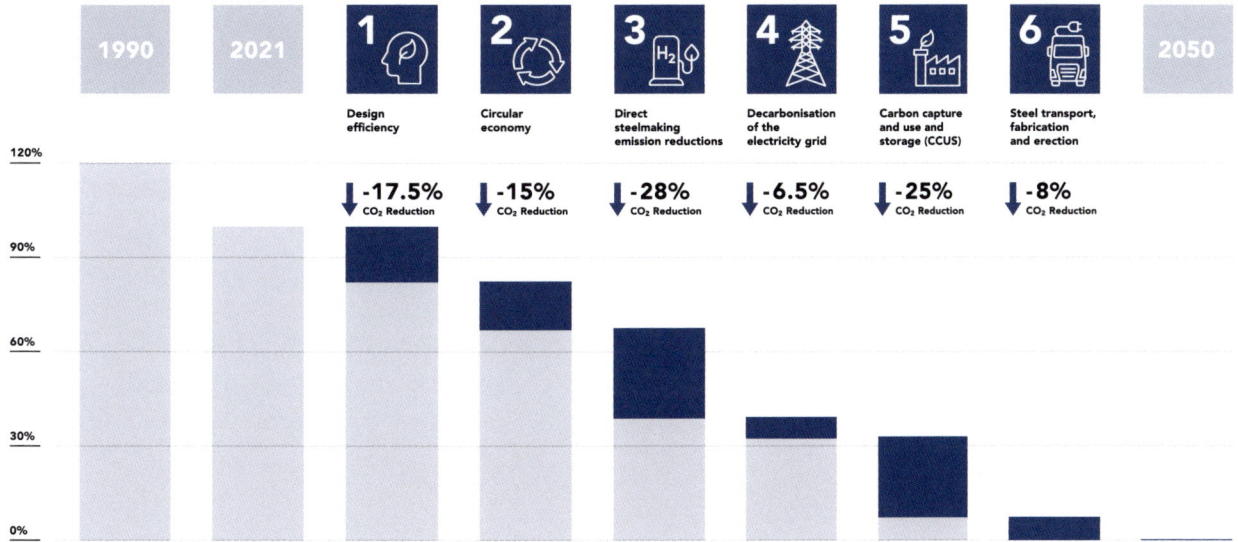

QUESTIONS
FOR PROJECT TEAMS AND SUPPLIERS

☐ Is it possible to reuse any of the steel being taken from the existing building that we are about to deconstruct/refurbish?

☐ Can we use reclaimed structural steel efficiently in this new design? (If an equivalent design with new steel would be 30% lighter, we should instead use that.) And if so, how soon can we start talking about a source for the reclaimed steel?

☐ Is the structural form efficient for steel? Are the spans the right length? Have we minimised the use of complex transfer structures and load paths?

☐ Are the structural zones sufficiently deep to facilitate efficient beam depths?

☐ Could cellular beams or beams with web openings be used to integrate structure and services into the same depth?

☐ How can we enable time for the design team and contractors to work together to further optimise steel section sizes and avoid over-rationalisation?

☐ What's the best time to engage with the contractor to review for buildability?

☐ Is this design going to last? Do the details ensure the shedding of water, are the coatings correct, and have we enabled future building adaption and deconstruction (without significantly increasing upfront carbon)?

FUTURE TRENDS

Steel has a vital role to play in the transition to a zero-carbon global economy and the industrial and infrastructure revolution that is required to reach net zero by 2050. Specific examples include wind turbines, solar harvesting and other renewable energy technologies, including tidal, hydrogen production facilities and networks, nuclear power and biomass plants.

One way for the steel sector to decarbonise would be to make all new steel from recycled scrap, using renewable energy. However, global demand for steel currently exceeds the supply of scrap by a factor of approximately 3, and while the amount of scrap available will increase by 2050, the demand for new steel will increase, too. The International Energy Agency predicts growth of 35% over 2019 levels by 2050.[25]

As the global population stabilises (expected to be between 10 and 12 billion people in the early part of the 22nd century[26]) and global development slows, steel scrap supply may eventually match demand for new steel, as is already the case in some richer economies. However, given the urgency of the climate crisis, and the fact that scrap steel is unlikely to match the demand for new steel globally in this century, the focus today must be on decarbonising primary steelmaking.[27]

While the ultimate goal is to achieve zero-emissions structural steel, a number of key steps are being taken to transition towards this. These include retrofitting existing BF-BOF facilities to increase efficiency, and transitioning to biomass instead of coke as the main source of fuel.

Greater emissions savings are anticipated from increasing the use of direct reduced iron (DRI) processes to produce iron that can be fed into the EAF process. DRI-EAF is already in use around the world, though it accounts for only 100 million tonnes of steel today, about 5% of global production. This uses natural gas within the DRI furnace as opposed to the coke used in a blast furnace, and so the emissions from DRI-EAF production are around 50% lower than BF-BOF, while still producing primary steel.

What steel sourcing credentials should we be insisting on? BES 6001 sustainably certified? ResponsibleSteel aligned?

Has the design of our steel structure complied with the requirements and practices of the National Structural Steelwork Specification, Annex J (Sustainability Specification)?

Research is also underway to use green hydrogen (produced using renewable energy) in place of natural gas in DRI-EAF steelmaking. A consortium called HYBRIT, based in Sweden, produced a small amount of steel in 2021 using this process, and fossil-free steel is expected to be commercially available in 2026.[28] While affordable green hydrogen will be in short supply (and high demand) in the short term, with several other sectors also looking to it in their decarbonisation plans, this is likely to provide one route to decarbonising primary steelmaking.

While carbon capture, utilisation and storage (CCUS) is presently technologically premature and yet to be proven economically, it is also being relied upon by the steel industry (and many other hard-to-decarbonise industries, such as fertiliser and cement) as a way to capture residual emissions that cannot be tackled through the other approaches listed here.

KEY TAKEAWAYS

- Efficient configurations and designs are key to reducing emissions in new steel-framed buildings – it is only through efficient use of material that the designer can reduce the upfront embodied carbon impact of a new design.

- While scrap-based electric arc furnace (EAF) steelmaking has a carbon footprint of around 20% of basic oxygen furnace (BOF-BF) steelmaking, this does not simply mean that the designer should specify EAF steel to reduce emissions of their project, as doing so will not reduce global emissions due to a globally constrained scrap supply.

- The inherent recyclability and reusability of structural steel means that it has excellent circular economy credentials. In the UK,

structural steel products are virtually 100% recovered for either reuse or recycling.

- Designers do not need to take any specific steps to ensure steel recycling because the economic value of scrap steel means that it is efficiently recovered and recycled through a well-established and efficient supply chain.

- The reuse of reclaimed steel elements to create 'new' steel frames is gaining momentum as there are few technical barriers to reuse. Reclaiming and reusing structural steel can save around 95% of the embodied carbon compared to using new steel. This approach is likely to increase in developed countries like the UK.

STONE

Alex Lynes, Pierre Bidaud and Steve Webb

Number of active UK quarries

1,850[1]

Amount of stone quarried in the UK annually

250 million tonnes[2]

Embodied carbon of European stone blocks, pavers, beams and columns

50–250 $kgCO_2e/m^3$ [3]

Embodied carbon of European stone cladding, tiles and decoration

150-600 $kgCO_2e/m^3$ [4]

Stone is readily available across the globe just by digging into the ground. In modern times, stone is extracted from the earth in quarries by large machinery and then worked into a wide range of products. As new materials and technologies developed, stone evolved from a common, everyday building material to a luxury. Due to the current focus on the climate emergency, stone is emerging as an alternative to more carbon-intensive materials.

Stone is particularly strong in compression, though weaker in tension. It is dense, hard and does not corrode, melt or burn. It is capable of being split, cut or shaped by hand tools and is very hardwearing.

While stone is stronger and more durable than concrete or brick, it is a natural material and therefore subject to variability, which must be considered when specifying. Uses of stone vary widely, with some types of stone more suited to external use and others to delicate sculptural work or affordable block walling.

Stone is most commonly used in blocks or slabs. Historically, blocks were either dry laid or mortared together to form walls, columns, vaults and lintels. Almost 90% of stone quarried now in the UK is crushed and used for aggregates or fill materials. Currently only stone valued for its aesthetic appeal (less than 1% of UK production) is cut into slabs or tiles or used for historical restoration, but stone can be used for much more.

APPLICATIONS

- Foundations
- Structure
- External and internal finishes

PROS

- Dense and durable
- Low carbon, depending on processing and transport
- Typically stronger than brick and concrete
- Broad range of colours and textures available
- Natural material with no off-gassing
- Infinitely reusable

CONS

- Heavy, so can be costly to construct and transport
- Thin tiles can be brittle
- Expensive if heavily selected or worked

Stone is potentially the oldest building material and the master stonemasons of old are the forefathers of both the architectural and engineering professions. More permanent and stronger than timber, stone was the primary material to build large structures, influencing the look of buildings to align with its characteristics. The genesis of much architecture stems from the practices of building with stone.

Nowadays, stone is used to make buildings appear monumental and prestigious but very often stone is just the finishing layer rather than the structural building material. Cladding and internal tiling are more common than load-bearing applications, and stone is considered expensive to use as solid blocks.

Stone is now rightly being rediscovered as a low-carbon alternative to both concrete and masonry. Designers, architects and engineers are rediscovering past techniques, as well as developing new construction methods using modern manufacturing machinery. Stone is a natural, variable material that requires a deeper understanding than other more uniform, modern materials, but that should not prevent utilising its benefits, especially with the target of a low-carbon future.

Figure 15.1 Approximate distribution of uses of stone quarried in the UK. Almost 90% is crushed for aggregate, and less than 1% is currently cut into blocks as 'dimensioned stone'.[5]

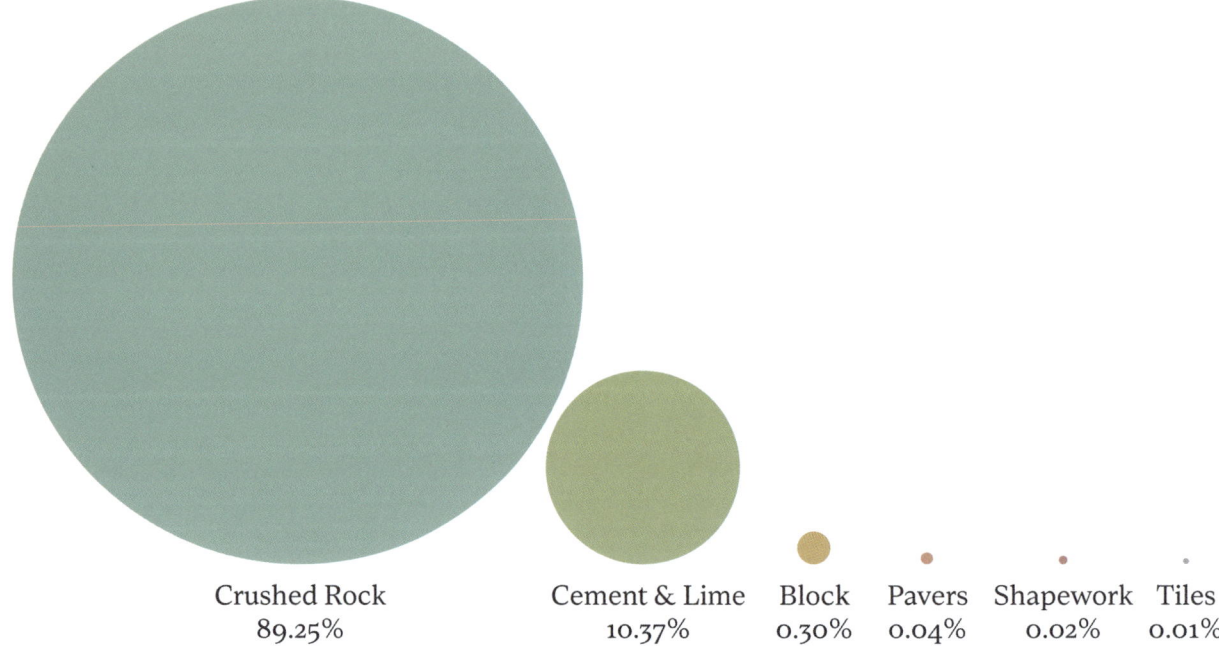

Crushed Rock 89.25%   Cement & Lime 10.37%   Block 0.30%   Pavers 0.04%   Shapework 0.02%   Tiles 0.01%

QUARRYING: UNDERSTANDING DIFFERENT STONE TYPES

Not all stone is created equal. Tremendous variation exists between different geological classifications, different quarries and even within a single bed of a quarry. Stone is a natural material, influenced by its location and creation conditions, much like timber or wine.

As a start, a basic geological overview of the main stone types is needed:

- **Basalt** Volcanic lava that formed at the planet's surface. Basalt is slightly weaker and more crystalline than granite. Typically, it has a darker colour and finer grain because it formed in a relatively short period. It can also be melted and processed into fibres and used in polymers to make fabrics, bars and plate products.
- **Granite** Magma that forms deep underground. Granite is strong and hard with no bedding planes but can have temperature stress fissures or joints, depending on how it formed. Larger crystalline grains are visible, especially when polished.
- **Limestone** Formed from shelled creature deposits on riverbeds. It has fine particles, but because it forms over long periods, its properties can vary and be influenced by the water. Limestone has a wide range of strengths and textures. Porosity can also vary, determined by the type of shelled creature deposited. Limestone has medium hardness and therefore is easy to cut and shape but can exhibit wear.
- **Marble** Heated and compressed limestone. It is generally denser, harder, less porous and more brittle than other stone, so can be hard to work. It is normally a little stronger than limestone and can be finished to a high polish.
- **Sandstone** Sand compressed over time. It varies from large particles loosely compacted to finer particles forming a dense and strong stone. Particles are silicates and so are very hard, making them difficult to shape and finish. Sandstone is very variable in strength – it can be barely held together or very hard and able to scratch glass.

EXTRACTION

There are as many types of quarries as there are different types of geological conditions and so it is very hard to sum up the make-up of the ground beneath our feet. For example, limestone quarries, due to the sedimentation of the debris, the tectonic movement and the layering of other material, such as clay, schists or gas, are made up of a succession of layers, called bedding layers. These beds will have different visual aspects, colours, textures and technical properties. As extraction moves in the quarry, horizontally and vertically, different beds are exposed. The height of a bed can vary from 300mm to 8 or 10m.

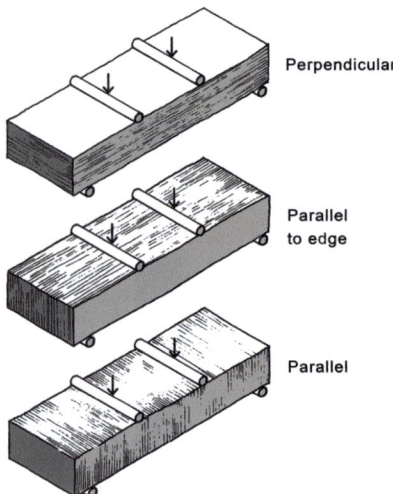

Perpendicular

Parallel
to edge

Parallel

Figure 15.2 Stone bedding layer direction relative to imposed load. Stone often has different strengths when loaded in different directions, with parallel-to-bedding layers (bottom image) typically being the weakest direction. It is important to understand both the variation between the directions and the orientation that the stone will be loaded in use to use it effectively.

The quarry's layering influences the extraction strategy, which is tailored to be the most efficient in terms of both the energy needed to get the stone out of the quarry and the yield of each block. Because stone is a solid material, the aim is to use natural fractures, such as a weak layer or fault, to save a long cut. Once these weak points are identified, several types of extraction can be used: drilling and splitting, wire sawing or by large hydraulic chainsaw. The deeper and more horizontal a bed is, the easier the extraction. Size of blocks does not always depend on quarry conditions but also on the tools and methodology used. New technologies enable large blocks to be extracted.

The variations in a stone quarry depend on the sediments – for example, type of shells, organic matter or other minerals – the speed of its formation and the different compression stages. Colours are determined during the stone's formation by sipping or leaking of liquid or gas containing iron, bitumen or manganese into the stone.

The main issue with natural stone in modern construction is its unpredictability. Traditionally, the way stone was selected was based on years of experience of the quarry master and local stonemason, who relied on visual observations and sounding the material with a hammer. Since the early 1900s, more destructive tests have been used that help source reliable stone. More recently, non-destructive tests have been developed with some success to get a better understanding of the material without relying on lengthy and expensive destructive tests.

Until stone cladding became popular in the 1980s, there was very little visual selection for load-bearing masonry. Typically, stone was triaged depending on its strength, suitability and mechanical properties. Quarry masters would have known the appropriate use for each bed: simple ashlar from the soft bed for walls, and hard limestone for plinths, windowsills and cornices.

APPLICATIONS
Stone can be used in many different ways, as illustrated in Figure 16.3. It is useful to consider the priorities for a specific project. The choice between load-bearing and non-load-bearing stone can have a big impact on form, whereas the choice between site work and prefabrication will have a greater impact on look and buildability.

Solutions in the centre of the diagram will likely be less efficient (more expensive) as they do not make good use of stone's inherent properties.

Walls and cladding

One of the most effective (and oldest) uses of stone is for constructing walls. The compressive strength of stone normally greatly exceeds brick or block, and with precision cutting and thin mortar joints, the full strength can be most effectively utilised. Generally, a solid stone block wall will be thinner than the equivalent blockwork one.

When stone is used for aesthetic effect, it becomes cladding. The thickness can be reduced but additional support may be required, which often increases the overall complexity, cost and carbon. Cladding can also be bonded to a substrate as tiles, either to floors or walls. Size is driven by availability and buildability constraints, but large slabs may need additional thickness if used as floor tiles to prevent cracking.

Figure 15.3 This stone-use decision matrix shows the types of stone to use for different project drivers and applications: prefabrication versus site work and load-bearing versus non-load-bearing. Options in the central circle should be avoided due to inefficient material use.

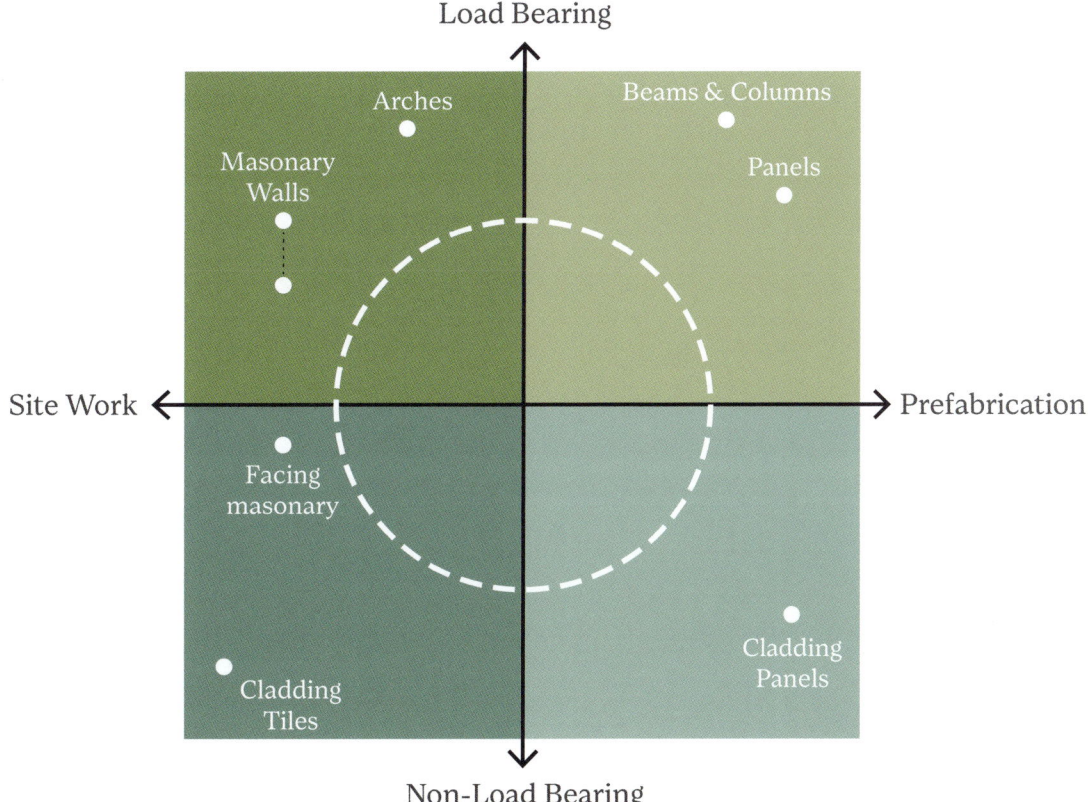

Figure 15.4 Padre Pio Pilgrimage Church, San Giovanni Rotondo, Italy, Renzo Piano Building Workshop, 2004. A modern stone arch structure with steel cables running through 22 arches, arranged in a radial pattern, creates a light-filled space. The overlapping arches form an impressive exposed structure and support the lightweight roof, using minimal material to span a vast space.

Arches and vaults

Arches and vaults are the most classic stone forms, using stone's high compressive strength to span efficiently over a space. The principles of these structures are well understood but they are underused in modern design, partly due to the perceived complexity of building anything curved and partly because additional structural depth is required compared to flat-spanning structures. Arches and vaults should be factored into a design from an early stage to ensure that form follows function and the geometry is dictated by the structural requirements or they lose their efficiency rapidly. When proposing complex forms, buildability must be carefully considered and repetition is often desirable.

Beams and columns

Unreinforced stone lintels are solid blocks of stone spanning a short gap. As stone typically has twice the strength of concrete, the size of the lintel can be reduced to a practical thickness. The span achievable is normally governed by the size of block that can be quarried or the weight of block that can be lifted.

Beams can be either reinforced or tensioned and preassembled or the blocks assembled on site, though the structural principles remain the same. Reinforced stone beams are typically similar in size to concrete beams, while tensioned beams can be smaller and are normally limited by bar spacing at the anchorages. As beams are typically straight, bars rather than cables are normally used.

Stone also works well for columns, either simply built with narrow mortar joints and a lintel or with beams on top or reinforced with bars. If the maximum block size permits, columns can even be single elements.

Spanning floor or roof structures can also be made from stone, as monolithic single slabs supported by walls or beams or for longer spans reinforced (or tensioned) planks made up of multiple blocks. Construction details can be identical to concrete precast planks, with support from walls or beams, and a structural screed can be added to the top to level and provide a continuous floor.

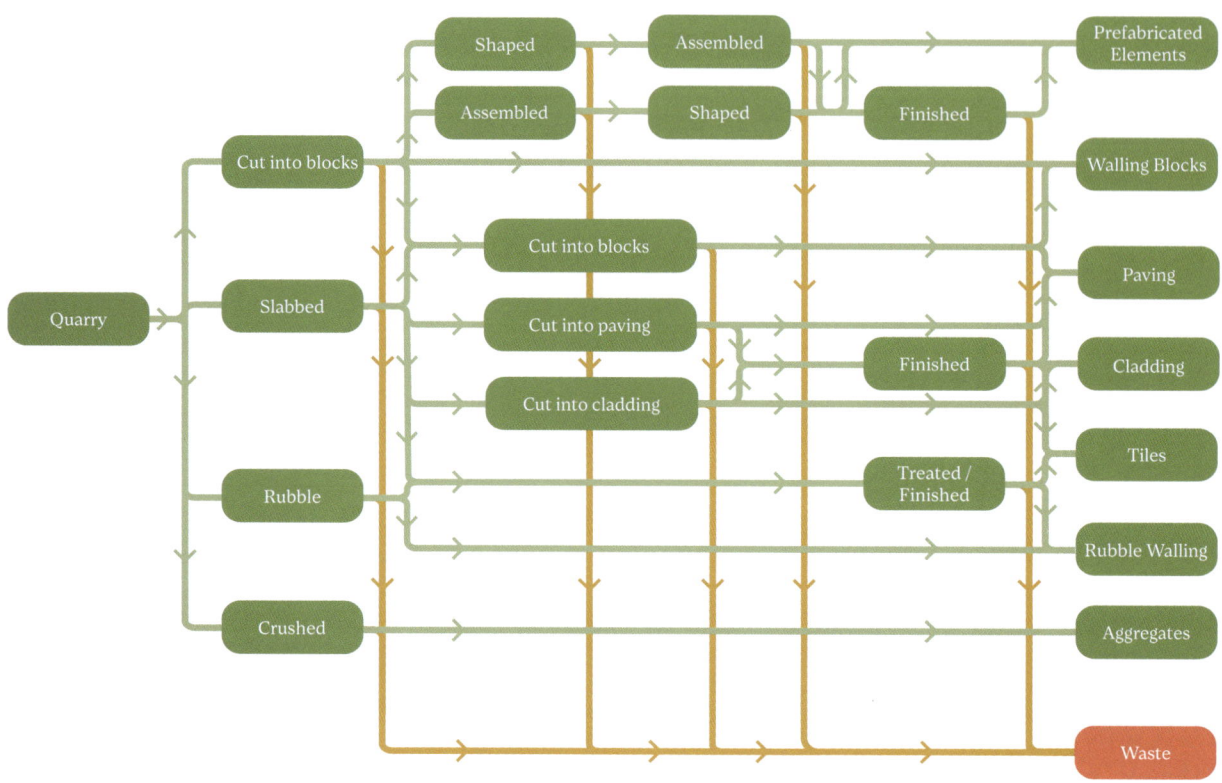

Figure 15.5 Diagram showing a selection of possible uses and manufacturing routes for stone, from quarry to site. Each process can be done in the same location or the material can be transported between different locations or companies, adding carbon and cost.

MANUFACTURE

The shaping of stone depends on its hardness and as such the most appropriate tools will differ. The craftsman's knowledge of stone cutting has evolved, along with technological development in steel and diamond cutting; from manual tools of percussion (hammer and chisel), to grinding with sand, to the use of compressed air for disc and hammer, and portable electrical tools.

Once extracted, large blocks of stone, usually three cubic metres, are processed via a primary gang saw, wire or circular saw, slicing the stone in thickness and then processing it through a secondary saw to cut the stone to its final shape. Depending on the level of detail required, shaping is then undertaken by a skilled stonemason or a CNC machine.

Manual finishing and shaping is still very much in use, especially for work that necessitates fine detailing, intricate connections or reproduction of classical details.

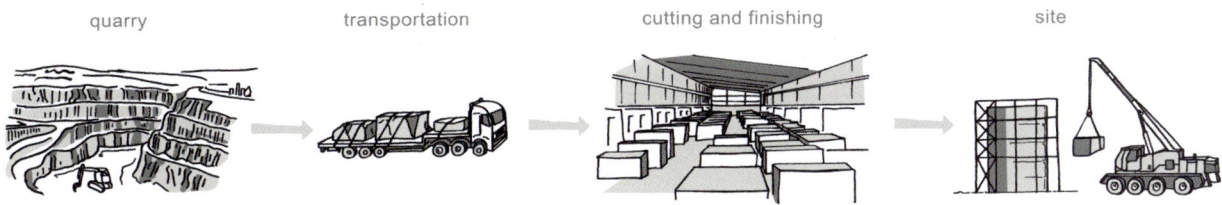

quarry   transportation   cutting and finishing   site

Figure 15.6 Stone extraction and production diagram, from stone quarry to site.

CNC

CNC (computer numerical controlled) machines have revolutionised stone-cutting technology because any robot arm can shape stone simply, reducing the skill required. The only restriction is that it demands significant power and an effective dust suppression system. CNC machines are particularly effective when repetition is required, or for roughing out intricate or large shapes such as a sculpture. The high precision of CNC machines means they can simplify complex connections, similar to wood interfaces. CNC wire machines also reduce waste, making it possible to use 100% of the stone.

Assembly

Unlike other trades like joinery or metalwork, stonemasons do not typically preassemble their work off site. Sometimes, a dry lay is used to check connections between stones but mostly pieces are sent to site and assembled traditionally on a bed of mortar.

The prefabrication of stone – the off-site building of larger components – has emerged in the last two decades, although until recently this was very rare. Prefabrication requires a method

Figure 15.7 Graph showing embodied carbon for the main uses of stone (kgCO₂e/m³). Numbers are approximate, for LCA modules A1 to A3, for limestone or sandstone. Embodied carbon for marble, granite and basalt is typically 50% higher due to the increased energy needed for cutting and transport. Shaped work and cladding can vary significantly depending on the amount of shaping and finishes.[6]

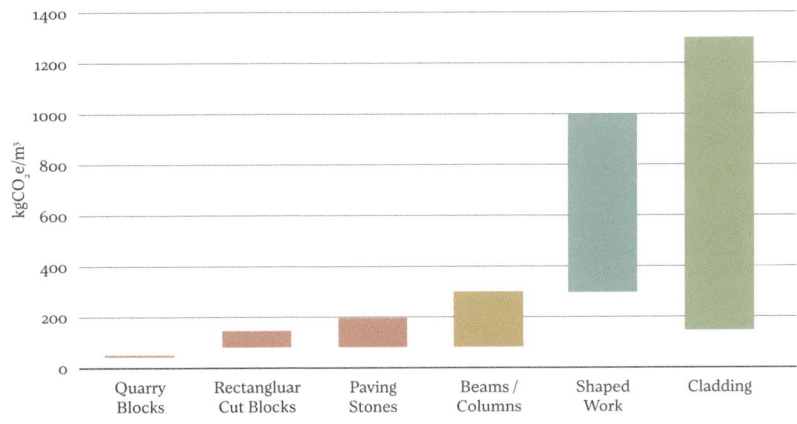

of fixing stone blocks together so they can be lifted as a single component. This is typically done with steel dowels, bars or cables but can also be done with carbon or basalt fibre or concrete backing.

Fitting is one of the areas where stone construction has been left behind. The industry still relies extensively on highly skilled labour that is expensive and rare. Pre-tensioned stone sections were used at the top of the towers of the Sagrada Família in Barcelona to speed up construction and in future, prefabrication of larger elements could further reduce work on site.

Finishing

Finishes differ from one stone to another, from the heavy, clean chiselling of sandstone to highly polished marbles. The most basic technique leaves traces of tools marks, the most efficient way to finish stone. However, as taste developed for more elegant finishes, chiselling became more refined and scripted, and sanded finishes became more popular.

Finishing is a key determinant of stone's visual appearance. In the past, specific textures were sometimes used for a particular reason, such as stopping water ingress or helping water flow down a wall. Tooling was often used to reinforce the contrast of light and shadow on a façade. Light catching on the tooling creates a deeper relief.

CONSTRUCTION
Lifting

In contrast with in situ concrete, stone is delivered as dry blocks that are assembled on site. Each block, therefore, requires lifting into place. Small blocks can be lifted by hand but there will be a lot of lifts. Larger blocks require fewer lifts but each will take more time and is likely to require a crane. The relationship between the method of lifting and the type of construction requires a careful balance that should be considered to ensure an efficient construction sequence.

In modern construction, where there is often access to a crane, stone is well suited to prefabrication into larger elements. This speeds up construction and allows better control of stone elements off site in the workshop.

Support

Where stone is not load-bearing, or where it is used in combination with other elements, it requires support. For flooring, this is a subfloor system, often a screed, or alternatively plyboard or other types of board that provide a stable base. Stone wall cladding, tiles or panels require both vertical and lateral support back to structure. This can be a subframe, typically steel, with all panels supported, or

Figure 15.8 15 Clerkenwell Close, London, Groupwork and Webb Yates, 2017. Large, unfinished limestone blocks from France form a load-bearing façade that supports the building's concrete floor slabs. The large size of the blocks reduced construction time on site. To reduce finishing and for visual effect, the stone blocks have been left raw.

stacked blocks supported at the base and restrained laterally with wall ties, as in masonry wall cladding systems. Stone can normally be thinner than masonry and requires fewer movement joints as there is negligible long-term movement. As stone panels get thinner, support becomes more complex, requiring pins tied back to the supporting frame.

Stone used as cladding panels is at its least effective. Often the main expense and carbon cost is in the supporting frame rather than the stone itself. A frequent value engineering conundrum is that stone is made thinner to save cost but the frame cost increases more than the saving.

DURABILITY

If stone is porous, it will often weather faster and be subject to freezing damage. It can also discolour as lichens and moss can grow. For this reason, more porous stone is often not used in contact with the ground. The porosity of stone depends on its type and grain size – limestones particularly can be very porous and should be detailed carefully or sealed if used externally.

Some stones are susceptible to attack from certain chemicals. For example, soft limestones can be partially dissolved by acids such as some cleaning agents or de-icing salts.

END OF LIFE

The key to ensuring a future use for stone is designing a building that can be easily dismantled. For stone, the main consideration is mortar type. The mortar should always be weaker than the stone so it can be easily removed. Lime mortars are particularly suitable.

Disassembled blocks can easily be reassembled with very little loss of quality. Stone can also be recut or refinished on or off site. The smaller and more fragile the element, the more difficult it is to reuse. Tiles are often too thin and bonded too tightly to substrates to be efficiently removed, but thicker paving slabs can be.

Unfortunately, much of the stone from demolished buildings is simply crushed for use as hardcore or aggregate. This is still a better use than landfill and is less wasteful than using fresh crushed rock from a quarry, but this should change as more buildings are dismantled rather than demolished.

Figure 15.9 Approximate sources of cost and carbon in stone construction. The main drivers of cost in stone construction are the choice of stone type, any shaping required and construction on site. Embodied carbon is predominantly from transport and quarrying.

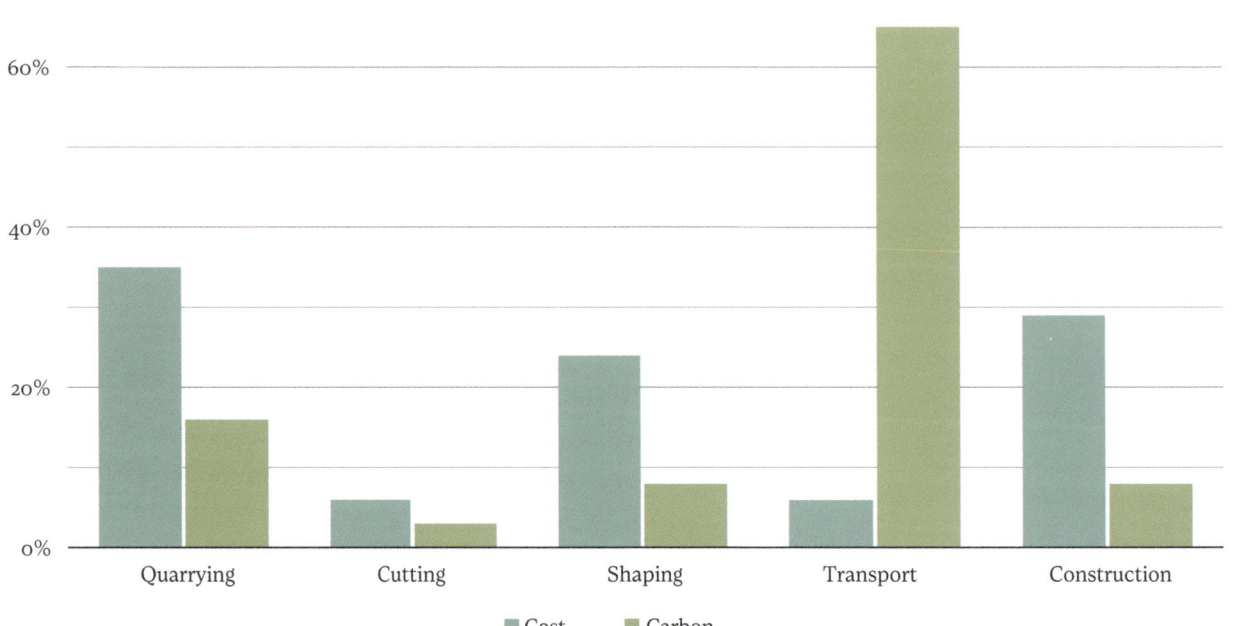

Figure 15.10 Prefabricated stone beam, The Stonemasonry Company. This 9m-long by 1.2m-deep transfer 'beam' is made up of French limestone blocks secured with tensioned steel bars. The extended block with holes will be connected to a second perpendicular beam.

DESIGNING FOR LOWER IMPACTS
Extraction and manufacturing

Quarries exist all over the world in all shapes and sizes. The challenge is extracting blocks efficiently and the technique varies depending on the type of stone, intended use and available technology. Understanding how a specific stone is quarried is crucial to ensuring it is being used in the most efficient way. If most blocks from a quarry are only 400mm high, then blocks that measure 450mm will be hard to find.

Standardising machine-cut items

The more complex the shape, the more cuts required, and the more expensive. When stone is cut to size, it is significantly quicker to cut the same dimension multiple times, therefore standardisation is key to an efficient proposal. By contrast, if the stone is being hand-worked, then agreeing generous tolerances will save time by eliminating additional close control.

QUESTIONS
FOR PROJECT TEAMS
AND SUPPLIERS

☐ Where is the quarry?

☐ What are the typical and
    maximum block sizes available?

☐ What stone property and testing
    data is available? (For example,
    compression and flexural
    strengths, with freeze–thaw
    data if the stone is to be used
    externally. Density, porosity
    and chemical attach data is also
    useful.)

☐ Where will the stone be cut and
    shaped?

☐ Is it suitable for the proposed use?

☐ Where will the stone be finished
    and what surface finishes are
    available?

Finishing

Finishing is always an extra process. Flat shapes can be finished more easily than curved ones and the higher polish levels are exponentially more expensive and carbon intensive. It is easy to spend more on the finish than the price of the stone.

Local is low carbon

Transport represents the largest portion of stone's embodied carbon, from the quarry to the stonemason's workshop and from there to site. **Local stone, cut and shaped locally, can dramatically reduce embodied carbon and multiply the benefit of using stone.**

SUSTAINABILITY

Dust from cutting and shaping stone is one of the main hazards of working with stone. Fine dust is prevalent throughout the process and can lead to respiratory illnesses.

Water is often used for cleaning and cutting but most modern quarries now use dry cutting. The water used in cutting is filtered and reused in a closed process so water use in modern stone manufacture is relatively low.

One of the main criticisms levelled at stone extraction is the scar it leaves on the landscape at the quarry. The regulation of quarries is overseen by the Environment Agency in the UK, who administer permits and review environmental impacts such as dust, waste, water use, emissions and noise, but not what happens when a quarry stops operating. Examples of old quarries being reused after they have yielded their usable stone are increasing. Much of the stone to build Edinburgh came from a nearby quarry that is now an underground car park.

FUTURE TRENDS
Better quarrying and transport

In the UK, stone is often quarried using more intensive methods than other countries as the quarries mainly focus on extracting rock for aggregates rather than blocks. As a result, most stone products for construction are imported. By modernising its quarries, the UK could compete with European imports on cost, and with reduced travel distances. When coupled with a switch to electric vehicle transport and the decarbonising of the UK's electricity grid, the embodied carbon of UK stone for UK projects could be near zero.

Better testing and certification

Stone is a variable material and is not typically used structurally so its mechanical properties are normally secondary to its look. This makes it difficult to specify and select as the quarries provide limited data and no guarantees. By standardising and regulating the strength of stone in a similar way to timber, it would be much easier to find the right stone for the job and ensure that supply meets project requirements.

Composite/hybrid products

Combining stone and timber is a compelling way forward for low-carbon design.
Stone can be used for its mass and compressive strength, while timber is used for its tensile strength, light weight and carbon sequestering properties. Structures that use tensioned stone beams with CLT floors, or glulam timber beams with stone slabs, are very effective. In the second option, if timber beams can be connected compositely to stone planks, the properties of both materials can be maximised for a very efficient structural system.

One of the challenges of building with stone is the available block size, both as quarried and for lifting on site. Often the blocks or elements are too large to lift by hand but below the lifting limit to justify a site crane. One approach is to prefabricate panels made from multiple blocks using a thin fibre-reinforced concrete backing to hold them together. This backing will be a similar strength as the stone and so can also be used structurally in combination with the stone.

KEY TAKEAWAYS

- Stone supplies are plentiful almost everywhere across the globe.

- Modern methods of analysis and manufacturing mean that stone can be used in ways that it wasn't before. Stone can be analysed with computer software, it can be cut to precision with industrial diamonds, and it can be post-tensioned with high-tensile steel. It can be turned into columns, beams and slabs.

- Stone can play an important role in replacing concrete and steel and could form the backbone of a decarbonised building industry.

- It is important to use the appropriate porosity of stone for a given application and to exploit stone's structural capacity in the most effective way.

- Consideration must be given to how stone will be built on site: block size and lifting method.

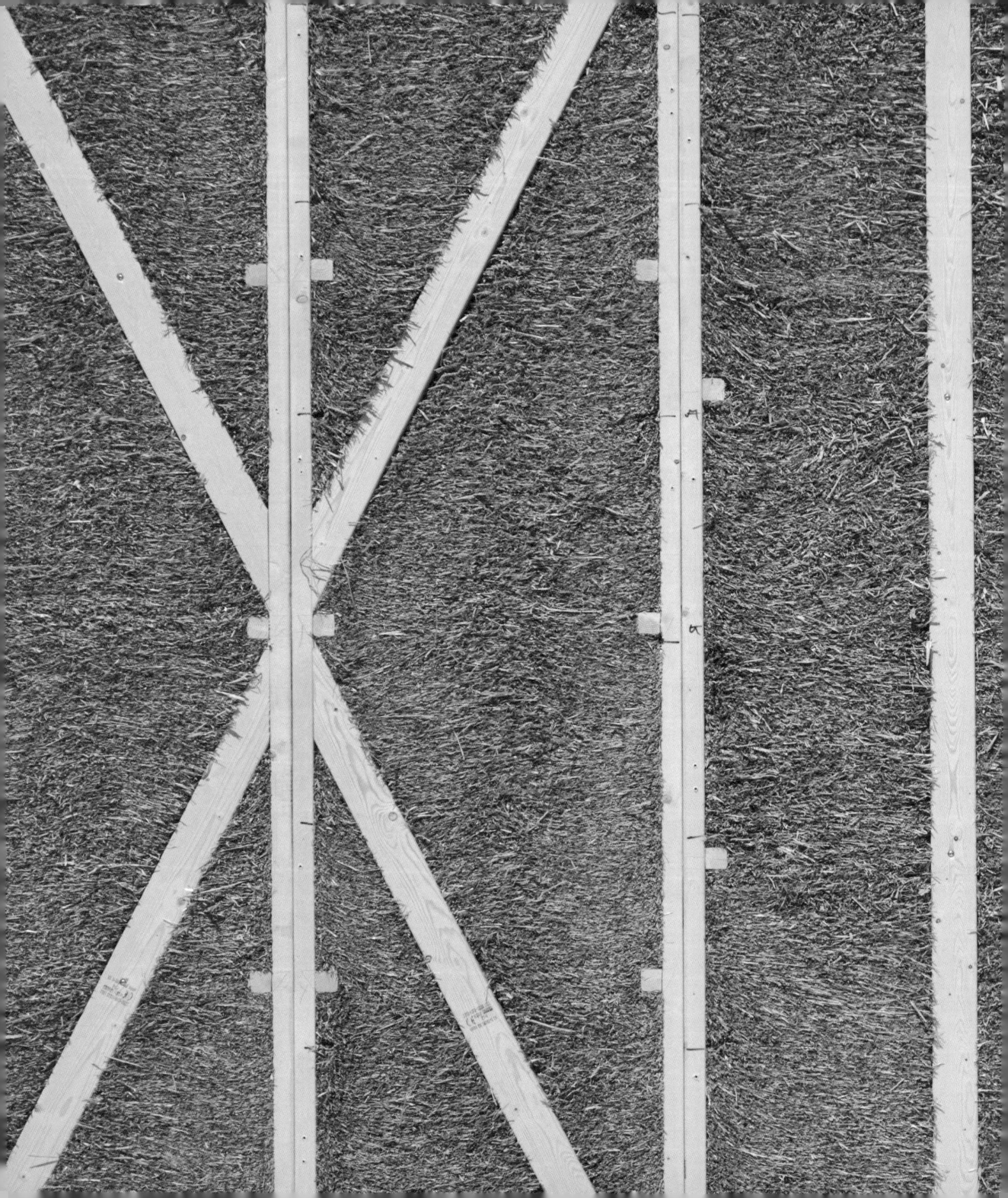

STRAW

Barbara Jones

Annual UK production

5.5 million
tonnes/year[1]

Thermal mass (straw wall)

c206 kJ/m^2K[2]

Embodied Carbon of UK straw

12.97 $kgCO_2e$[3]

Nominal size of a small trimmed straw bale

1,000 x 450 x 345 mm

Lambda value for straw

0.065 W/mK[4]

Sequestered carbon

129 $kgCO_2e/m^3$ [5]

Straw is an annually renewable natural material that sequesters carbon and can be used for construction without any change to current agricultural practices. The average annual UK wheat yield between 2010 and 2020 was 5.5 million tonnes per year of which at least 2 million tonnes per year had no useful purpose and was ploughed back in.[6]

Straw has been used in construction for thousands of years as an insulator and fibrous additive to earth-based structural elements (cob), as covering/infill methods (thatch, wattle and daub) and in flooring and plasters. Straw can be used as a structural element on its own, as an infill insulator within a framework and, more recently, as a prefabricated timber-and-straw structural panel system. It is likely that there will be a UK factory building prefabricated timber-and-straw panels within five years. The amount of straw produced annually could easily support such a factory.

Despite common misconceptions, well-designed compressed straw construction does not present a fire risk.

Straw construction is likely to increase substantially in the UK from now on, as it has done in France in the past decade. This is due to the climate emergency focusing attention on embodied carbon and the fuel crisis focusing attention on operational carbon; increased awareness due to ACAN's Natural Materials group, the Alliance of Sustainable Building Products and others; and the availability of factory-based straw panel production.

APPLICATIONS

- Monolithic walls
- Prefabricated wall systems

PROS

- Plentiful supply
- Modular prefabricated timber-and-straw panels can be erected extremely quickly
- Easily meets building regulations/warranties for buildings up to 11m high
- Annually renewable and carbon sequestering
- Naturally fire resistant
- Longevity 100+ years

CONS

- Must be kept dry during construction
- Small bales not always locally available in the correct quality
- Thick walls, approximately 400mm minimum
- Cannot be used below ground

BUILDING WITH STRAW - A TIMELINE	
1921	The oldest straw-bale house known in Europe is a two-storey straw-and-timber-frame house in Montargis near Paris, built in 1921 and still inhabited.
1994	The first load-bearing straw building in the UK was built.
1996	The first straw building with planning permission and building regulation approval was built in Crickhowell, Wales in 1996 using load-bearing jumbo bales.[7]
2002	A UK company began making structural CLT timber panels filled with bales of straw.[8]
2008	Structural twin stud timber-framed panels with straw insulation (not bales) began to be manufactured as prefabricated panels for MMC (modern methods of construction) in a Slovakian factory.
2009	The world's first social housing built of straw was two pairs of load-bearing semidetached houses in Lincolnshire built for North Kesteven Council.[9]
2016	Prefabricated timber-and-straw panels were introduced to the UK in 2016 as a certified Passivhaus product.[10]

GLOSSARY

Load-bearing (Nebraska) The original method of building with straw which does not require a framework.

Mini-hestons The large rectangular 'jumbo' bales often seen stacked high in fields after harvest.

Round bales The large circular bales commonly baled by farmers.

Straw applications discussed in this chapter fall into two categories: straw in small bales and straw in prefabricated straw panels.

WHAT IS STRAW?

Straw is the hollow dead plant stems of a grain crop, the inert stalk that remains after the grain has been harvested. It has a similar chemical make-up to wood and is not to be confused with hay or grasses, which contain leaves, flowers, pollen and are cut green. The majority of straw in the UK is wheat, but barley is also available and, to a lesser extent, oats or rye. Straw does not decompose easily, especially rye which contains a natural fungicide.

Despite so much straw having no use each year, it is officially considered a co-product rather than waste from the grain industry. Currently the only straw that is actually grown for construction are the long-straw varieties used for thatching, such as the modern hybrid wheat/rye Triticale.[11]

Straw requires little processing as it is baled into small bales or much larger rectangular bales in the field after harvest and drying. The black plastic-wrapped round bales often seen in the field actually contain hay for silage, not straw, as straw would rot down in this condition. Instead, the round bales of straw are wrapped in a light netting. It is then generally stored for use in a nearby barn. Small bales can be used as is, after a quality check, and large round bales are fed into a machine in the factory and compressed to produce panels. It is usually possible to source good-quality construction bales fairly close to the building site/factory.

MATERIAL SOURCING

Currently there is not enough straw used in construction for it to enter the UK straw annual statistics. Stated uses of straw include home use, bedding, biomass fuel and feed. Home use does not include the vast amount that is ploughed back into the field as there is no other way to dispose of it since burning straw was banned. Government statistics show that the average annual UK wheat yield between 2010 and 2020 was 5.5 million tonnes per year.[13] Most UK straw is grown on the east coast, in Yorkshire and East Anglia, but it is grown in most areas to some extent.

Figure 16.1 UK wheat straw average production 2010–2020 by end use (not including straw ploughed back in).[12] Source: John Butler www. sustainablebuildconsultancy.com

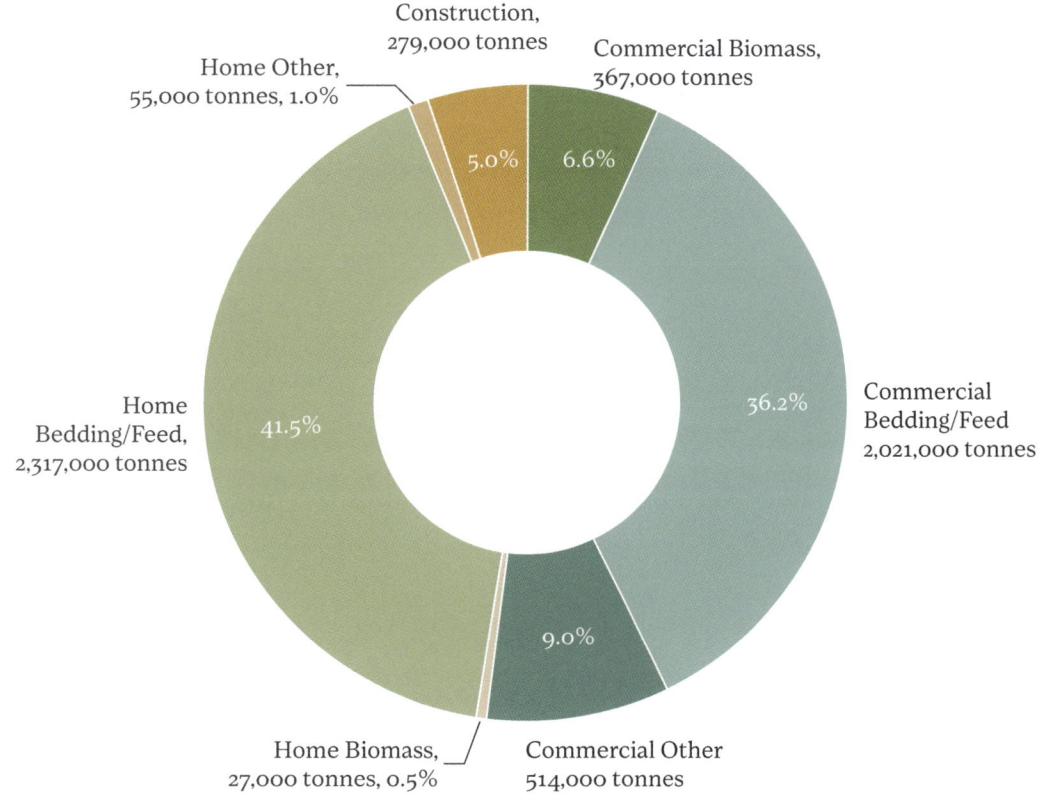

Construction, 279,000 tonnes

Commercial Biomass, 367,000 tonnes

Home Other, 55,000 tonnes, 1.0%

5.0%

6.6%

Commercial Bedding/Feed 2,021,000 tonnes

36.2%

Home Bedding/Feed, 2,317,000 tonnes

41.5%

9.0%

Commercial Other 514,000 tonnes

Home Biomass, 27,000 tonnes, 0.5%

Type of straw	Construction method	Storey height	Comments
Small bales	Load-bearing/ Nebraska	2–3	Could be up to 4 storeys but never been tried/tested
Small bales	Infill/ framework	5–6	Depends on the engineering of the frame and limits set by building regulations (currently restricted to buildings below 11m in height and more than 1m from the boundary of another property)
Factory-packed from round bales	Prefabricated MMC	5–6	The timber frame is engineered with appropriate bracing; height limits are set by UK building regulations (currently restricted to buildings below 11m in height and more than 1m from the boundary of another property)
Factory-packed from round bales	Prefabricated MMC	12	Sweden only: the timber frame is engineered with appropriate bracing; Swedish building regulations have different methods of assessing fire risk than in the UK

Table 16.1 Uses of straw in construction.

Levels of building with straw in the UK show a small annual increase each year. It is estimated that there are around 500 straw-bale buildings in the UK and in the region of 30 being built each year.[14]

If only 5% of the total annual production of UK wheat straw was used for building houses instead of being ploughed back in, almost 40,000 homes could be built every year without planting extra straw.[15]

APPLICATIONS AND USE OF STRAW
Straw can be and has been used for many building types, including schools, offices, museums, council housing, refurbishment, large and small community centres, commercial, residential and self-build.[16] It cannot be used underground.

DESIGNING WITH STRAW
Small bales
The following guidance applies to designing with small bales.
- Design for a good hat, a good pair of boots and a good overcoat. This means ideally a roof overhang of 450mm, plinth/stem wall of 300–450mm and 30mm of lime plaster or rain screen cladding/ ventilated façade.
- Use nominal bale sizes of 1m length by 450mm width by 345mm depth (size after compression) unless actual bale sizes are known.

Figure 16.2 La Ferme du Rail, Buttes-Chaumont, Paris, Grand Huit, 2019. Located on an urban farm with a strong community ethos, this 20-unit residential building is insulated with locally sourced straw and clad with untreated chestnut posts.

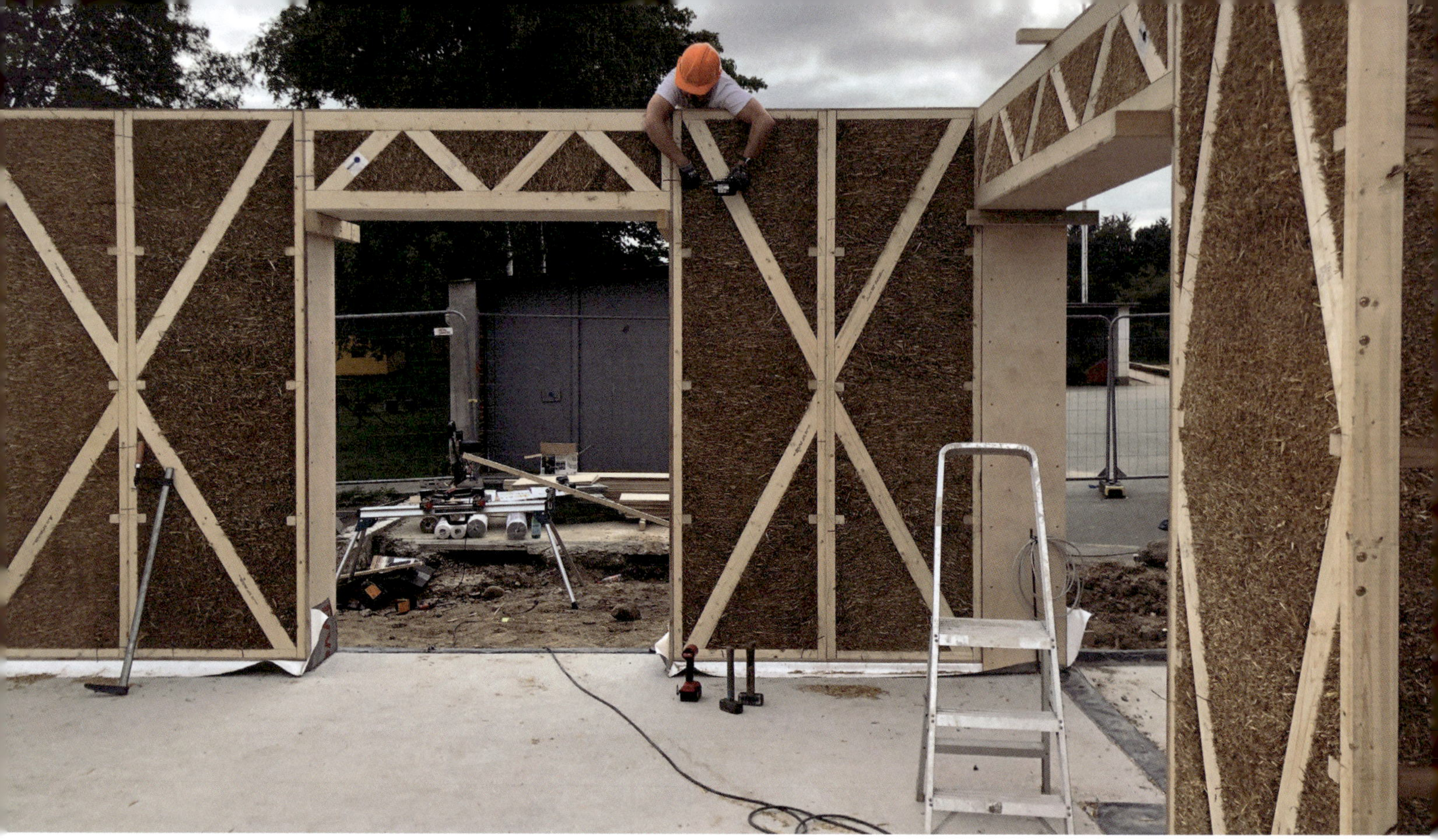

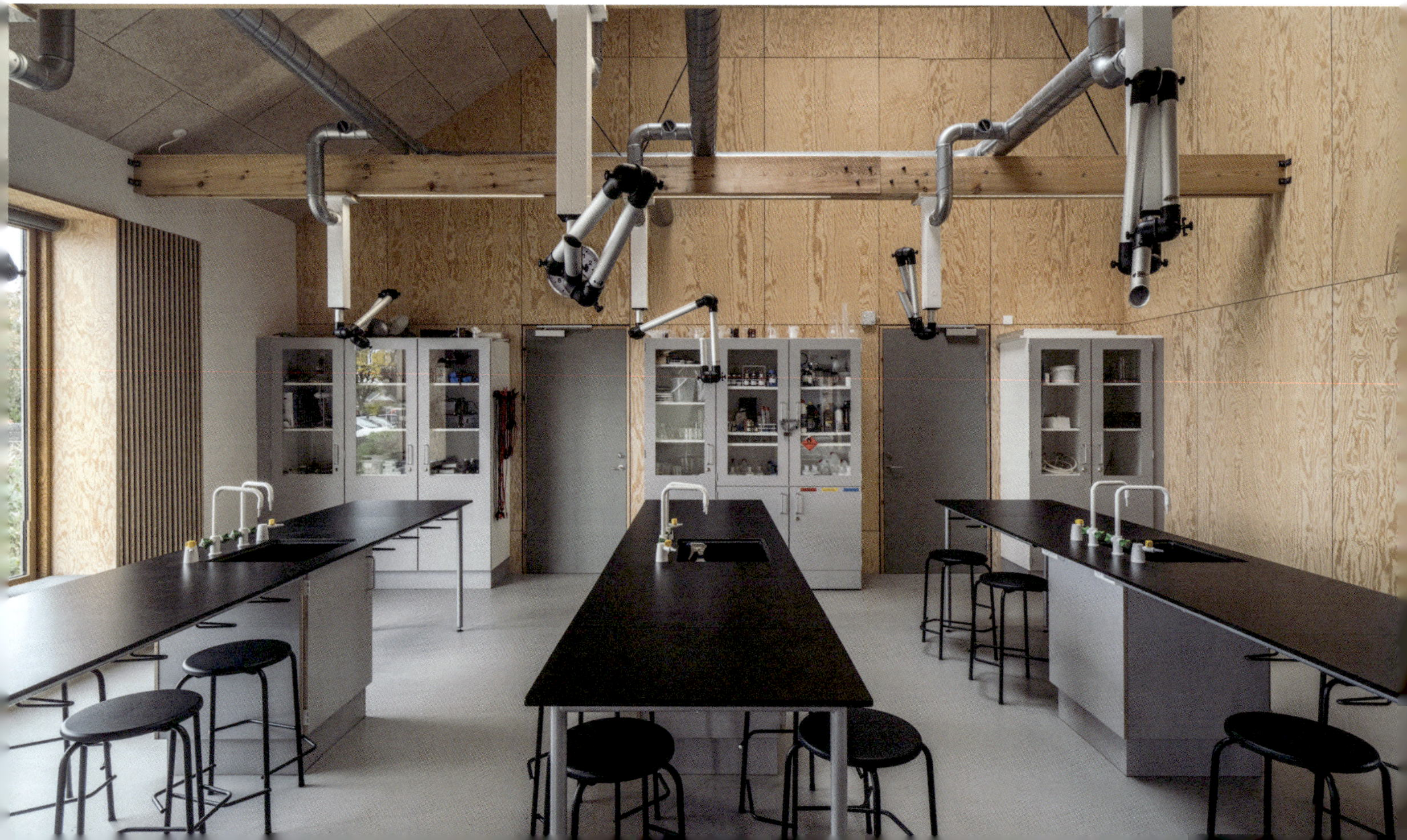

Figures 16.3a and b Feldballe School, Rønde, Denmark, Henning Larsen, 2022. Slovakian prefabricated panels of compressed straw by EcoCocon form the external walls of this 250m² school building in rural Denmark.

- Design for bale heights after compression; the designer must understand how compression will be achieved and design it in.
- Draw accurate sections that show bale sizes and size of timber in beams/wall plates, especially at planning stage, otherwise roof and floor heights may change by a significant amount on construction.
- Ensure the dimensions of structural openings and the distance between framing posts correspond to bale sizes to avoid extra customising and waste.
- Keep the design as simple and as uncomplicated as possible for speed, affordability and ease of construction.

Prefabricated panels

The following guidance applies to designing with prefabricated panels.

- Ensure the design wall thickness corresponds to the size of the panel plus surface finishes. These vary by manufacturer: 450/400/300mm.
- The external finish could be 60mm wood fibre board with either 3–10mm of lime plaster or a ventilated façade, or a 30mm lime plaster coat.
- The internal finish is typically 25–30mm of clay/lime plaster or other breathable board finishes with or without service voids.

With some manufacturers, the actual sizes of panels/lintels/sills and the number of braces required is engineered by the company itself according to the specifications on loading and weather conditions, once the design is finalised. Other manufacturers provide standard panels and the architect needs to design with these in mind.

See further details about construction in the *Technical Guide: Straw Construction in the UK*.[17]

THERMAL CONDUCTIVITY AND THERMAL MASS

Straw does not actively contribute any carbon emissions during use, and it can also help reduce operational energy carbon emissions by its thermal efficiency. In addition, using straw as insulation significantly reduces summer overheating and adds to indoor comfort in winter due to the thermal mass and decrement delay of the wall system.

Heavy materials such as concrete, brick and clay are often chosen for thermal mass, but dense straw also has significant mass, which is increased by adding a lime or clay plaster. The typical thermal mass of different materials is as follows:

- **High thermal mass (e.g. solid masonry)** c500KJ/m²K[18]
- **Low thermal mass (e.g. lightweight timber frame)** c55KJ/m²K
- **Thermal mass of straw walls** c206KJ/m²K (before the addition of plaster/ render)[19]

Due to differing research methods, the Lambda value for straw may vary by country:

- **Austria** 0.050W/mK
- **Slovakian (e.g. EcoCocon straw panels)** 0.065W/mK
- **UK straw EPD (Environmental Product Declaration)** 0.052W/mK (based on pan-European results)
- **UK *Technical Guide*** 0.065W/mK

At the time of writing, the University of Bath is undertaking research on straw bales made to fit the testing machine, which will give a more accurate figure.[20] It is fair to say that for *all* insulations, a more realistic and honest figure for actual performance is achieved by carrying out post-occupancy testing.

COMBUSTIBILITY RATING

Straw has suffered from negative stereotypes regarding fire that bear no relation to reality. Fire test results from many countries show that a well-designed compressed straw system (load-bearing, infill and prefab) does not present a fire risk.

It is also important to note the following.

- Straw contains up to 4% silica, a naturally occurring fire retardant.
- Compressed straw contains little air; in a fire, the carbonised outer coating protects against further combustion (the same as for large-dimension timber).
- The insulating properties of straw protect against self-ignition within the body of the panel/bale.
- At the end of fire tests, the integrity of the wall is still intact.

The density of straw in straw-bale construction can provide a near airless environment so that fire ratings are compliant with building regulations. When installed to a high standard, its compactness, lack of air gaps, conductivity, specific heat capacity, density and thickness prevent the spread of heat and therefore combustion beyond the wall. The application of clay and lime plasters seals the bales and further increases fire resistance.

Fire tests for both small bales and prefabricated panels show results that are better than the required levels to meet UK building regulations, which require the Reaction to Fire performance of the external surface of walls to be Class B-s3, d2 (BS EN 13501-1-2018) or better. In fact, for straw they are much better, being Class B-s1, do. From this evidence, designers can specify straw walls with confidence that occupants are within and beyond the correct safety margins.

The minimum fire resistance required by building regulations ranges from 30 to 120 minutes and straw achieves between 120 and 135 minutes without failure. In comparison, the BRE carried out fire resistance tests on PUR (polyurethane rigid foam) or EPS (expanded polystyrene) thermally insulated structural insulation panels (SIPs) with 18mm orientated strand board (OSB) facings and concluded the average result was 60 minutes.[21]

EMBODIED CARBON

It's a biological fact that plant-based materials store more carbon than any other type of material – plants take in carbon dioxide during their life cycle and release oxygen. Both straw and timber have this quality, storing 106 $kgCO_2e/m^3$ of straw – a major advantage of straw is that it is annually renewable. This means that not only does it sequester carbon, but it does this again and again each year, building up a store of carbon in each crop very quickly, much more quickly than timber, which requires 30 to 40 years' growth for structural timber.

The whole-life carbon impact of straw from land use and land use change is $0.02kgCO_2/m^3$. Across the whole life cycle, the carbon emissions of $1m^3$ of UK straw are:
- **A–C** $14.12kgCO_2e/m^3$
- **A1–A3** $12.97kgCO_2e/m^3$

The UK Straw EPD gives the environmental impacts for $1m^3$ of straw at $100kg/m^3$ density, enabling straw to be used as a generic material within a building life-cycle assessment (LCA), regardless of the construction method used. The separate EPD for the EcoCocon straw panels can be used when these are part of the building envelope.

LONGEVITY

When protected from excessive moisture, usually by being bound in clay or plastered/rendered by clay or lime, straw can last indefinitely. Straw used by the ancient Egyptians in mudbricks survives today. Clay and lime plasters are a common covering for contemporary straw construction; and lime render, in particular, is very durable. Modern building methods using straw as structure and insulation began in the late 19th century and some load-bearing straw houses are still inhabited from this era in the USA (over 100 years later). It is expected that with good design, straw construction can last several hundred years – as heritage buildings do. The LCA for EcoCocon straw panels and for small bales gives their life expectancy as 75 years.

END OF LIFE

As set out in the UK straw EPD, it is assumed that 50% of straw (small bales) can be reused and 50% is incinerated. End-of-construction-use straw can be used as mulch, animal bedding, biomass fuel or composted.

The Slovakian prefabricated panels can also be deconstructed and reused. At the end of construction life, the timber and straw can be separated, the straw being used as above and the timber also, if chipped. However, unlike the UK straw EPD, the EPD for the Slovakian panels states that 95% of the straw and 80% of the timber is incinerated, which has a net overall benefit as this then replaces fossil fuels for heating.

COMPARISON WITH FRANCE

It is useful to compare the UK development of straw buildings with that of France, which has 1.4 million tonnes of straw available annually after all other uses, such as for biomass, bedding and mulching, are removed. French projections anticipate that straw construction will continue to grow and consume almost 70% of production.[22] France introduced new environmental regulations in 2022 that require new buildings from 2031 to achieve a 52% reduction in embodied carbon emissions over today's standard. Modelling of different fabric choices shows that the use of plant-based materials (timber and straw) will be required in order to meet this strict criteria.[23]

So why is building with straw gaining traction in France so much faster than in the UK? Over the last decade, the French have been massively outstripping the UK in terms of the number of straw buildings constructed, the infrastructure to deliver good-quality construction bales, and the establishment of experienced straw construction companies. In 2023, France built more than 700 straw buildings and the country has more than 6,000 straw buildings in total.[24] France also has several training centres for straw-bale building, generating approximately 1 million euros annually, which indicates the seriousness with which the construction industry takes this building method.[25] The French straw sector has jumped from training five individuals in 2011, to 559 in 2022. Moreover, of the 95 public buildings built in France in 2022 to Passivhaus standards, over 70% were built with straw.[26] France has more than 450 companies building with straw, with an annual turnover of 114 million euros.[27]

A major reason is that the politics and availability of land in the UK is very different than in France. In the UK, it is extremely difficult for small builders (who are the usual pioneers of a new technique) to find land that is affordable on which to build. This is the opposite in France, where there is good availability of cheaper building plots. It is also difficult for small developers to break the monopoly of larger land-banking developers and try out new ideas. The UK is dominated by eight very large national construction companies, while France has predominantly regional companies.

If pioneers are stifled, it is very difficult for a new industry to grow. Despite this drawback, numerous straw-bale houses were built in the UK and Ireland in the early days between 1994 and 2006, with full legal approvals for planning and building regulations. The UK never had to wait, as the French did, for regulations to be written before straw building could take off. Our regulatory system, based on guidance rather than prescriptive rules, allows innovation, and the essence of good practice was established with the *Building with Straw Bales* manual as long ago as 2002.[28] Nevertheless, the sector has suffered from comments that reduce straw construction to 'hobbit' and 'hippy' houses.

Figure 16.4 Student housing, Vaugirard, Paris, NZI Architectes, 2021. Retrofit of an existing office building into student accommodation. The envelope is comprised of timber-clad panels of compressed straw bales, manufactured off site. The straw was sourced from the Île-de-France.

QUESTIONS
FOR PROJECT TEAMS AND SUPPLIERS

Small bale designs

☐ Are construction-grade bales available locally?

☐ Will a reputable natural builder be involved in the specialist aspects of construction to ensure quality and durability?

☐ Are lime-based plasters, free of fossil-fuel additives such as polyester and plastic fibres, being used?

Prefabricated designs

☐ Is there access to site for an articulated lorry? If not, can an alternative unloading site be arranged with smaller transport to site?

☐ Will the other insulations used for the roof, floor, etc. match the sustainability criteria of the straw/timber panels and be both natural and breathable?

It is the UK construction industry that is lacking in vision to design and build with straw, rather than straw building techniques not being appropriate.

CONCLUSION

No other structural building material matches the environmental credentials of straw. It sequesters carbon as part of its life cycle, which is then bound up within the building for its lifespan, which can potentially be more than 100 years. Straw is annually renewable so can be grown year on year indefinitely, producing very fast-renewing carbon sequestration. It can be used for construction without any change to current agricultural practices as it is a co-product of grain production, produced surplus to requirements by several million tonnes annually.

Straw is plentiful enough to build all or nearly all the houses the UK needs every year. Straw is now available as a prefabricated modular panel made with twin timber studs that allows the walls of a three-bedroom, two-storey house to be built in three to four days. Designers only need to specify the wall width (typically 400mm plus finishes) and the supplier works out sizes and engineering. It is a very simple way to build, comprising only five layers: external lime render or cladding, wood fibre board, vapour open membrane, the panel, internal clay plaster or breathable board with finish. Building with small bales is also a viable option, often chosen by self-builders and community groups for economic reasons.

Straw buildings have been shown in all fire tests to meet or exceed the requirements of UK building regulations, which currently allow straw buildings up to 11m in height. In Sweden, where fire risk is assessed differently, a 12-storey straw-panel residential building is underway at the time of writing.[29]

Straw building in the UK has suffered from lack of awareness, negative stereotyping and serious lack of funding. A viable commercial product has recently emerged: the advent of straw panels, now available in the UK, is likely to be a game-changer.

KEY TAKEAWAYS

- A plentiful supply of UK-grown, annually renewable straw already exists as a co-product of grain production.

- Just 5% of current straw production could supply almost 40,000 houses per year.

- Fire test results demonstrate that well-designed compressed straw systems present no fire risk.

- Prefabricated timber-and-straw panels open new possibilities for straw construction.

TIMBER

Marlene Cramer and Gabriele Tamagnone

Amount of timber used in the UK annually

11,189,000 tonnes [1]

Percentage of UK timber products that are imported

65% [2]

Embodied Carbon of wood panel products

130-360 $kgCO_2e/m^3$ [3]

Amount of CO_2e locked away in timber products

650-1100 $kgCO_2e/m^3$ [4]

Wood waste per year in the UK

4.5 million tonnes [5]

Embodied carbon of mass timber

100-220 $kgCO_2e/m^3$ [6]

Forest cover comprises 13.3% of the UK's total area, almost evenly divided between conifers and broadleaves.[7] UK forest cover is well below the European average of 46% (ranging from 74% in Finland and 0.5% in Iceland).[8] Almost 75% of UK forests are privately owned and not necessarily managed for timber production; even so, 64% of timber production comes from private forests. A felling licence is required for harvesting trees, and almost all felling licences require the woodland to be restocked with new trees. In 2021 in the UK, 140km² of land was restocked after harvesting in the years prior, plus an additional 133km² of new forest, equal to 1.5% of the UK's total forest area, was created.[9] The UK's own harvest is small compared to the volume of imported products, the UK being the second-largest net importer of forest products after China.

The UK imports more wood products than it produces and has been reliant on timber import for centuries. The local harvest comprises mostly of softwoods, and more than 85% of the hardwood harvest is burnt for energy production.[10] Around one quarter of the UK's softwood harvest is used in construction products, including structural timber, flooring, doors, windows and detailing. Often, timber used in construction products has lower embodied carbon than comparable building materials, because it can be extracted and processed using relatively little energy and its low weight contributes to lowering emissions.[11]

APPLICATIONS

- Structure
- Façade cladding
- Finishes: flooring, walls, ceilings
- Furniture
- Window frames
- Doors

PROS

- A renewable resource sequestering CO2 from the atmosphere and locking it away within the wood
- A versatile material, able to be used throughout the building
- Creates a natural built environment that promotes biophilic responses from occupants
- Can be readily integrated into circular economy plans through careful cascading of products

CONS

- Fire performance needs careful consideration during design and construction
- A majority of timber products are imported
- Sustainability and timber quality are dependent on good forest management
- There are fewer certifications and warranties available for timber products compared to other materials that have been more commonly used over the past 30 years

GLOSSARY

Biogenic carbon Carbon associated with natural processes, rather than industrial processes. In the case of timber, carbon is sequestered during tree growth.

Board Planar engineered wood product that is made from chips, strands or fibres, e.g. chipboard and fibreboard.

Clearfelling Removing the trees in an area all at once.

Continuous cover forestry Forest management system in which trees of different ages co-exist and are continuously harvested and replaced as they mature.

Engineered wood products (EWPs) Secondary wood products made by combining primary wood products. A permanent bond between them is achieved, for example, by gluing, nailing or embedding particles in a matrix.

Lamella (plural lamellae) Primary product usually between 6mm and 45mm thick and 45mm to 300mm wide. Length. Length varies depending on use.

Mass timber A family of engineered wood products made from layers of timber, glued or fastened together into elements usually of 100mm thickness/height or more, e.g., cross-laminated timber and dowel-laminated timber.

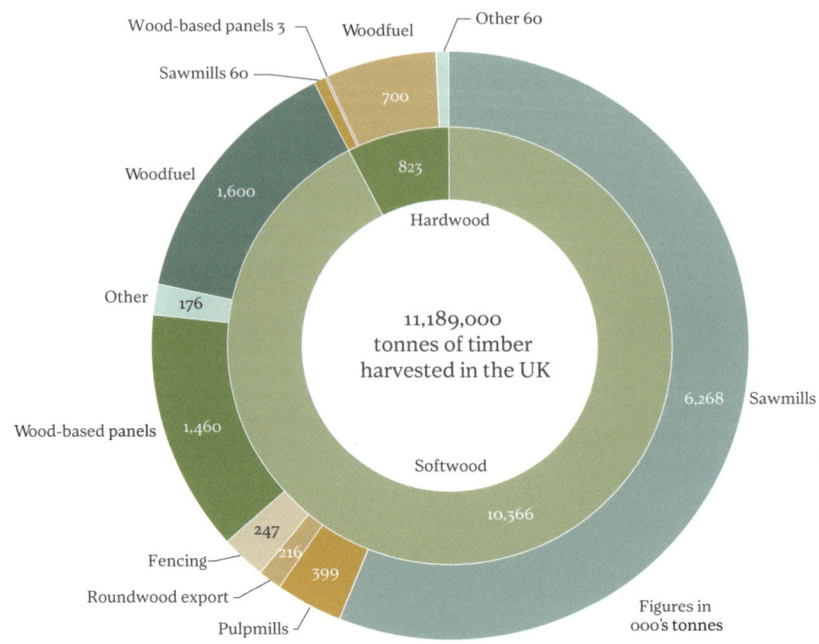

Figure 17.1 Uses for timber in the UK (annual figures), showing the vast amount of timber resources currently dedicated to energy production that could be diverted to higher-level uses, including construction, particularly engineered wood products.[12]

THE RAW MATERIAL

Wood and forests are inseparably linked, which means that material sourcing of timber does not start with harvesting, but decades prior, with planting the tree. During its life, the tree has been part of a forest that has formed a habitat, food source, recreation space, flood prevention and erosion protection, and all these functions need to be balanced with timber production. Contrary to popular belief, legal timber production is not usually a driver for deforestation; converting forests to agricultural land is much more problematic globally. The forest area in Europe and the UK is increasing by an average 0.2% per annum (2010–2020), although the global forest area is still shrinking.[13]

UK forestry and timber production

Only around 13% of the UK's land area is covered in forest. Even though a significant majority of timber consumed in the UK is imported, the local production does make an important contribution

Minor species Tree species which are not among the major commercial species grown for timber production. Because little is known about their properties, they are often overlooked for use in building products.

Primary (wood) product Timber products after the first stage of processing, e.g. sawn timber and veneer.

Recovered wood Wood waste that has been salvaged for reuse, recycling or energy recovery.

Roundwood Wood in the form of stems or branches, normally with a diameter larger than 7cm.

Thinnings Roundwood that is harvested before it has reached target diameter as part of forest management to create space and access to light for other trees. Thinnings are usually small stems and can have poor form.

Table 17.1 Forest sustainability principles.

to meeting our wood needs, given the production of sawn timber more than tripled since the 1960s, while the manufacturing of board products increased 15-fold.[14]

Although more than half of the UK's forests are broadleaved, 90% of the UK's commercial timber harvest is softwood. Intensely managed conifer plantations are the primary source of UK timber, with Sitka spruce being the UK's main commercial species, making up more than 50% of the UK's conifer forests.[15]

The dominant management approach for conifer plantations in the UK is to plant even-aged stands and undertake thinning cycles before clearfelling and restocking the land with new trees. Recent forestry guidance has shifted towards mixed species planting and management techniques that balance economic, environmental and social objectives. Management systems, like continuous cover forestry, especially in mixed stands with more than one species, balance these different objectives and can improve biodiversity and forest resilience.

The ins and outs of forestry certification

The UK Woodland Assurance Standard (UKWAS) forms the basis for independent certification schemes that help timber users identify sustainably grown timber products.[16] In 2021, 44% of all UK forests were certified, mostly by the Forest Stewardship Council (FSC) or the Programme for the Endorsement of Forest Certification (PEFC), which certify approximately 365 million hectares globally

Forest sustainability principles	Example duties
Legal compliance	Conforming with local laws
Management planning	Setting socially, economically and ecologically viable management objectives Restoring stands to pre-harvesting or more natural conditions
Woodland operations	Minimising fertiliser, biological control agent and pesticide use Minimising pollution
Natural, historical and cultural environment	Conserving and enhancing areas of special conservation value
People, communities and workers	Protecting legal and customary rights Providing employment and training opportunities

(9%), with around 85 million hectares holding both certifications. Forest sustainability is not only about planting as many trees as we fell, but certified forests also need to be managed according to a long-term management strategy that maintains biodiversity, ensures minimum impact from forest operations and retains the long-term function of soil and water, next to social and economic sustainability objectives.

The UK imports vast amounts of timber, and FSC and PEFC certifications also include chain-of-custody controls, which allow tracking of wood products to their origin and assure their sustainable production. Tropical timber can also be imported under a UK FLEGT (Forest Law Enforcement, Governance and Trade) licence which provides assurances of sustainability and legality.

Importing tropical timber can support tropical forests and local communities. Making forests a profitable source of income through

Figure 17.2 UK-sourced and imported timber, split by end uses of timber. Data based on Timber Trade Federation 2021 sourcing data[17] and Timber Utilisation Statistics.[18]

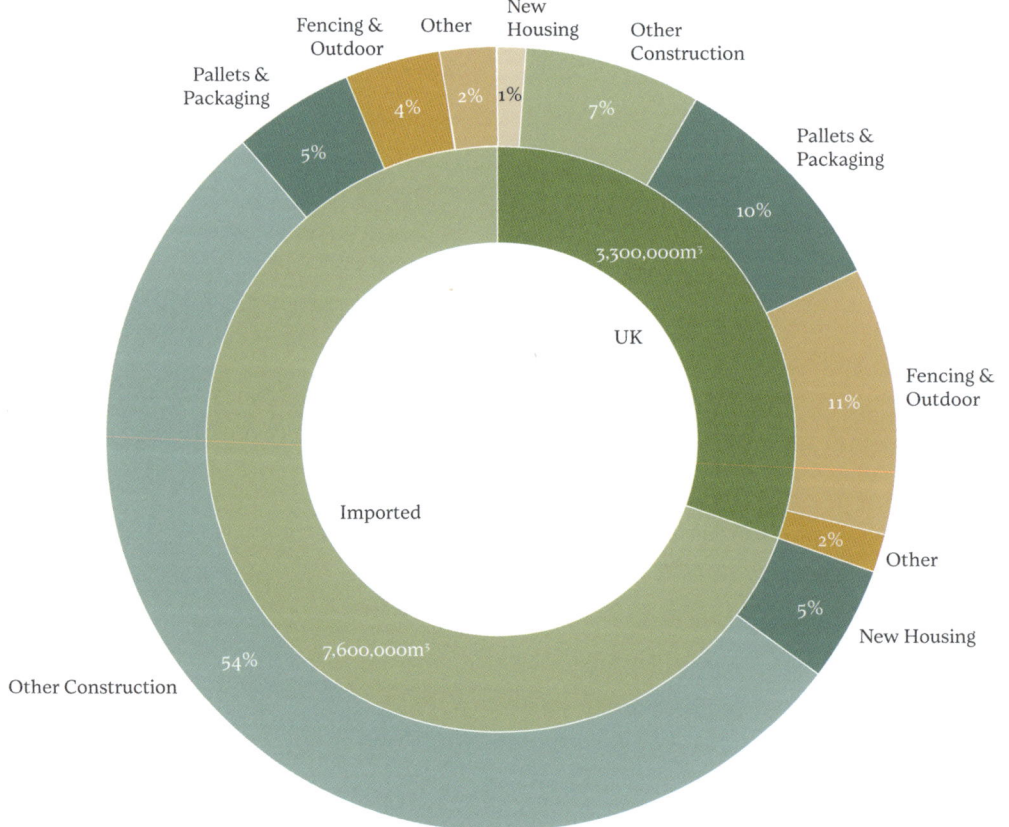

timber production is one of the best ways to encourage forest preservation and creation and to support forest communities. It helps prevent deforestation for more profitable land uses like agriculture, which is the largest driver of forest loss in the Global South. Using certain tropical timbers, especially in applications with very high demands to resistance against decay, such as marine and freshwater construction, can be the most sustainable choice.

It is always advisable to seek further information on the forest conditions of a particular timber resource from a trusted merchant, especially when procuring unusual resources or large volumes. Merchants who are members of Timber Development UK (TDUK) sign a responsible sourcing policy which helps ensure timber is sustainably forested.

Engaging with the timber supply chain also allows designers to source timber via alternative pathways, because not all sustainable timber is certified. Partner Forest Program's *Sustainable Wood for Cities* guide sets out detailed guidance for inclusive procurement and demystifies the certification process.[19]

Carbon sequestration

The big advantage of timber is that trees sequester and store carbon from the atmosphere. During their growth, trees transform CO_2 from the atmosphere into wood. Around half of the wood's dry mass is carbon, and a kilogram of wood contains as much carbon as $1.83kgCO_2$.[20] In a forest, wood decays, which releases carbon back into the atmosphere. Trees are harvested when their growth slows down and carbon sequestering action decreases, making replanting the most effective carbon use of that area. One strategy for carbon capture is planting and replanting fast-growing trees in short harvesting cycles. The benefits of carbon sequestration are only maximised if we don't release it back into the atmosphere immediately. Successive growth and harvesting of trees for products, and their long lifespan in the built environment, could draw down and store significant amounts of carbon and contribute to mitigating climate change, if adopted on a big enough scale.[21]

EXTRACTION OF RAW MATERIALS

The life cycle of timber really begins with the felling of the tree (which accounts for less than 2% of carbon emissions for timber[22]). Trees are typically felled using large harvesting machines that de-limb stems and cut them to size. The branches, treetops and stumps could remain in the forest and help refertilise the soil when they decompose, but, in recent times, these are commonly removed and chipped for board products and biomass, or used in the kilns that dry the wood. At this stage, the timber could be used in log construction, but more commonly it is processed into sawn timber or other products.

Figure 17.3 Lea Bridge Library Pavilion, Leyton, London, Studio Weave, 2021. The pavilion uses street trees that have been felled within London to create a mixed palette of different timbers, based on what was available locally.

MANUFACTURING

After harvesting, timber is taken to sawmills to be processed into different products (accounting for 10% of carbon emissions for timber[23]). Log-processing sawmills, as well as many engineered wood product (EWP) manufacturing facilities, are located close to the forest. Logs are sorted into diameter classes, debarked and milled to improve their round shape. Sawn timber comprises products like beams, boards or battens that can be further processed into joinery, cladding, decking, etc. Some sawmills outside the UK produce veneer, thin sheets of wood that are used decoratively, in structural products or in common panel products like plywood. Veneers are sawn or peeled from logs that have been steamed at high moisture and temperature to soften the wood. Most sawn timber, rough sawn or planed, is used directly (e.g. in construction), but some sawn timber and residues are further manufactured into a variety of secondary products (e.g. board products or mass timber).

Whether in the production of sawn timbers or veneers, offcuts and residues are inevitable, but these are valuable co-products. A variety of residues, from thin stems to board ends, can be chipped into particles of varying size that are used in board products like OSB (oriented strand board) and chipboard, or as animal bedding. Another option is to reduce the residues into even smaller particles – fibre bundles – and use them in the production of fibreboard or fibre insulation. Sawdust and other residues are often burned onsite for energy production.

DRYING

Freshly harvested 'green' timber contains a lot of water and is usually dried before it is sold. The simplest method for this is air drying, but more commonly drying kilns are used to accelerate the process. When timber is dried below around 30% moisture content, it begins to shrink, typically losing around 5% of its volume when it is dried to 15% moisture content. Different products and manufacturing processes require different moisture content, hence the target moisture content of each drying process depends on further processing or final use.

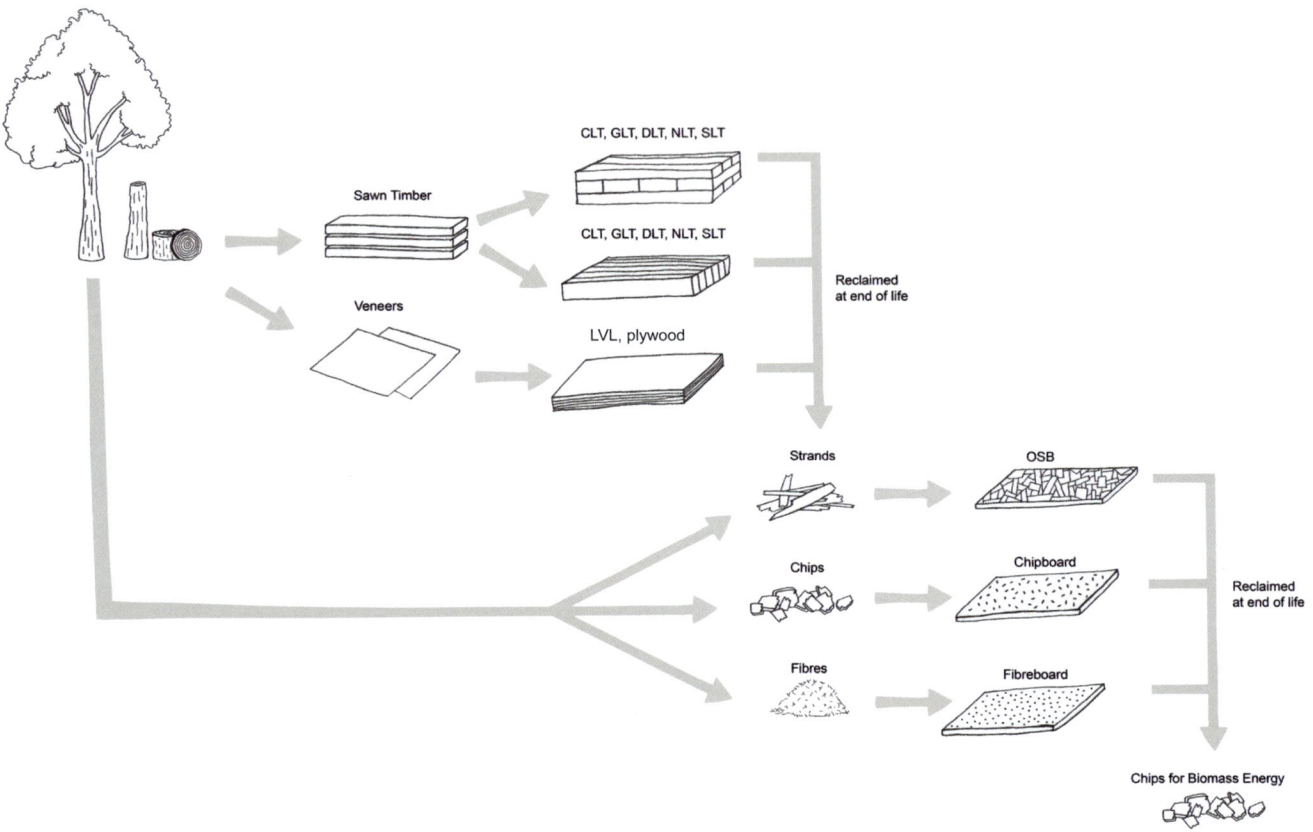

Kiln drying can become a very costly part of the sawmilling process, not only because it is energy intensive and responsible for a further 2.5% of global timber industry emissions, in addition to the 11% of emissions due to timber manufacturing in general.[24]

MODIFICATION

Sawn timber and veneers can be thermally or chemically modified to permanently alter characteristics. Modification mostly aims to improve the timber's dimensional stability in contact with moisture and its resistance to wood-decaying organisms.

Typical modified wood products are acetylated wood, furfurylated wood and thermally treated timber. Chemically modified timber is not currently manufactured in the UK and inevitably has longer transport distances.

Figure 17.4 Wood processing – primary wood products like boards and veneers can be used directly or processed into engineered wood products (EWPs). At the end of life of the timber products, they can be cascaded into new panel products.

STRENGTH GRADING

Sawn timber can be strength graded to be used as structural timber or in structural wood products. This can be done as machine grading or visual grading, and both try to reject the worst pieces of timber to raise the properties of the remaining population.[25] The timber assigned to a strength class, collectively, meets the thresholds set for three primary design properties: strength, stiffness and density.

Timber users most often encounter structural timber graded to a strength class listed in EN 338. There are three groups of strength classes:
- **C** includes softwoods and similar hardwoods tested in edgewise bending[26]
- **T** is based on tension tests
- **D** includes hardwoods assessed through edgewise bending tests.

Each class features a set of predefined values of strength, stiffness and density, which can be used to perform structural calculations. This simplifies the design and specification process, as the wood species or origin of the timber don't need to be specified.

However, only certain species and origin combinations can be assigned to specific strength classes, reducing choice. For example, you cannot buy visually graded C16 spruce from the UK because the visual grades for this resource correspond to C14 and C18. It is important to remember that no one strength class is better than another, it is just a question of choosing a strength class that fulfils the design requirements. This can also be a strength class that is not listed in EN 338, like C16+ or TR26. Bespoke strength classes can also be developed for individual construction projects, specific building products or reused elements. Manufacturers can also declare timber properties directly, without the use of strength classes, provided standardised assessing procedures are being used.

TRANSPORT

Around half the carbon emissions of all timber products used in the UK result from transport (55%). Since the UK imports large amounts of sawn timber and board products, most of these emissions result from maritime transport (30%) and road transport outside the UK (49%).[27] Sourcing local products can therefore drastically reduce the carbon footprint of wood products, making it even more important to bring more of the UK timber resource into material-efficient uses.

SAWN WOOD LIMITATIONS
AND ENGINEERED WOOD PRODUCTS

Each primary product can be used to manufacture so-called engineered wood products (EWPs). EWPs have been devised to overcome some intrinsic limitations of sawn timber, as can be seen in Table 18.2.

PRODUCTS

To make timber products, the sawn timber is further processed to enhance its properties, and this additional manufacturing step accounts for around 28% of timber carbon emissions.[28] Sawn timber makes more than 60% of wood products consumed in the UK. It is commonly used where the enhancement of performance by EWPs is not needed. Sawn timber is typically lower carbon than EWPs, although a careful comparison between timber products should be undertaken as higher emissions from EWPs could be balanced by their optimised use of resources.

Mass timber is a term for engineered wood products that use solid wood or veneers to form large panels or beams, often used in structural applications. Examples are cross-laminated timber (CLT), glued-laminated timber (GLT or glulam) and laminated veneer lumber (LVL). Mass timber products have revolutionised the way we build with wood. They offer greatly enhanced structural strength, high precision, factory-production and rapid on-site assembly. Less than 1% of wood products consumed in the UK are mass timber products, but their increasing use opens possibilities to build more mid-rise and commercial buildings in timber.

Board products are EWPs that are usually made from smaller particles like strands, wafers, chips or fibres. Offcuts, low-quality stems and recovered wood are often used in their manufacturing, and they are sometimes seen as low-quality materials;

Solid timber	EWPs
Affected by knots and other defects	Defects segmented and relocated: more homogeneous and higher mechanical properties
Different characteristics in different directions (anisotropic material)	Some products are manufactured to create elements with similar characteristics in at least two directions
Geometry constrained by the shape of the tree	Can assume virtually any shape or size
One grade, determined characteristics	Can combine different grades to optimise the use of material; different EWPs can be coupled to create composite EWPs

Table 17.2 Comparison of solid timber versus engineered wood products (EWPs).

Figure 17.5 Black & White Building, London, Waugh Thistleton Architects, 2022. Thermally modified tulipwood louvres provide shading on this six-storey office building with a CLT core and LVL frame.

however, board products are an excellent example of circular timber use and can be very cost effective. They are used as building products and even structurally, and in these applications have the same life expectancy as solid wood.

Composite products, such as I-joists, structural insulated panels and metal-web joists, combine different EWPs and other materials. They enable the different benefits of timber products and other building materials to be maximised. Determining which product offers optimal performance depends on specific project requirements, and designing for reduced impacts requires an increasingly nuanced understanding of hybrid material choices. An overview of EWPs is given in Table 18.3.

GLUE

Most EWPs contain glue, which has its own impact and should be considered when assessing product sustainability.

The type of glue impacts a product's moisture resistance, fire behaviour, aesthetics and emissions, and the performance of a wood product regarding these aspects is normally declared by manufacturers. Timber products have acquired a poor reputation for containing formaldehyde resins; they often still do today, but they are safer.

Product	Materials	Use	Impact
Cross-laminated timber (CLT)	• Lamellae of sawn timber • Odd number of layers • Each layer is orthogonal to adjacent layers • Glued	• Walls • Floor slabs • Infill in frame systems (both timber and non-timber) • Roofs	Little UK production, but project-based manufacturing is possible More than 30,000m³ of CLT and GLT are mainly imported from Austria
Glued-laminated timber (GLT or glulam)	• Lamellae of sawn timber • All layers are parallel • Glued	• Columns • Beams • Portal frame • Joists	Good reuse and recycling potential, subject to testing[29]
Dowel-laminated timber (DLT)	• Lamellae of sawn timber • Wood dowels instead of glue	• Columns • Beams • Walls • Floor slabs	No widespread use Very good reuse and recycling potential, subject to testing
Nail-laminated timber (NLT)	• Lamellae of sawn timber • Like CLT or GLT, but with nails instead of glue	• Columns • Beams • Walls • Floor slabs	No widespread use Medium-low reuse and recycling potential
Laminated veneer lumber (LVL)	• Veneer layers • Glued	• Beams • Floors • Roofs • Joists • Composite I-Joists • Sheeting	No UK production More than 450,000m³ softwood plywood imported mainly from Brazil 1 million m³ hardwood plywood imported mainly from China
Plywood	• Odd number of veneer layers, laid orthogonally • Glued • Available in different thicknesses and species	• Sheeting – both structural and decorative	
Oriented strand board (OSB)	• Strands • Wafers • Flakes • Glue	• Sheeting • Shear-loaded elements of composite products (e.g. I-joist)	More than 2.5 million m³ is produced in the UK More than 450,000m³ OSB and nearly 700,000m³ chipboard is imported, mainly from Europe
Chipboard	• Chips and sawdust • Glue	• Flooring, kitchen tops, etc. • Substrate for composite products (e.g. laminated flooring)	Produced from recovered wood and residues Can be recycled again
Fibreboard, including MDF, HDF	• Fibres • Glue	• Flooring, skirting, kitchen fronts • Shear-loaded elements of composite products (e.g. I-joist).	Nearly 800,000m³ is produced in the UK Nearly 900,000m³ is imported Produced from recovered wood and residues Cannot currently be recycled

Table 17.3 Characteristics, uses and impacts of the main engineered wood products (EWPs).

The presence of glue directly impacts a product's end of life. Glues decrease the reuse potential of wood products because the connection is not reversible and the effects of glue ageing are not well defined. Incineration of glued wood products is also likely to lead to more harmful emissions.

New research aims to either produce glues and resins from natural resources, such as soy flour adhesives, or create glues from the recycling process of other products. To date, few are cost effective.

LONGEVITY
Durability and decay

The lifespan of timber products can be very variable. Timber that makes it into buildings can have a life of several centuries, but this requires careful planning, material choice, detailing and maintenance.

All building materials, not just wood, are subject to decay. The longevity of any product is decided by two factors: the material's resistance and its exposure to decaying agents. Wood has a good resistance against chemicals, corrosion and heat, but, as a natural material, it is susceptible to biological decay. The most important factor in preventing biological decay is moisture, and the most important consideration for designers should be keeping timber dry, wherever possible. This can be achieved by following the 4-D hierarchy: deflection, drainage, drying and durability – four principles that should be considered as extensions of one another, rather than alternatives (see also exemplary design guidance for timber frame buildings[30] and for CLT structures[31]).

Some woods have natural durability, with an inherent resistance against insects, fungi and marine borers. The resistance of wood against fungi is categorised in durability classes, according to EN 350.[32] Natural durability varies between wood species, but also depends on origin, growing conditions of the tree, and the portion of the stem - sapwood is never resistant to decay.

Whether the natural durability of a wood product is sufficient to prevent decay depends on the circumstances in which it is used and its moisture content. Timber is constantly absorbing and desorbing moisture from the surrounding air, and in heated indoor environments normally reaches a moisture content between 8 and 13%, but moisture contents in more humid environments can be above 20%, which is when timber becomes at risk of fungal decay. To facilitate design choices, EN 335 and EN 460 provide further guidance on durability classes and ratings.[33]

Figure 17.6 The Gramophone Works, London, Studio Rhe, 2022. Lightweight timber enables the extension of the 1919 reinforced concrete building from two to six storeys without reinforcing existing foundations. The hybrid extension sits on a steel transfer deck on top of the original roof slab and is comprised of a glulam frame with CLT floors. In terms of whole-life carbon, the most impactful design decision is retention of the existing concrete frame because it is already there, but the modern timber construction elevates the original structure, giving the building a new identity as a creative office space and riverside café.

Treatments and modifications

Where the natural durability of timber is not enough to protect it from decay, even after implementing best design, it can be treated to increase its resistance. Many treatment, modification and coating options exist (see Table 18.4), and all are linked to certain cost and environmental impacts. Treatments should therefore be used as much as necessary but as little as possible. Whether treatment is necessary also depends on the risk that is involved should the product fail, and on the expected service life.[34] Structural and inaccessible components are therefore often treated with preservatives.

Wood degrades when exposed to UV light, environmental moisture changes and loading. These can lead to swelling and shrinking of wood. Wood does not swell and shrink equally in all three fibre directions, with movement much greater across the grain than along the fibres, so that these processes can also lead to deformations. A degree of moisture movement needs to be considered in design and timber needs to have the correct moisture content at the time of construction to avoid these problems. During construction, the moisture content must be checked to ensure the application of coatings or treatments is done at the right moisture content, so water is not trapped inside the material.

Timber products that are less exposed, for example concealed or protected structural timber elements, can have an indefinite service life, in theory. These products often reach the end of their life because of changes in building use or accidental damage.

Accidental damage – fire

Timber products' reaction to fire has been studied extensively in the last few decades, and well-documented charring rates allow designers to design timber structures with confidence, even when accounting for an unpredictable action like fire.

Wood's performance in fire can be understood in terms of 'reaction' and 'resistance'. The reaction is the contribution of a material to the spread of fire and is classified by EN 13501 in seven groups – timber products are typically rated as D or E, meaning they contribute to a fire. When exposed wood-based products need to be used in an area where a grade D or higher is required, flame-retardant treatments should be considered. Treatments can be applied at the timber factory/sawmill (typically impregnation or incorporation during manufacture) or during construction (surface coatings). While both approaches improve the resistance to flame, treatments applied during manufacture of the product can guarantee a higher standard and reliability than on-site applications, but will affect the underlying properties of the timber, unlike surface coatings. Not all wood-based materials are suitable for post-manufacture impregnation (e.g. OSB or MDF), and may need recertification following any additional treatment.

Resistance is the ability of the system to tolerate fire before it collapses or is exceedingly deteriorated. This means it is not possible to assess this characteristic based on the single element, but it is necessary to account for assemblies and to consider how parts are connected or separated. However, the charring rate (the speed at which timber burns) is reasonably predictable. Solid hardwood and softwood reduce their thickness (charred layer) by 30 to 40mm per hour of exposure. EWPs require more complicated evaluations, as the exposed layers delaminate and fall off when the glue is softening. An accurate study of the geometry and composition of the spaces should always be undertaken, but new concepts often require expensive research and testing.

Treatment option	Advantages	Disadvantages
Preservative treatment: biocide impregnation	• Industrial process (controlled, efficient, effective). • Level of treatment can be adjusted to requirements. • Can be used in wood products like OSB. • Little maintenance required. • Simultaneous fire-retardant treatment possible.	• Use of potentially harmful chemicals (especially in heavy-duty applications). • Reuse might be limited. • Incineration at end of life might release harmful substances. • Only permeable wood species can be treated. • Strength and stiffness can be affected (see EN 15228[36]).
Modification: acetylation furfurylation thermal modification	• Very good resistance against most wood-decaying organisms. • Reduced moisture uptake and movement (swelling and shrinking). • Little maintenance required. • Use of harmless chemicals, sometimes made from waste products (furfuryl alcohol). • Mimics the look of tropical wood (thermal modification, furfurylation)	• High cost. • High energy use. • Mechanical properties might be reduced. • Potential corrosion of metals (acetylation). • Discolouration under UV light.
Coatings: paints varnishes stains oils waxes	• Wide range of material choices and aesthetics (with wide range of protective properties against different degradation agents). • Natural products available (e.g. vegetable oil). • Can be used in addition to other treatments. • Can increase resistance to decay, UV-light, abrasion, indentation.	• Coatings increase resistance by shielding and do not change wood properties. • May contain potentially harmful chemicals and give off VOCs. • Reuse and recycling options might be limited. • Level of protection strongly depends on flawless application. • Coatings need to be repeated at intervals.

Table 17.4 Treatment options for timber products to improve durability.

END OF LIFE

Around 4.5 million tonnes of wood waste arise in the UK annually.[37] Most of it is chipped for recycling or incineration, even though waste wood often has great reuse potential. Ideally, wood could be cascaded into new products after its first life, before finally being incinerated or composted. These end-of-life treatments release the stored carbon back into the atmosphere, but they can provide a valuable function at the same time.[38]

Cascading wood means reusing it multiple times, with minimal processing, in products with decreasing demands of timber quality and particle size. This way, a wood product could be used again and again, keeping the carbon locked away.

Timber cascading requires waste wood to be in a good condition, but demolition timber is not normally suitable for reuse, because it gets damaged during demolition.[39] Currently, questions of time and profitability tip the scale towards demolition as the method of choice. In future, buildings could be deconstructed and their parts reused, but this needs to be incorporated into the design. Even small changes towards design for deconstruction and reuse (DfDR) can improve the reuse potential of timber components and building assemblies.[40]

New products utilising recovered wood are being developed steadily, but the future should see more reuse and cascading. Researchers at University College London (UCL) have developed cross-laminated secondary timber (CLST), reusing Victorian roof timbers in new CLT.

DESIGN FOR LOWER IMPACTS

Choosing between different wood products that come with different environmental implications is one way designers and architects can influence the environmental impact of buildings. Even in the simplest case, when specifying solid timber, the design can influence material efficiency and transport distances, as well as forest sustainability.

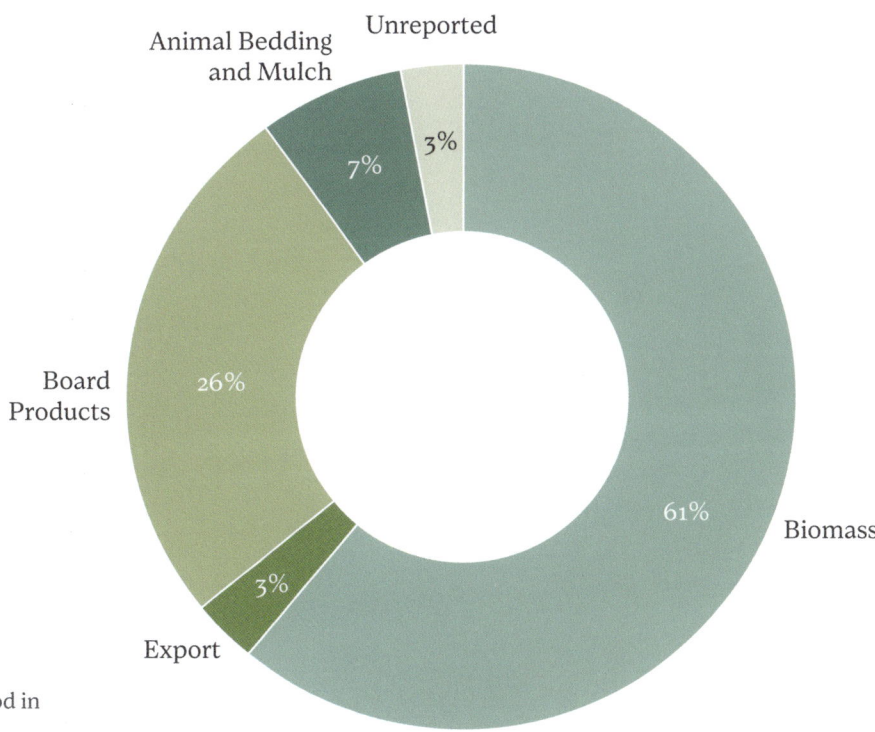

Figure 17.7 Use of recovered wood in the UK.[41]

Design for timber

Timber cannot span the same distances as steel and has a lower stiffness than concrete. Design iterations for any structural timber product should start with sensible configurations for the chosen material – for example, solid softwood floor joists can accommodate spans up to around 6m for loads between 1 and 5kN/m² while being economical, and glulam systems under the same conditions might be used for spans between 6 and 10m.[42]

The following sections present the three key design principles which support the greater use of local materials.

Use what you need

Specify timber by performance (e.g. a durability class or strength) rather than by a species or product, to enable a broader range of timber choices. Similar designs can often be achieved with two different approaches: either using less materials, but specifying higher strength classes, or using more materials but lower strength classes.

Figures 17.10a and b Sara Kulturhus, Skellefteå, Sweden, White Arkitekter, 2021. A mixed-use cultural centre and hotel, the 20-storey building is prefabricated and built with timber sourced from regional forests. It is one of the tallest all-timber-structure buildings, including timber cores.

Design out waste

Currently, biomass for energy production represents a large portion of UK timber consumption, and in 2021 biomass for energy consumed 15% of the UK's softwood harvest, 85% of the UK's hardwood harvest, and 61% of waste wood.[43] It would be better for the environment if we found alternative uses for UK timber.

Currently, using these resources structurally is only possible on a small scale because we lack strength grading assignments for most hardwood species, many softwood species and recovered wood. Even so, designers could specify materials that are currently considered waste. This may mean working with manufacturers to assess (structural) properties on a piece-by-piece basis and working out designs that can use the available resources. **The creative use of such unusual resources can lead to fascinating designs**, as demonstrated, for example, by the Massachusetts Institute of Technology's (MIT's) wooden pavilion using joints made from discarded tree forks.[44]

Design for the future

Material choices can also influence the longevity (life-cycle assessment modules B1 to B5) of timber products, as well as reusability at their end of life (C1 to C4, D). Designers can have a major influence on the environmental impact of timber products by designing for a long service life and multiple uses.

WIDER IMPACTS

Timber production affects the forest area and the forest types we create, which in turn affects the carbon flows between soil, vegetation and the atmosphere. How increased use of forest products will affect ecosystems, the climate and local communities cannot be answered precisely, because the answer is usually, 'It depends.' The following questions should be considered.

- Is increased timber demand leading to more forest creation or to more forest exploitation, including illegal logging and transformation of natural ecosystems to plantations? Either can be the case, depending on regional context. The best solution we have is certification.
- Is timber production linked to economic growth or exploitation of local communities? Again, either can be the case, as above.
- Does commercial forestry cause 'biodiversity deserts', or does it do good for nature? Replacing natural forests with monocultures does lead to a decrease in biodiversity, but this shouldn't happen in certified forests. Planting productive forests that comprise mainly of one tree species to replace other land uses or restock recently harvested woodland might not be the ideal choice for maximising biodiversity, but even these forests make important habitats, more so under recent standards, while providing the vast majority of the timber we need.[45]
- Is it better to grow a lot of timber quickly and store away carbon or to grow trees for longer, to create benefits to timber quality and ecosystem services? Balancing different objectives is the role of forestry management. Using the *Sustainable Wood for Cities* procurement guide can help match timber selection to the objectives of a particular project and balance competing considerations.[46]
- Should burning waste wood be counted as carbon neutral despite the fact that carbon is released when it is burned? It is counted as carbon neutral or even negative, as we are displacing grid energy, but we are still releasing carbon. If we have a 100% renewable energy from grid in the future, the better option might be to landfill wood and keep the carbon locked away.[47]

Timber brings a number of benefits to our built environment that are not easily tangible. Wood used in indoor environments can alleviate stress, improve occupants' health and contribute to a well-balanced room climate. Even though these benefits might not be directly linked to timber's carbon emissions, they might be a major factor in a product's longevity. Buildings that are appreciated by the public are less likely to be demolished

QUESTIONS
FOR PROJECT TEAMS
AND SUPPLIERS

☐ Are the legality and sustainability of the timber product assured, either by independent certification or by working with trusted suppliers?

☐ Have all the implications result of timber selection been considered, including transport distances or recycling potential?

☐ Is the environmental exposure appropriate for the product and the intended design life?

☐ Have the 4-D principles for addressing moisture – deflection, drainage, drying and durability - been considered?

☐ Has the natural movement of the material been considered, especially for exposed elements, such as cladding?

☐ Have fire regulations been followed when sizes of members, materials, and preservatives were specified?

☐ Has timber been considered from the outset of the design, ensuring that geometries suitable for other construction materials have not been inappropriately adapted?

☐ Can the same design be achieved either by using less material or specifying lower material quality, such as a lower strength class?

☐ Has the material been stored and handled properly on site?

☐ Has the design considered reuse and cascading potential?

prematurely, and products are more likely to be actively maintained if they are perceived as valuable. Timber has a vast advantage in these respects, especially if construction projects are regionally or historically linked to the resource.

FUTURE TRENDS

One challenge for the future will be realising the ambition of using more timber in construction while using the same amount of timber overall. An even bigger impact will be achieved by bringing a wider range of bio-based materials into use, which requires research and innovation. Creative use of timber, including cork, together with hemp, straw and lime, can upgrade existing buildings, and be used for new buildings when needed.

Much home-grown timber is used in applications with low demands regarding material quality, such as fencing and packaging, but reclaimed timber could be cascaded to meet the material needs for these less demanding uses. While this shift lies beyond the remit of built environment practitioners, designers can leverage their sphere of influence by retrofitting and designing pioneering timber buildings that point new ways forward.

Since the conception of mass timber in the 1990s, higher and higher timber buildings have been realised. We have an opportunity to realise more high-rise buildings (greater than 18m) in timber, though this represents a tiny percentage of UK construction and requires overcoming challenges with insurance and regulations. The UK's low-rise housing sector presents an enormous opportunity for a shift to timber within existing codes and regulations. Other sectors are also well suited to timber's shorter spans, such as mid-rise and commercial buildings including schools, doctors' surgeries and care homes.

Innovative wood products

Innovations in timber construction can be grouped in three categories:
- bringing a variety of resources into use via basic properties research
- improving products to make them more material efficient or more resistant
- decarbonising building products by reducing the use of fossil resources.

Secondary EWPs are made from recovered timber as a form of reuse or recycling. Recycling timber in board products, like chipboard, is already common practice, but reusing wood in mass timber products is only at the proof-of-concept stage.[48] Mass timber products also have potential to use other resources with low mechanical properties or higher variability, like homegrown hardwoods. Components can incorporate layers of different grades (usually higher grades for outer layers, and lower grades for inner layers) to optimise the use of wood, with higher strength where it is needed. Research on these topics is underway in the UK, which could result in local production of these innovative products.

New research aims to either produce glues and resins from natural resources, such as soy flour adhesives,[49] or create glues from the recycling process of other products.[50] The use of mechanical fasteners (e.g. dowels in DLT, nails in NLT) or self-bonding particles (binderless MDF, binderless chipboard) are other promising areas of innovation that could reduce reliance on glues and improve the end-of-life scenarios for EWPs.

KEY TAKEAWAYS

- We are probably growing enough wood, we are just using too much. Treating timber as a finite, valuable material and allocating resources to be used to their full potential is essential.

- The ways in which timber consumption affects forests and local communities are hard to quantify. It is important to engage with the supply chain or at least specify certified timber to have a high chance of making things better, rather than worse.

- Up to half the carbon emissions of the timber supply chain result from transport, so specifying local materials can greatly reduce emissions. Materials should be specified according to actual requirements, and attention should be paid to locally available supply.

- Timber inside structures can have a service life of centuries, keeping the sequestered carbon locked up.

- Recovered timber has a high potential for being cascaded into EWPs before it is incinerated for energy production. Demand for secondary timber products can be created by specifying them in your design and ensure the reusability of components by implementing DfDR.

- The design should be matched to the available materials, not the materials matched to the design. Timber has specific span limitations depending on the selected product, so it is important to make sure specific characteristics are accounted for from the very beginning.

ZINC

Nick Hodges

Durability

60–100+ years

Embodied carbon

20,000-34,000 kgCO$_2$e/m^3 [1]

Amount of refined zinc produced globally in 2018

13.2 million tonnes, approximately[2]

Melting point

420°C

Zinc has long been in use in buildings as part of brass alloys, but it has been used as a material in its own right for only around 200 years. While zinc is common in construction, it is predominantly through its use in galvanising steel and iron. When used as a metal, it has characteristics of lightness and rigidity, and is used in roofs, façade cladding and rainwater goods.

Zinc as a pure metal is lightweight but relatively brittle. It is typically alloyed in small quantities with other metals to provide malleability and flexibility. In early uses this was copper, while in modern constructions zinc is now also alloyed with titanium. The proportion of zinc in the alloys is typically greater than 99%. Zinc is infinitely recyclable without loss of performance. Zinc has high energy consumption in mining and processing, particularly through its use of electricity in extracting zinc from its ores, although this is now providing potential for carbon reduction by using low-carbon energy. It can be readily and repeatedly recycled, and around 80% of zinc is reclaimed, including from galvanising.[3] For zinc sheets used in construction, this increases to up to 95%.[4]

APPLICATIONS

- Cladding
- Roofing
- Rainwater goods
- Galvanised protection of steel

PROS

- Long life
- High recycled content
- Excellent recyclability at end of life
- Robust
- Relatively low environmental risks in production

CONS

- High energy use in extraction, processing and production
- Risk of bimetallic corrosion and affected by some tannins from timber

Compared to other metals, such as copper, iron and lead, zinc has only relatively recently been used as an individual metal. Zinc ores were used, with copper, to make brass for many centuries, but the first smelting of metallic zinc is thought to have begun in 14th-century India.

Within 50 years of its isolation as a pure metal in 1746, zinc was being rolled into sheets and considered for use in construction. It became a characteristic roofing material of the renovation and replanning of Paris by Baron Haussmann in the mid-19th century. Zinc has continued to be popular as a roofing and cladding material for its lightness, stiffness and longevity. Through the discovery of hot-dip galvanising, zinc has also become an important anticorrosion treatment for steel and iron.

CHARACTERISTICS

Zinc is typically a bright silver colour as a natural material, which quickly dulls to silver-grey and eventually patinates through oxidisation to a grey-blue. As a coated or self-finished material, zinc has good longevity and resistance to corrosion – as an alloy there are examples of it installed for over 100 years. It has been in use as a roof or cladding material and rainwater goods for several centuries, although can suffer increased corrosion in industrial and coastal environments.

Brasses and coppers offer the greatest risk of bimetallic or galvanic corrosion with zinc and care should be taken with copper, and copper alloys, in contact with other metals.

AVAILABILITY AND PRODUCTION

Zinc is the fourth most widely used metal in the world. Much of this production is used in construction, with approximately half in anticorrosion galvanised coatings for steel and iron and a further third used in modern metal alloy production, including brass, bronze and zinc aluminium. While zinc has excellent recyclability, post-consumer sources are expected to provide around 50 to 60% of the expected demand for zinc by 2050.

Zinc is also used in battery production, and white powder zinc oxide is used to create white paint and as a topical treatment, usually as creams or ointments, such as calamine. In some formations, it is used as an antibacterial agent.

As a proportion of total material use, only a relatively small amount of zinc is used in its own right, as a cladding material, for example. Recycled sources of zinc contribute around one-third of total zinc production and around 80% of zinc is reclaimed, including from galvanised steel. When used in construction in sheet form, this increases to up to 95%. Zinc is recyclable even in its alloyed condition, but use of coatings and finishes can complicate the recycling process.

Zinc production is global, with China, Peru and Australia currently accounting for over 50% of worldwide production.[5] Zinc ore contains up to 15% zinc, and extraction is an energy-intensive process that has historically focused on heat energy to smelt raw material. Since the 1980s, production has been predominantly through electrolysis, which uses sulphuric acid in electrolytic cells and uses approximately 3,900kWh of electrical power per tonne of zinc produced. Zinc produced in this way still requires further heat processing to create ingots of high purity metal, or zinc containing alloys.

MATERIAL SOURCING

Zinc ingots are processed into rolls or sheets for use in construction, typically as façades, roofs and cladding made from thin sheets (0.7mm thick), with lightweight fixing and connections. Sheets and tiles are cut from larger zinc rolls produced through the primary manufacturing process. For the UK, most commercially available zinc products are based in mainland Europe and require transportation to site.

Zinc can be pressed into more rigid forms as part of a whole roofing or cladding cassette system, or fixed directly to substrates with clips and standing-seam connections. It can be finished as either a natural material which will patinate through weathering or be coated. These additional processes involve chemical treatments such as pickling, which can add further environmental risks if not well managed.

CONSTRUCTION METHODS

Standing-seam installations are most common on roofs, which are recommended to be laid at a minimum 5 degrees and up to vertical. The seams are typically at 450mm-width centres (maximum approximately 600mm), supported at the seams with stainless-steel clips which are fixed into the substrate. Standing-seam installations can be laid as a level surface or on curved substrates for barrel or vault roofs.

Roofs are typically designed to be ventilated, but should always be designed to avoid interstitial condensation that can damage the roof structure. Ideally, zinc should be laid onto a breather membrane or other geotextile to reduce the impact of rain 'drumming' noise on the metal.

Zinc façades can also be installed as interlocking flatlock panels or shingle tiles, often in a diamond pattern. Tiles are fixed to a rigid substrate with stainless-steel clips and installed from the bottom up. These installations are able to form three-dimensional curved forms and can be quite sculptural.

An important point to bear in mind with all zinc applications is that it has bimetallic corrosion risks when used in contact with copper, brass, steel, cast iron and chromed surfaces. It also has risks of damage when used with some timbers with an acidic

QUESTIONS
FOR PROJECT TEAMS
AND SUPPLIERS

☐ What are the optimum sheet or roll sizes for the most efficient use of material?

☐ Where was the zinc produced and what is its carbon factor?

☐ Is a low-carbon zinc available and possible for the project?

☐ How should the material be fitted in a way that will last and reduce risks of substrate failure or bimetallic corrosion?

☐ How can the metal be fixed in a way that makes it easy to disassemble?

tannin (such as oak and red cedar), cements and plasters, and bitumen roofs – which should not run off onto a zinc surface.

CIRCULARITY AND RECYCLABILITY

Zinc is readily recycled and around 95% of zinc used in construction is recycled, making a significant contribution to the total amount of zinc from recycled sources. Deconstructing zinc assemblies requires disconnecting zinc sheets or tiles from the substrate and separating the stainless-steel fixing clips, essentially in a reverse of the installation process. In a well-organised demolition process, this should be a relatively straightforward disassembly.

DESIGNING FOR LOWER IMPACTS

Zinc has high embodied carbon per kilogram produced, principally as a result of electrical energy used for extraction and heat energy in processing. The most effective way to reduce this is to specify 100% recycled material, and, if possible, from a manufacturer close to site to reduce carbon associated with transport.

Zinc has excellent longevity and designs should maximise material life through robust details with well-ventilated substrates. Site waste can be minimised by working with suppliers and subcontractors to identify the most efficient shingle and roll sizes. Where possible, zinc should be used in a way that makes it easy to deconstruct, through easy-to-remove fixings and avoiding composite building elements. Although current recycled zinc stock cannot meet the global demand, by increasingly asking for recycled zinc, designers will be informing the supply chain of its importance.

WIDER SUSTAINABILITY IMPACTS

While zinc is an essential element for human health, large concentrations of zinc can cause health problems, although this would be through direct consumption rather than other forms of exposure. Production of zinc does not generally produce directly toxic zinc-containing by-products, but does produce other chemical wastes, such as sulphur dioxide and cadmium.

Zinc can be an aquatic pollutant, and rainwater run-off should be managed to avoid entering any aquatic ecosystems. Where zinc is joined using lead-based solder, rainwater run-off can include lead which is highly toxic, so alternative soldering compounds should be used.

Figure 18.1 Fenton House, Dundee, Kirsty Maguire Architect, 2019. Fully demountable and recyclable zinc cladding encloses the timber structure of this Passivhaus project. A whole-life carbon analysis of Fenton House calculated its embodied carbon at 496.8kgCO$_2$e/m^2 GIA (gross internal area), with zinc representing just over 11% of the total.

FUTURE TRENDS

Technological developments in zinc have led to a range of coloured, pre-patinated zinc finishes – in a range of greys through to a near-black. These are developed as a factory finish and described as 'self-healing', maintenance-free finishes, while remaining fully recyclable and solderable.

Manufacturers are now offering EcoZinc with a stated 50% reduction in embodied carbon, which is achieved through extraction and processing using lower-carbon energy sources.

KEY TAKEAWAYS

- Zinc has great potential for circularity and when used in construction can be sourced as 100% recycled material.

- Zinc is a metal with excellent longevity, evidenced by buildings using zinc in installations that have lasted over a century.

- While zinc has a character of weathered blue-grey, a range of patinated colours are now available.

- Low-carbon zinc is now available for construction projects, making use of low-carbon electricity in production.

- Using zinc efficiently and from recycled sources can mitigate some of its carbon impacts, and it should be installed in a way that is easily demountable to support recycling at the end of the building's life.

FUTURE INNOVATIONS AND TRENDS

SOPHIE THOMAS

Innovation is a much-overused word, but when talking about future trends in construction materials it is highly relevant. If we look beyond the current discussions covered in the previous chapters surrounding engineering efficiency, manufacturing and energy improvement, we catch glimpses of material innovations that in 10, maybe 20, years' time will be accelerating the birth of a new era in the built environment. Right now, these materials and technologies are still in the minds of budding green chemists (the unicorns of the biotech sector), or they are a first prototype growing in a biotech start-up test tube.

These are the ideas to watch. They have the potential to disrupt the incumbents, change the brief and reimagine the outcome – if we let them.

The construction industry knows what big challenges it faces. The previous pages in this book demonstrate that for a long time there has been a heavy reliance on a core palette of materials with heavy eco-rucksacks and carbon burdens. And at the end of life, archaic and inadequate demolition systems bulldoze and crush buildings into contaminated piles, resulting in enormous losses of potentially reusable material.

Like every challenge, these bring opportunities to explore innovative processes and new materials. Technologies like AI (artificial intelligence), IoT (internet of things), smart sensors and 3D printing are already disrupting the way things are built, monitored and adapted. A focus on embodied energy in building design has put the spotlight on materials and the need for better products. Knowing that we can already see the seeds of future materials taking root, the question becomes, 'Where should we look for inspiration?'

It is an enormous subject with an expansive range of exploration and research going on around the world. All types of skill sets are involved, including scientists, designers

Figure 19.1 Cutaway view of a Stanford Torus Space Colony, designed by NASA, 1975. This proposed design for a space habitat was capable of housing 10,000 to 140,000 permanent residents.

and start-up companies – sitting in different types of universities and research labs – all driven by the desire to help build a more efficient, more adaptable, circular and low-carbon world.

Following is a selection of early-stage innovations that have the potential to respond to these crucial challenges. There are so many others out there and this list is certainly not exclusive but aims to demonstrate some of the opportunities being explored.

Each innovation is set out as a potential answer to a specific challenge.

Figure 19.2 The *Cyphochilus* beetle.

CHALLENGE: RESOURCE SECURITY – PAINTS AND COATINGS

Of the 9 million tonnes of globally mined titanium mineral per year, approximately 93% is used in making things white through titanium dioxide.[1] The extraction and processing associated with creating this ubiquitous mineral are environmentally harmful, polluting ground water and damaging fragile coastal areas.[2] Half the global supply of this mineral is supplied by the Asia Pacific region and Europe holds no reserves.[3] The use of titanium dioxide is quite extraordinary in its breadth – think of anything that is white – from paint to plastic, skimmed milk to toothpaste. Recently, concerns around nano particle toxicity have pushed the EU to ban the pigment in food.[4]

Opportunity: One day, could we eliminate our need for paints and coatings by looking to nature for inspiration?

Biomimicry research into the brilliant-white *Cyphochilus* beetle has discovered that, through its covering of perfectly spaced micro scales, all light wavelengths are scattered, creating its pure-white look.[5] Tech start-up Impossible Materials has used this idea to create a cellulose-based white pigment technology that mimics the beetle's exoskeleton structure without additional minerals.[6] This research could disrupt the paint and specialist coating markets that are so reliant on titanium dioxide[7] with a global market set to grow to around $27.9 billion by 2027.[8]

CHALLENGE: MATERIAL TOXICITY – PAINTS AND COATINGS

Paints and coatings have a significant dual role in the construction sector, being both protective and decorative. The global market is huge, valued at $164 billion in 2022.[9] The component materials of paint include pigment, a binder and a solvent. There are also additional additives in paint that include biocides to stop mould, surfactants to help the even spread of paint, and driers to accelerate paint drying time.[10] Paint and coatings chemistry comes with significant upstream and downstream environmental and health impacts, including toxic additives, heavy metals, polluting processes, health hazards and untraceable production. There are also issues at end of life – unused paint is treated as hazardous, requiring special disposal.

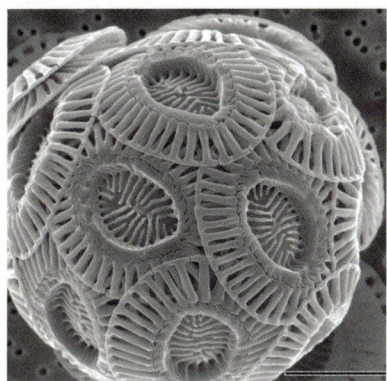

Figure 19.3 A scanning electron micrograph of a single coccolithophore cell.

Opportunity: Can we eliminate the need to add chemical-based pigments by looking at how nature creates colour through nano-sized structures?

Biomimicry research into the iridescent blue of the morpho butterfly's wings demonstrated how light-interacting nano structures on scales reflected and deflected the colour spectrum in a way that created a beautiful, iridescent colour even though the wings were translucent.[11] Cypris Materials has developed a structural colour nanotechnology that creates a colour palette free of pigments and dyes.[12] This technology could disrupt the paints and coatings industry and bring new colour opportunities literally into the structural design of objects without adding more materials.

CHALLENGE: CARBON INTENSIVE MATERIALS – CONCRETE

Other than water, concrete is the most used material on the planet, representing about 50% of all human-made materials by mass. This major construction material accounts for 8% of the global annual greenhouse gas emissions.[13] Cement makes up to 10 to 15% of concrete and making it a carbon-intensive process. A large part of this is in the extraction and burning of quarried limestone. This mineral is blasted from quarries, then heated to temperatures of up to 1,450°C to make it into clinker. This requires lots of energy and can be up to 40% of the carbon footprint of concrete.[14]

Opportunity: Can we find alternative, low-carbon ways to create limestone using algae?

The CU Boulder Engineers at Algal Resources Collection at the University of North Carolina Wilmington (UNCW), with others, have created a concrete that utilises micro-algae as a resource.[15] These cloudy white micro-algae, called coccolithophores, sequester and store CO_2 in mineral form through photosynthesis by creating calcium carbonate micro-shells around the cells.[16] The concrete mix contains cement made from biogenic limestone produced by the micro-algae. It has the potential to be carbon negative by reducing the embedded carbon of concrete and locking in the carbon sequestered by the algae.

CHALLENGE: CARBON INTENSIVE MATERIALS – STEEL

Steel is one of the most important construction engineering materials globally, but the industry has a big decarbonising challenge. Some 1.95 billion tonnes of steel was produced in 2021, with the total direct emissions equalling 2.6 billion tonnes CO_2e, representing between 3 and 4% of global CO2e emissions.[17] Over the years there have been many pushes to find alternatives to steel from composites with glass fibre and carbon. Although incredibly lightweight and malleable, such composites are complex and hard to recycle once they come to the end of life.

Opportunity : Can we find laighter way to bring structural strength through molecular structure?

Chemical engineers at the Massachusetts Institute of Technology (MIT) have recently made a breakthrough with a novel polymerisation process, creating a new material that is stronger than steel, as light as plastic and cheap to manufacture.[18] It is a simple, two-dimensional polymer that self assembles into sheets, giving it great strength without the need to increase the material mass. This new material structure has the potential to accelerate new material use where strength is required – the research team says it could even be used to build a bridge.

CHALLENGE: CARBON INTENSIVE PRODUCTION – 3D PRINTING

Large-scale 3D printing in construction could radically change the way we build, but most printers use concrete, furthering the carbon footprint of the construction sector. Even though concrete 3D-printing technology is still in its infancy, it is valued at $56.4 million and is growing fast. With large-scale construction 3D printers mainly relying on concrete as a printing medium, there has been little or no carbon benefit from the technology until now.

Opportunity Are there better 3D-print materials with lower carbon impacts?

The University of Maine Advanced Structures and Composites Centre has built the first 100% bio-based 3D-printed house.[19] The BioHome3D was printed on the world's largest polymer printer and researchers used scrap lumber, sawdust and construction debris to make 'wood flour' as the basic building material. New 3D-print materials have the potential to build faster, cheaper, more sustainable housing using one material, bringing speed to the process and reducing the carbon impact.

CHALLENGE: MATERIAL LOSS – FOOD WASTE

If food waste were a country, it would be the third-largest emitter of greenhouse gases (after China and the USA). Over one third of all food produced goes to waste, ranging from waste from the field to leftover dinners.[20] Some of this waste stream is collected and converted into gas through composting and anaerobic digestion systems (and UK law will soon require all food waste to be collected from households). Companies are waking up to this potential feedstock full of raw materials and are now beginning to divert this waste into making new materials.

Opportunity: Can we replace carbon-intense materials by creating alternatives from other bio-rich waste streams?

Using manufacturing techniques from the pressed wood board industries, Japanese tech start-up Fabula has been freeze-drying and pulverising food scraps into powder which is then compressed into boards.[21] Using the biological make-up of food waste meant that no adhesives were required to bind the material. In strength tests, some vegetables showed great promise – Chinese cabbage waste creates a board that has a bending strength four times stronger than concrete.[22]

CONCLUSION

Material innovation can take decades to accelerate into an off-the-shelf reality. It took more than 20 years to establish the commercial use of carbon fibre in aeroplanes and a further 30 years before it became the primary material for the Boeing 787.[23] But the future isn't what it used to be. We are at a moment in time where good ideas with accelerated action are urgently required.

Deep Tech like AI may come to our rescue, helping to take the risk out of material development and fast-tracking new recipes that can drop into our existing manufacturing processes. Robots are already helping us to accelerate prototyping and construction in a way we would not have imagined in past decades. We have a long way to go to reduce this industry's carbon, and in the material choices there are solutions coming.

The concepts discussed briefly here are just some ideas for new materials that are being thought up by scientists and start-ups wanting to help create a better, liveable and regenerative world. We must start somewhere and where better place than a Chinese cabbage or a white beetle?

CONCLUSION

Figure 20.1 Bloqs Meridian Water, Enfield, London, 5th Studio, 2022. Part retrofit and part newbuild, this reuse of a former vehicle testing facility refitted existing large-scale window assemblies with recyclable polycarbonate sheets that flood the workshop space with daylight. Aluminium was specified mill-finished to reduce embodied energy in manufacturing, with mechanic fixings for easy disassembly.

Every design decision impacts our planetary boundaries and how we use materials is no exception. In this book, we advocate the use of the right amount of the right material in the right place.

For a designer, this requires navigating complex supply chains and balancing multiple design parameters which often clash. While this book focuses primarily on understanding the carbon impacts of material choices, the wider ecological issues and health impacts are equally important.

Our hope is that by gathering together expert insights on the environmental impacts of the building materials of today – and the emerging materials of tomorrow – readers will be better equipped to exercise these choices. Through connecting with the material supply chain, designers will have a greater understanding of where the true impacts of a material occur and will be able to make informed decisions on efficient use of a material.

With every client brief, these questions must be asked: is a new building necessary? Is there an existing building that can meet the brief?

Designers must look upstream into environmental resource flows and downstream into communities on the ground impacted by material specification, sometimes in another part of the globe. This requires a broad grasp of ecological issues as well as a deep understanding of supply chains. Cities4Forests' exemplary work to build networks between municipal procurement and forest communities (both nearby and remote) does just that.[1]

Architecture must move away from a primary focus on eye candy to embrace an ecologically sound aesthetic. Buildings must be beautiful so that people will cherish and look after them, but they must be based on truly sustainable material choices. This requires constant focus and scrutiny throughout the design process, breaking out of the current design and construction processes.

DOUGHNUT ECONOMICS MEETS EPDS

Balancing the various impacts and benefits of a material can be incredibly daunting. An approach developed by Architects Climate Action Network (ACAN) together with Kate Raworth's Doughnut Economics Action Lab (DEAL) offers a framework for considering material choices through a lens of planetary boundaries, to help designers make choices that contribute to rather than compromise the planet's long-term resilience and ultimate survival.[2] This is not just another tool or a refresh of life-cycle analysis, but rather a lens through which to assess materials by considering both the local and global impacts of specification as well as their social implications.

This process starts with examining the Environmental Product Declaration (EPD) for a particular material (or product) and mapping its impacts on both people and the planet across the life-cycle stages. The second step digs deeper by mapping a material or product on the DEAL diagram to identify how they impact the planetary boundaries. Through this lens, building designers can begin to break down the complex relationship between our material use and our planetary impacts in an intelligible format. The final part of the exercise involves identifying ways to transform construction processes which exceed planetary limits and to promote social equity.

A NEW APPROACH TO MATERIALS

As we move forward in the 21st century, designers must leverage their agency to challenge business-as-usual. Designers can expand their sphere of influence by thinking beyond a project's boundaries and demonstrate that it is possible to build differently. It's okay to start small.

Figure 20.2 ACAN's mapping of a product onto the Doughnut Economic Action Lab diagram, highlighting areas of planetary and social impact in a semiquantitative way to help visualise the true impact of a product. This could be attempted for any product, and the identification of any knowledge gaps can be just as enlightening as where there is more certainty.

Collaborate – find the experts and the changemakers and get them on your team.

Question – no one person can know all the answers, and by asking the questions we can create a dialogue around material use with the design team, supply chain and contractors, raising awareness across the whole industry.

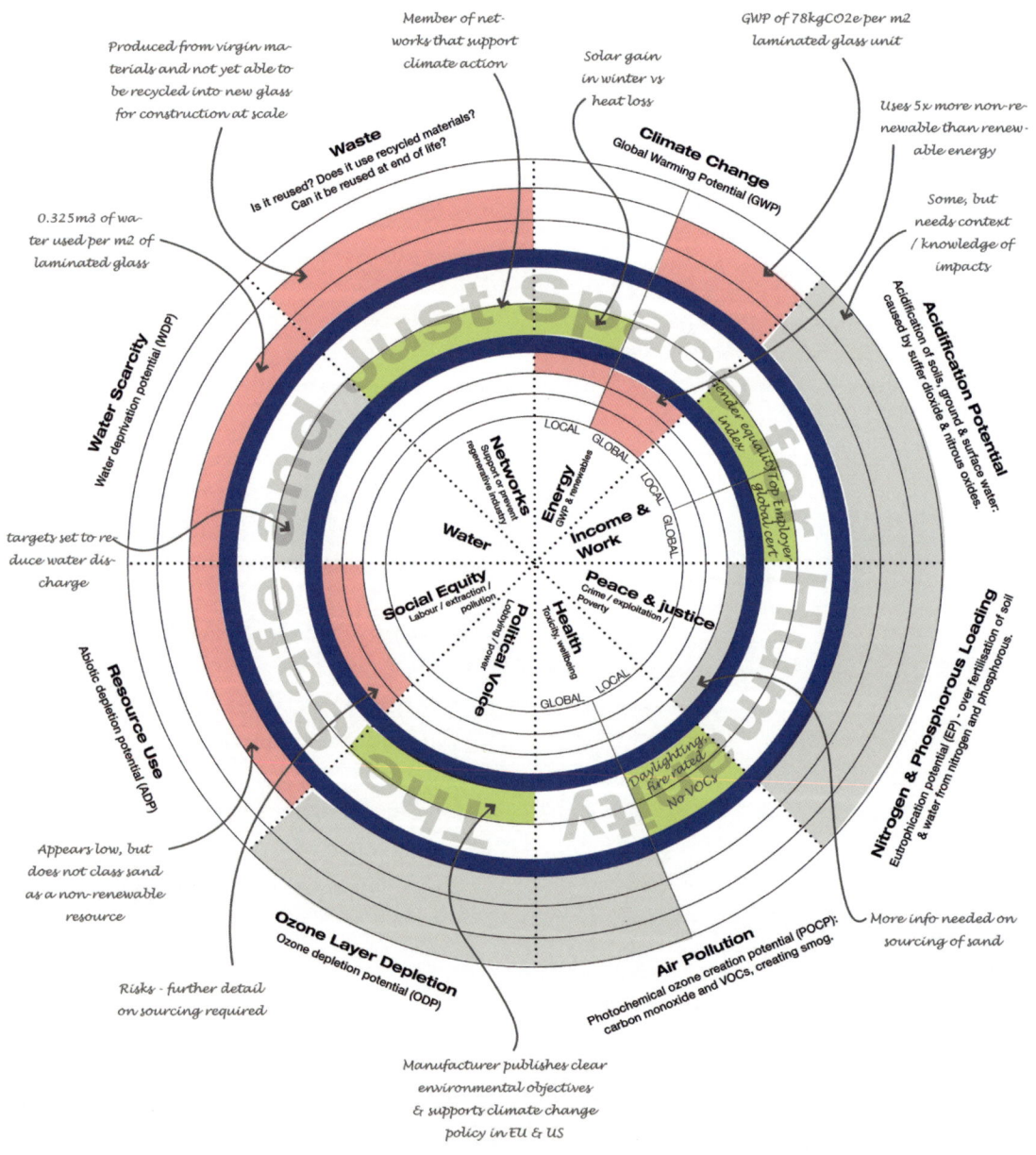

Figure 20.3 Le Magasin Électrique, a building on the campus of the LUMA Arles museum, Arles, by Assemble and BC Architects, 2023. Former train sheds have been adapted into a centre for Atelier LUMA, a collaborative of researchers working on the exploration of bio-regional materials. Rammed-earth walls topped with compressed-earth bricks made from excavated site soil form internal partitions. Different types of thermal and acoustic insulation were prototyped using rice straw, a by-product of the Camargue rice industry, and sunflower stems, a waste stream from making oil.

Use less – use of concrete, steel and other materials which have significant carbon impacts and draw on finite resources must be carefully considered and be limited to essential applications.

End of life – we must break the cycle of resource depletion and enable our buildings to be true material banks for future generations, designing-in simple deconstruction methods to maximise the value of the materials in our buildings.

Regenerative materials – the potential for reclaimed and bio-based renewable materials must be explored and built into the design programme.

Use local – a future of eco-regionalism where harvest maps, like those being developed in France as part of the *Cartographie Nationale des Ressources Locale* or in the Camargue by Atelier LUMA, identify local material resources so that more local sourcing of construction materials becomes the norm.

The material suppliers and product manufacturers of today's industry must be challenged to adapt working practices and develop products that comply with planetary boundaries. This is good for their business – and good for the planet. Use the questions in this book as a starting point for change.

And finally, push for material innovation on every project. The time is now.

HOW TO READ AN EPD

Throughout this book, the processes of how we make and use materials are discussed in terms of their key impacts on the planet. Sometimes it is enough to have a qualitative understanding of these impacts, but in light of the climate emergency, scrutiny of materials choices has sparked a demand for more rigorous quantitative assessment. More widespread use of Life-Cycle Assessment (LCA) has furthered this demand for metrics that can provide a definitive, quantitative understanding of material impacts.

This is where an Environmental Product Declaration (EPD) can be useful. EPDs are one of the few tools available that can be used to compare material impacts across a consistent set of parameters.

This short guide explains how to extract relevant information from an EPD to help guide materials choices. EPDs are complex technical documents, often running to 20 pages in length for a single product, and it can be daunting to interpret the numerous metrics and tables.

First EPDs are explained, and this is followed by a series of tips on how to read and decipher an EPD.

WHAT IS AN EPD?

An Environmental Product Declaration is a report that sets out the environmental impact of a product or material, following a standardised process. This process is governed by two key standards: ISO 14025:2006, which covers the overarching principles for creating an EPD, and BS EN 15804, a standard that defines the Product Category Rules (PCRs). The PCR guidance is a set of consistent assumptions and rules that are applied to broad product types, such as concrete or thermal insulation products, ensuring that the results from one product EPD are directly comparable to another within the same sector.

A key requirement of ISO 14025 is that each EPD is independently verified. This provides quality assurance to ensure the PCR has been applied fully, the calculations truly reflect the processes used, and that the reporting is aligned with the standards. It is always important to check that an EPD complies with the relevant standards.

EPDs that are not verified or do not follow the full standard can lead to misleading conclusions.

EPDs are optional documents, prepared by product manufacturers on a voluntary basis. Because each one costs over £5,000, depending on the complexity of the product being assessed, EPDs are prohibitively expensive for small manufacturers. In France, any company making a claim on their product's sustainability credentials must have an EPD to support their claims, leading to a significant number of French product EPDs in recent years.

In 2023, there were an estimated 130,000 EPDs available globally, with over 40,000 following the European standard of EN 15804.[1] This is a minute number compared to the number of materials and products available, but numbers of EPDs have proliferated in the last decade and are increasing steadily.

How are EPDs calculated?

An EPD contains measurements of the material and energy flows during the life of a product, beginning with the extraction of the raw materials, through transport and conversion to a product, and finally its end of life. The impact for each material or energy flow is then assessed and quantified, using peer-reviewed standard databases (such as EcoInvent or GaBi) that convert them into eight key indicators required by an EPD. Some EPDs analyse material production processes in great detail, while others remain high level.

Most EPDs focus on a specific product by a manufacturer, but there are also generic product EPDs, usually developed by industry associations. These can be useful at the early stage of a project when a manufacturer is not known, giving guidance on estimated impacts when changes are simpler to make.

What do EPDs assess?

All EPDs assess environmental impacts across eight key indicators:

- **Global warming potential (GWP)** The greenhouse gas impact, measured in $kgCO_2e$, including both emissions and removal. In terms of climate change, this is the metric to look for in assessing material impacts, notably within embodied carbon assessments. Recent EPDs now also include the impact of land use and land use change (LULUC), such as farming or forestry.
- **Ozone depletion potential (ODP)** The impact on the high altitude ozone layer, such as from CFCs.
- **Acidification potential (AP)** The impact of acidification of water and the earth, such as from direct pollution or causing acid rain.

- **Eutrophication potential (EP)** The impact of over-fertilisation of water and earth that can lead to imbalances in biomass, such as algae blooms.
- **Photochemical ozone creation potential (POCP)** The impact of chemicals that can lead to ground-level ozone, which is poisonous.
- **Abiotic depletion potential: elements (ADPE)** The impact of consuming minerals and non-renewable resources.
- **Abiotic depletion potential: fossil (ADPF)** The impact of consuming fossil-based resources, e.g. oil.
- **Resource and waste data** The flows in and out of energy, fuels, water and waste.

Each of these aspects is quantified with a specific unit that allows a product's environmental impact to be assessed and compared with other products. These impacts can be particularly useful if considering the impact on planetary boundaries or other broader metrics, but are typically used only by experts.

HOW TO READ AN EPD

EPDs follow the same LCA modules noted in the primer (A1, A2, A3, etc...). For most products, modules A1–A3 are typically the most important to understand as these are particular to the product, while the other modules are dependent on the site location, construction methods, maintenance schedules, building use, etc.

Not every module is needed for every product, for example B4 and B5 (replacement and refurbishment), as the EPD is for the life of the product, not the life of the building. There will always be a table within the EPD that notes what is and isn't included. See Figure B.1 below.

Production				Construction		Use							End of Life				Loads & Benefits
Raw Material	Transport Supply	Manufacturing	Production Total	Transport	Installation	Use	Maintenance	Repair	Replacement	Refurbishment	Operational Energy	Operational Water	Deconstruction	Transport	Waste Treatement	Disposal	Reuse / Recovery / Recycling
A1	A2	A3	A1-A3	A4	A5	B1	B2	B3	B4	B5	B6	B7	C1	C2	C3	C4	D
X	X	X	X	X	X	MND	MND	MND	MND	MND	NR	NR	X	X	X	X	X

X = included   **NR** = Not Relevant   **MND** = Module Not Declared

Figure B.1 Each EPD will have a table which describes which LCA modules are included in the analysis, which are not relevant, and which are not declared.

Check the units!

Before reading any data from an EPD, it is essential to understand the quantity of material being measured. This is the most common source of error from any interpretation of data from an EPD. Next to the table it will always state the 'functional unit', which is explained earlier on in the EPD. For most products, this is typically linked to how it is sold, e.g. m2 for flooring or per door. Some products are measured by volume or mass, and will need conversion to a practical unit before use. Note, some products have less intuitive units which will need some thought regarding how to convert them, such as PV panels, which are commonly measured by the amount of energy generated.

For any product, check what is and isn't included with the unit. For example, one façade product may contain all the steelwork to hang it, whereas another may include just the rain screen cladding, making direct comparison very difficult. What is included within the product will be clearly described within the EPD, so do check!

Understanding scientific notation

For many of us, scientific notation was last seen at school and it can seem incredibly confusing. However, it is straightforward once the basic principles are understood, with the 'E' shorthand for the exponent, and the final digits representing the power of 10 to be applied, i.e:

$2.75E\text{-}02 = 2.75 \times 10^{-2} = 0.0275$

$2.75E\text{-}01 = 2.75 \times 10^{-1} = 0.275$

$2.75E\text{+}00 = 2.75 \times 10^{0} = 2.75$

$2.75E\text{+}01 = 2.75 \times 10^{1} = 27.5$

$2.75E\text{+}02 = 2.75 \times 10^{2} = 275$

$2.75E\text{+}03 = 2.75 \times 10^{3} = 2750$

Reading the data

Once the functional unit of a product is understood, it's time to get stuck into the data. For this example, the focus will be on global warming potential (GWP), but the principles apply to each of the impacts within an EPD.

There may be many tables and multiple products within the same EPD, so find the correct product table and identify the GWP line. Many EPDs will contain emissions for each of the LCA modules, but the focus should be on the A1–A3 column. This represents the emissions from the creation of the product, whereas the other modules all contain estimates based on typical product usage. These other LCA modules should be treated with caution if used within any analysis.

A majority of current available EPDs include biogenic carbon as part of the A1–A3 total, notably where GWP figures are given as negative values. This can give misleading results, particularly if reporting only the upfront carbon emissions of a project, where the biogenic carbon is typically excluded from the calculation. To find the emissions without the biogenic carbon, add the GWP value for C3 to the A1–A3 total. The C3 emissions represent the release of the biogenic carbon through combustion of the product at the end of its life, and will provide a good estimate of the biogenic carbon within the product included in the A1–A3 total.

Note, newer EPDs use an updated format and split out the A1–A3 emissions by fossil fuel, biogenic, and land use and land use change (LULUC), so an emissions-only total can be formed from adding just fossil fuel and LULUC together.

	Production				Construction		Use							End of Life				Loads & Benefits
	Raw Material	Transport Supply	Manufacturing	Production Total	Transport	Installation	Use	Maintenance	Repair	Replacement	Refurbishment	Operational Energy	Operational Water	Deconstruction	Transport	Waste Treatement	Disposal	Reuse / Recovery / Recycling
	A1	A2	A3	A1-A3	A4	A5	B1	B2	B3	B4	B5	B6	B7	C1	C2	C3	C4	D
Global Warming Potential (GWP), kgCO₂e	-7.29E+02	1.40E+01	8.10E+00	-7.07E+02	3.84E+01	1.47E+01	0.00E+00	0.00E+00	0.00E+00	0.00E+00	0.00E+00	0.00E+00	0.00E+00	5.42E-01	2.04E+00	7.62E+02	0.00E+00	-8.14E+02
Ozone Depletion Potential (ODP), kgCFC11e	2.44E-06	3.90E-06	2.01E-06	8.35E-06	9.58E-06	2.82E-06	0.00E+00	0.00E+00	0.00E+00	0.00E+00	0.00E+00	0.00E+00	0.00E+00	1.19E-07	5.10E-07	0.00E+00	0.00E+00	-7.73E-06
Acidification Potential (AP), molH+e	1.45E-01	5.89E-02	1.97E-01	4.01E-01	1.30E-01	7.17E-02	0.00E+00	0.00E+00	0.00E+00	0.00E+00	0.00E+00	0.00E+00	0.00E+00	5.55E-03	6.59E-03	0.00E+00	0.00E+00	-3.89E-01
Eutrophication Potential (EP), kgPO₄e	3.53E-03	2.70E-03	4.30E-03	1.05E-02	3.17E-03	2.22E-03	0.00E+00	0.00E+00	0.00E+00	0.00E+00	0.00E+00	0.00E+00	0.00E+00	3.75E-05	1.67E-04	0.00E+00	0.00E+00	-1.03E-02
Photochemical Ozone Creation Potential (POCP), kgNMVOCe	2.26E-01	4.71E-02	9.09E-02	3.64E-01	1.17E-01	8.02E-02	0.00E+00	0.00E+00	0.00E+00	0.00E+00	0.00E+00	0.00E+00	0.00E+00	7.20E-03	6.05E-03	0.00E+00	0.00E+00	-3.51E-01
Abiotic Depletion Potential, Elements (ADPE), kgSbe	1.07E-04	2.75E-05	1.99E-06	1.54E-04	7.40E-05	2.61E-05	0.00E+00	0.00E+00	0.00E+00	0.00E+00	0.00E+00	0.00E+00	0.00E+00	2.69E-07	3.96E-06	0.00E+00	0.00E+00	-1.50E-04
Abiotic Depletion Potential, Fossil (ADPF), MJ	5.44E+02	2.62E+02	1.12E+02	9.18E+02	4.71E+00	1.90E+02	0.00E+00	0.00E+00	0.00E+00	0.00E+00	0.00E+00	0.00E+00	0.00E+00	7.68E+00	3.32E+01	0.00E+00	0.00E+00	-8.78E+02

Figure B.2 Example of the output figures within an EPD, showing the key global warming potential (GWP) row and the A1–A3 production column, where the key embodied carbon data can be found.

REFERENCES

Introduction

1. Optioneering is an iterative process of developing alternative designs to compare and assess which is best.
2. Roynon J, *Embodied Carbon: Structural Sensitivity Study*, Buro Happold, April 2020, https://www.istructe.org/resources/case-study/embodied-carbon-structural-sensitivity-study (accessed 15 August 2023).
3. https://materialepyramiden.dk (accessed 19 August 2023).
4. Global warming potential (GWP) is a scale for measuring the climate effects of different gases relative to CO2 and it compares the capacity of a greenhouse gas to trap heat in the atmosphere. The GWP of CO2 is 1.
5. FCBS CARBON is a free software tool to enable analysis of embodied carbon; see https://portal.fcbstudios.com/fcbscarbon (accessed 16 August 2023).
6. 'Planetary boundaries' is a term that describes the limits of impacts of human activities on the earth, beyond which the planet can no longer self-regulate. The planetary boundaries framework was developed in 2009 by scientists at the Stockholm Resilience Centre and the Australian National Institute and includes nine thresholds: climate change (CO2 concentration in the atmosphere), ocean acidification, ozone depletion, disruption of the nitrogen cycle, global freshwater use, land system change, chemical pollution, biodiversity loss and atmospheric aerosol load.

Primer

1. https://portal.fcbstudios.com/fcbscarbon (accessed 16 August 2023).
2. https://www.hawkinsbrown.com/sub-services/hbert-emissions-reduction-tool (accessed 16 August 2023).
3. https://buildingtransparency.org/ec3 (accessed 16 August 2023).
4. Such as within the University of Bath and Circular Ecology's free Inventory of Carbon and Energy, https://circularecology.com/embodied-carbon-footprint-database.html (accessed 16 August 2023).
5. https://www.architectsjournal.co.uk/news/introducing-retrofirst-a-new-aj-campaign-championing-reuse-in-the-built-environment (accessed 16 August 2023).
6. Roynon J, *Embodied Carbon: Structural Sensitivity Study*, Buro Happold, April 2020, https://www.istructe.org/resources/case-study/embodied-carbon-structural-sensitivity-study (accessed 15 August 2023).

7. https://www.rics.org/profession-standards/rics-standards-and-guidance/sector-standards/building-surveying-standards/whole-life-carbon-assessment-for-the-built-environment (accessed 16 August 2023).
8. https://transparency.perkinswill.com (accessed 16 August 2023).
9. https://declare.living-future.org (accessed 16 August 2023).
10. https://natureplus-institute.eu/?lang=en (accessed 16 August 2023).
11. https://www.designforfreedom.org (accessed 16 August 2023).

Aluminium

1. Building Transparency EC3 European Aluminium dataset: https://buildingtransparency.org/ec3/material-search (accessed 27 October 2023)
2. International Energy Agency Aluminium tracker: https://www.iea.org/energy-system/industry/aluminium (accessed 27 October 2023)
3. Email from Alvance to the author, August 2022, and International Aluminium Institute statistics – https://international-aluminium.org/statistics/primary-aluminium-production (accessed 20 July 2022).
4. Ibid.
5. International Aluminium Institute, 'IAI Material Flow Model – 2021 update', 2021, https://international-aluminium.org/resource/iai-material-flow-model-2021-update (accessed March 2022).
6. International Aluminium Institute, 'Aluminium sector greenhouse gas pathways to 2050'.
7. For more information on the history and development of aluminium see: Stacey M, *Aluminium Flexible and Light: Towards Sustainable Cities*, Cwningen Press, 2016.
8. For Julia Brainerd Hall's contribution, see Kass-Simon G, *Women of Science: Righting the Record*, Indiana University Press, 1993, pp 173–178.
9. Alfred R, 'April 2, 1889 Aluminum process foils steep prices', *Wired Magazine*, 4 May 2002, www.wired.com/2008/04/dayintech-0402 (accessed April 2019).
10. World Aluminium, 'Statistics', www.world-aluminium.org/statistics (accessed January 2015).
11. https://international-aluminium.org/new-iai-data-reveals-aluminium-industrys-continuous-improvement-in-energy-efficiency (accessed July 2022).

12. Data from Hydro Circal (formerly Hydro75R) EPD, https://www.hydro.com/globalassets/download-center/certificates/nepd-1841-768-hydro-75r-aluminium-extrusion-ingot.pdf (accessed August 2022). European average supplied by the International Aluminium Institute to the author.

13. Email from Alvance to the author, August 2022, and International Aluminium Institute statistics.

14. https://international-aluminium.org/new-iai-data-reveals-aluminium-industrys-continuous-improvement-in-energy-efficiency (accessed July 2022)

15. Stacey M, *Aluminium Flexible and Light*. Cwingen Press, 2016

16. Stacey M, *Aluminium: Sympathetic and Powerful*, Riverside Architectural Press, 2020, p 4.

17. Stacey M, *Aluminium Flexible and Light*.

18. Eurocode 9: Design of aluminium structures, European Commission, 2007, https://eurocodes.jrc.ec.europa.eu/EEurocodes/eurocode-9-design-aluminium-structures (accessed June 2023).

19. Stacey M, *Aluminium Recyclability and Recycling: Towards Sustainable Cities*, Cwningen Press, 2015, pp 230–245.

20. Adjaye D, *David Adjaye: Constructed Narratives*, Lars Müller, 2017.

21. Charlson A, *Counting Carbon: Practical Approaches to Life Cycle Assessment in Facade Engineering*, Institution of Civil Engineers, 2011, pp 1–13. GaBi software is supplied by Thinkstep.

22. Stacey M, *Aluminium and Durability: Towards Sustainable Cities*, Cwningen Press, 2014 (second edition 2015).

23. Stacey M and Bayliss C, 'Aluminium and durability: Reviewed by inspection and testing', *Materials Today: Proceedings 2*, 2015, pp 5088–5095.

24. Stacey M, *Aluminium Sympathetic and Powerful*, pp 236–259.

25. Carlisle S, Frielander E and Faircloth B, *Aluminium and Life Cycle Thinking: Towards Sustainable Cities*, Cwningen Press, 2015.

26. Qualicoat, 'Technical Information Sheet 1: Recommendation for cleaning coated aluminium', 2011, https://www.qualicoat.net/main/about-us/assured-quality/information-for-coaters.html (accessed August 2023).

27. Total solar reflectance (TSR) testing to ASTM G-173 by the Faculty of Engineering, University of Porto.

28. Macintrye HL and Heaviside C, 'Potential benefits of cool roofs in reducing heat-related mortality during heatwaves in a European city', *Environmental International* 127, 2019, pp 430–441.

29. Sandhu N, AkzoNobel, RIBA Approved CPD, https://www.ribacpd.com/akzo-nobel-powder-coatings/10065/fire-performance-of-polyester-powder-coating/410600/movie (accessed June 2022).

30. Das SK, Green JAS and Kaufman JG, 'Aluminum recycling: Economic and environmental benefits', *Light Metal Age*, February 2010, p 22, www.phinix.net/services/Recycling/Aluminum_Recycling_Economic.pdf (accessed February 2015).

31. European Aluminium Association/Organisation of European Aluminium Refiners and Remelters Recycling Division, *Aluminium Recycling in Europe: The Road to High Quality Products*, EAA/OEA, 2006, p 6, http://recycling.world-aluminium.org/uploads/media/fl000217.pdf (accessed April 2014).

32. International Aluminium Institute, 'Aluminium sector greenhouse gas pathways to 2050', September 2021, https://international-aluminium.org/resource/aluminium-sector-greenhouse-gas-pathways-to-2050-2021 (accessed May 2022).

33. Hydro Circal (formerly Hydro75R) EPD.

34. International Aluminium Institute, 'IAI Material Flow Model – 2021 update', 2021, https://international-aluminium.org/resource/iai-material-flow-model-2021-update (accessed March 2022).

35. Ibid.

36. Boin UMJ and van Houwelingen JA, *Collection of Aluminium from Buildings in Europe: A Study by Delft University of Technology*, EAA (European Aluminium Association), 2004, http://recycling.world-aluminium.org/uploads/media/_TUDelftBrochure2004.pdf (accessed April 2015).

37. Founded by Hartmann, Hueck, Gartner, Schüco and Wicona, www.a-u-f.com (accessed September 2019).

38. International Aluminium Institute, 'Aluminium sector greenhouse gas pathways to 2050'.

39. International Aluminium Institute, *Sustainable Bauxite Mining Guidelines*, Second Edition, 2022, p 4, https://international-aluminium.org/resource/sustainable-bauxite-mining-guidelines-second-edition-2022 (accessed September 2023).

40. www.hydro.com (accessed December 2022).

Bamboo

1. Calculated using ecoinvent v3.7 & IPCC2013, based on Zea Escamilla, E. Habert, G, 'Environmental Impacts of Bamboo-based Construction Materials Representing Global Production Diversity', *Journal of Cleaner Production*, Vol 69, 2014, pp.117-127

2. Ibid.

3. https://www.inbar.int (accessed 17 August 2023).

4. Zea Escamilla, Habert, *Journal of Cleaner Production*, pp.117-127

5. Bundi T, 'Carbon offsetting with bamboo-based social housing: A case study in the Philippines', MSc Research Thesis, ETH Zürich, 2022.

6. Eleftheriou, E., Lopez Muñoz, L. F., Habert, G., & Zea Escamilla, E. (2022). Parametric approach to simplified life cycle assessment of social housing projects. *Sustainability*, 14(12), 7409. https://www.mdpi.com/1681876 (accessed 17 August 2023).

Brick

1. BRE, BREG EN EPD No 000002 UK Clay Brick, 2019, https://www.brick.org.uk/uploads/downloads/breg-en-epd-000002-v4.pdf (accessed 9 June 2023).

2. British Geological Survey, 'Mineral planning factsheet – brick clay', 2022, https://nora.nerc.ac.uk/id/eprint/532490/1/Brick%20Clay%20Mineral%20Planning%20Factsheet.pdf (accessed 9 June 2023).

3. Ibid.

4. Baum E, 'Emissions from S Asian brick production and potential climate impact', 2015, https://cdn.cseindia.org/docs/aad2015/Baum%20Emission%20and%20climate%20S%20Asia%20bricks.pdf (accessed 9 June 2023).

5. British Geological Survey, 'Mineral planning factsheet – brick clay'. https://nora.nerc.ac.uk/id/eprint/532490/1/ Brick%20 Clay%20Mineral%20Planning%20Factsheet.pdf (accessed 9 June 2023).
6. Ibid.
7. Ibid.
8. Department for Business and Trade, 'Building materials and components statistics: May 2023', 2023, https://assets. publishing.service.gov.uk/government/uploads/system/uploads/ attachment_data/file/1160690/23-cs6-_Construction_ Building_ Materials_-_Tables_May_2023.xlsx (accessed 9 June 2023).
9. AAC Worldwide, 'The European calcium silicate masonry unit industry – building on its past, ready for the future', https://www.aac-worldwide.com/category/special/ the-european-calcium-silicate-masonry-unit-industry- building-on-its-past-ready-for-the-future-739 (accessed 9 June 2023). Construction Products Europe, 'Waste dossier', 2016, https://www.construction-products.eu/application/fil es/7015/2473/3831/20160608002359-2016_Waste_dossier.pdf (accessed 17 July 2023).
10. Sources: various EPDs and manufacturer data sheets; IStructE values: Gibbons OP and Orr JJ, *How to Calculate Embodied Carbon*, Second Edition, The Institution of Structural Engineers, 2022.
11. British Geological Survey, 'Mineral planning factsheet – brick clay', 2022. https://nora.nerc.ac.uk/id/eprint/532490/1/ Brick%20Clay%20Mineral%20Planning%20Factsheet.pdf (accessed 9 June 2023).
12. Ibid.
13. British Geological Survey, 'Mineral planning factsheet – natural hydraulic lime', 2005, https://nora.nerc.ac.uk/id/ eprint/534437/1/mpf_hydraulic_limes.pdf (accessed 17 July 2023).
14. 1British Geological Survey, 'Mineral planning factsheet – limestone', 2006, https://nora.nerc.ac.uk/id/eprint/534436/1/ mpf_limestone.pdf (accessed 9 June 2023).
15. British Geological Survey, 'Mineral planning factsheet – cement', 2014, https://nora.nerc.ac.uk/id/eprint/534425/1/ mpf_cement.pdf (accessed 9 June 2023).
16. British Geological Survey, 'Mineral planning factsheet – brick clay', 2022. https://nora.nerc.ac.uk/id/eprint/532490/1/ Brick%20Clay%20Mineral%20Planning%20Factsheet.pdf (accessed 9 June 2023).
17. Ibid.
18. Brick Development Association, *Sustainability Report 2021*, 2021, https://www.brick.org.uk/uploads/downloads/ Sustainability-Report-2021.pdf (accessed 9 June 2023).
19. Piddington J, Nicol S, Garrett H and Custard M, 'The housing stock of the United Kingdom', 2020, https://files.bregroup. com/bretrust/The-Housing-Stock-of-the-United-Kingdom_ Report_BRE-Trust.pdf (accessed 9 June 2023). Department for Communities and Local Government, 'English housing survey HOMES 2011', 2011, https://assets.publishing.service. gov.uk/government/uploads/system/uploads/attachment_ data/file/211324/EHS_HOMES_REPORT_2011.pdf (accessed 9 June 2023).
20. British Geological Survey, 'Mineral planning factsheet – brick clay', 2022.

21. UN, UN Comtrade Database, 2021, https://comtradeplus. un.org/TradeFlow?Frequency=A&Flows=M&Commod- ityCodes=690410&Partners=826&Reporters=all&peri- od=2021&AggregateBy=none&BreakdownMode=plus (accessed 9 June 2023).Burton E and Friedrich N, 'Net waste tool guide to reference data', Version 1.0, WRAP, 2008.
22. Burton E and Friedrich N, 'Net waste tool guide to reference data', Version 1.0, WRAP, 2008.
23. Environment Agency, 2021 Waste Data Interrogator, 2023, https://www.data.gov.uk/dataset/d8a12b93-03ef-4fbf-9a43- 1ca7a054479c/2021-waste-data-interrogator (accessed 9 June 2023).
24. Construction Resources and Waste Platform, 'Overview of demolition waste in the UK', BRE, 2008.
25. Skouteris G et al., 'Water footprint and water pinch analysis techniques for sustainable water management in the brick- manufacturing industry', Journal of Cleaner Production 172, 2018, pp 786–794.
26. Baum E, 'Emissions from S Asian brick production and potential climate impact'. https://cdn.cseindia.org/docs/ aad2015/Baum%20Emission%20%20and%20climate%20S%20 Asia%20bricks.pdf (accessed 9 June 2023).

Concrete

1. Green Construction Board, Low Carbon Concrete Routemap, Institution of Civil Engineers, 2022, https://www.ice.org.uk/ engineering-resources/briefing-sheets/ low-carbon-concrete- routemap (accessed 18 August 2023).
2. Ibid.
3. Ibid.
4. Ibid.
5. Ibid.
6. Values taken from Figure 1.1 of Green Construction Board, *Low Carbon Concrete Routemap*, Institution of Civil Engineers, 2022, https://www.ice.org.uk/engineering-resources/briefing-sheets/ low-carbon-concrete-routemap (accessed 18 August 2023).
7. Concrete gains strength rapidly in the first few days, the rate of strength gain then decreases. However, concrete continues to gain strength for many months or years. Designers usually specify the compressive strength which must be achieved within 28 days. Alternatively, the strength to be achieved at 56, 72 or 90 days can be specified. In some cases, this enables a reduction in the embodied carbon of the concrete.
8. Data from Green Construction Board, *Low Carbon Concrete Routemap*.
9. Concrete curing takes place immediately after the concrete is placed and finished; it involves keeping the moisture and temperature conditions of the element at specific levels. Proper curing is crucial to concrete strength development and durability.
10. See, for example, https://www.slimbreker.nl/smartcrusher. html (accessed 18 August 2023).
11. Green Construction Board, *Low Carbon Concrete Routemap*.
12. https://www.buildingsandcities.org/insights/news/embodied- carbon-concrete.html (accessed 18 August 2023).

13. Roynon J, *Embodied Carbon: Structural Sensitivity Study*, Buro Happold, April 2020, https://www.istructe.org/resources/case-study/embodied-carbon-structural-sensitivity-study (accessed 18 August 2023).

14. Cousins S, 'Thin vaulted floor slab could slash embodied carbon by 60%', RIBA Journal, 18 February 2022, https://www.ribaj.com/products/acorn-research-robot-built-thin-concrete-arches-embodied-carbon (accessed 18 August 2023).

15. See, for example, HiLo research and innovation unit for NEST, https://brg.ethz.ch/hilo (accessed 18 August 2023).

16. MPA, 'Fact Sheet 18: Embodied CO_2e of cement, additions and cementitious material', https://cement.mineralproducts.org/MPACement/media/Cement/Publications/Fact-Sheets/FS_18_Embodied_CO2e.pdf (accessed 6 September 2023).

17. See Green Construction Board, Low Carbon Concrete Routemap, Figure 4.2: Cradle-to-gate carbon by cement types. The Concrete Centre's *Guide to Specifying Sustainable Concrete* (2019) reported that SCMs account for around 18% of UK cement.

18. Scrivener K, Martirena F, Bishnoi S, Maity S, 'Calcined clay limestone cements (LC3)', Cement and Concrete Research 114, 2018, pp 49–56.

19. https://www.istructe.org/resources/guidance/efficient-use-of-ggbs-in-reducing-global-emissions/#:~:text=GGBS%20should%20continue%20to%20be,reducing%20global%20greenhouse%20gas%20emissions (accessed 21 October 2023)

20. Seratech is a concrete product which sequesters CO2, see https://www.seratechcement.com (accessed 22 August 2023).

21. See references and case studies here: Green Construction Board, *Low Carbon Concrete Routemap*; PAS 2080 Carbon management for buildings and infrastructure; Guidance Document for PAS 2080.

22. The Environment Agency (EA) has recently committed to default to low-carbon concretes when constructing its infrastructure projects, provided they meet performance requirements.

23. Embodied Biodiversity Report, Expedition; https://expedition.uk.com/wp-content/uploads/2023/11/231103_Embodied-Biodiversity_Report_Compressed.pdf (last accessed 24/11/2023)

24. https://www.sandstories.org (accessed 18 August 2023).

25. British Ready-Mixed Concrete Association, BRMCA Best Practice: Managing Concrete Plant Water and Wash Water, 2016, https://brmca.org.uk/documents/BRMCA_Best_Practice_Managing_Concrete_Plant_Water_and_Wash_Water_June_2016.pdf (accessed 18 August 2023).

26. https://www.seratechcement.com (accessed 7 September 2023). https://www.karbonite.co.uk (accessed 7 September 2023). https://www.caths.cam.ac.uk/cement (accessed 7 September 2023).

27. https://www.manchester.ac.uk/discover/news/greener-and-cheaper-graphenemanchester-solves-concretes-big-problem (accessed 7 September 2023).

28. https://www.archdaily.com/979145/3d-printing-with-low-carbon-concrete-reducing-co2-emissions-and-material-waste (accessed 7 September 2023).

29. https://block.arch.ethz.ch/brg/research/cable-net-and-fabric-formworks-for-concrete-shells (accessed 7 September 2023).

30. https://dbt.arch.ethz.ch/project/foamwork (accessed 7 September 2023).

31. https://www.slimbreker.nl/smartcrusher.html (accessed 7 September 2023).

32. https://www.concretecentre.com/Specification/Innovative-concrete/Bioreceptive-Concrete.aspx (accessed 7 September 2023). https://www.sciencedirect.com/science/article/pii/S2352710221004022 (accessed 7 September 2023).

Copper

1. KME TECU cladding products EPDs available from Okobaudat: https://www.oekobaudat.de/no_cache/en/database/search/daten/db2.html#bereich2 (accessed 27 October 2023)

2. Based on EPDs for the 'Copper' category on Building Transparency EC3 portal: https://buildingtransparency.org/ec3/epds (accessed 27 October 2023)

3. https://www.carbonchain.com/blog/understand-your-copper-emissions (accessed 11 September 2023).

4. https://www.statista.com/statistics/264626/copper-production-by-country (accessed 11 September 2023).

5. Clark, Shirley E., et al. "Roofing materials' contributions to storm-water runoff pollution." Journal of Irrigation and Drainage Engineering 134.5 (2008): 638-645. https://doi.org/10.1061/(ASCE)0733-9437(2008)134:5(638)

Cork

1. APCOR (Portuguese Cork Association), 'Cork Yearbook 2020', 2020, https://www.apcor.pt/en/portfolio-posts/apcor-year-book-2020 (accessed 4 April 2022).

2. https://www.amorimcorkinsulation.com/xms/files/EPD_Amorim_Cork_Insulation_Expanded_Insulation_Corkboard_-ICB-_2023.pdf (accessed 27 October 2023)

3. https://erfmi.com/wp-content/uploads/EPD-9-Cork-floor-tiles-according-to-EN-12104-6Cork-floor-tiles-according-to-EN-12104.pdf (accessed 27 October 2023) https://www.firstplanit.com/doc_images/product_epd/product_epd_749141.pdf (accessed 27 October 2023)

4. Ibid.

5. Knapic S et al., 'Cork as a building material: A review', *European Journal of Wood and Wood Products* 74(6), 2016, pp 775–791.

6. Ibid.

7. APCOR, 'Cork Yearbook 2020'.

8. Wilton O and Barnett Howland M, 'Cork: An historical overview of its use in building construction', *Construction History: Journal of the Construction History Society* 35(1), 2020, pp 1–22.

9. APCOR, 'Cork Yearbook 2020'.

10. ERFMI (European Resilient Flooring Manufacturers' Institute), 'Environmental Product Declaration (EPD) for cork floor tiles', 2019, https://erfmi.com/wp-content/uploads/EPD-9-Cork-floor-tiles-according-to-EN-12104-6Cork-floor-tiles-according-to-EN-12104.pdf (accessed 12 September 2023).

11. Smith JT, 'Process of treating cork', United States Patent Number US 456068 A, 1891.

12. Amorim Isolamentos, 'Expanded insulation corkboard datasheet', 2021, https://www.amorimcorkinsulation.com/xms/files/FICHAS_TECNICAS_2021/FT_Corkboard_EN_2021.pdf (accessed 20 July 2022).

13. Amorim Cork Insulation Expanded Insulation Corkboard (ICB) EPD: https://www.amorimcorkinsulation.com/xms/files/EPD_Amorim_Cork_Insulation_Expanded_Insulation_Corkboard_-ICB-_2023.pdf Last accessed 27/10/2023.

14. Amorim Isolamentos, 'MD Façade datasheet', 2021, https://www.amorimcorkinsulation.com/xms/files/FICHAS_TECNICAS_2021/FT_MDFacade_EN_2021.pdf (accessed 20 July 2022).

15. Ty-Mawr Lime, 'Expanded cork insulation system – external wall', 2022, https://www.lime.org.uk/applications/retrofit-insulation-systems-for-old-buildings/internal-wall-insulation-system.html (accessed 4 September 2022).

16. Permanent Delegation of Portugal to UNESCO, 'Montado, cultural landscape', 2017, https://whc.unesco.org/en/tentativelists/6210 (accessed 2 February 2019).

17. Wilton, O; Barnett Howland, M, 'Cork: an historical overview of its use in building construction', Construction History, 35 (1), 2020, pp. 1-22.

18. Wilton O and Barnett Howland M, Cork Construction, Bartlett Design Research Folios: The Bartlett School of Architecture, 2022, https://issuu.com/bartlettarchucl/docs/design-research-wilton-barnett-howland-cork-constr (accessed 16 February 2023).

19. Wilton O and Barnett Howland M, 'Cork construction kit', The Journal of Architecture 25(2), 2020, pp 138–165.

Earth

1. Fernandes J, Peixoto M, Mateus R and Gervásio H, 'Life-cycle analysis of environmental impacts of earthen materials in the Portuguese context: Rammed earth and compressed earth blocks', Journal of Cleaner Production 241, 2019.

2. Bennetts Associates, 'Environmental benefits of using site-excavated subsoil for making earth blocks for the building, Tribeca: The Apex', February 2023.

3. Maskell D et al., 'Determination of optimal plaster thickness for moisture buffering of indoor air', Building and Environment 130, 2018, pp 143–150.

Glass

1. https://glassforeurope.com/wp-content/uploads/2020/01/flat-glass-climate-neutral-europe.pdf (accessed 14 September 2023).

2. https://www.mordorintelligence.com/industry-reports/float-glass-market# (accessed 12 September 2023).

3. Hartwell R, Coult G and Overend M, 'Mapping the flat glass value-chain: A material flow analysis and energy balance of UK production', https://www.researchgate.net/publication/359074054_Mapping_the_flat_glass_value-chain_A_material_flow_analysis_and_energy_balance_of_UK_production (accessed 12 September 2023).

4. Circular Ecology, Inventory of Carbon and Energy (ICE) database, V3.0, 2019, https://circularecology.com/embodied-carbon-footprint-database.html (accessed 24 August 2023).

5. https://www.saint-gobain-glass.fr/fr/reseau-glass-recycling#enjeux (accessed 24 August 2023).

6. https://resource-recycling.com/recycling/2017/03/02/cullet-comparisons (accessed 19 August 2023).

7. https://www.saint-gobain-glass.co.uk/en-gb/saint-gobain-glossary (accessed 19 August 2023).

8. https://info.glass.com/what-is-the-float-glass-process/#:~:text=In%20the%20float%20glass%20process,to%20form%20a%20flat%20surface (accessed 19 August 2023).

9. Circular Ecology, Inventory of Carbon and Energy (ICE) database.

10. Hartwell R, Coult G and Overend M, 'Mapping the flat glass value-chain'. 2022; https://doi.org/10.21203/rs.3.rs-1401635/v1 (accessed 24/11/2023)

11. https://glassforeurope.com/wp-content/uploads/2020/01/flat-glass-climate-neutral-europe.pdf (accessed 14 September 2023).

12. https://www.mordorintelligence.com/industry-reports/float-glass-market# (accessed 12 September 2023).

13. AGC Glass Europe, https://www.agc-glass.eu/en/sustainability/environmental-footprint/carbon-footprint (accessed 19 August 2023).

14. AGC - Double vitrage de sécurité (Thermobel Stratobel, Thermobel Stratophone, ipasafe IGU, ipaphon IGU) - Composition de référence 4-16-|44.2 et 4|-16-44.2, 4-15-|44.2 et 4|-15-44.2 - Hors accessoires de pose (v.1.1),: https://www.base-inies.fr/iniesV4/dist/consultation.html?id=9112

15. Glass Alliance Europe, June 2019, https://www.glassallianceeurope.eu/images/cont/gae-position-paper-on-decarbonisation-june-2019_file.pdf (accessed 19 August 2023)

16. Ibid.

17. Saint-Gobain, A Guide for Improved Cullet Recycling, https://www.vetrotech.com/saint-gobain-guide-cullet-recycling (accessed 24 August 2023).

18. Deloitte, 'Resource efficient use of mixed wastes – improving management of construction and demolition waste', 2017, https://op.europa.eu/en/publication-detail/-/publication/78e42e6c-d8a6-11e7-a506-01aa75ed71a1/language-en (accessed 12 September 2023).

19. Percentage of material used in Saint-Gobain's float glass furnaces in 2018.

20. https://www.saint-gobain-glass.fr/fr/reseau-glass-recycling#enjeux (accessed 24 August 2023).

21. https://abcbirds.org/glass-collisions (accessed 21 August 2023).

22. Saint-Gobain, 'Saint-Gobain achieves the first zero-carbon production of flat glass in the world', 16 May 2022, https://www.saint-gobain.com/sites/saint-gobain.com/files/media/document/20220516_First%20zero-carbon%20production%20of%20flat%20glass_VA.pdf (accessed 21 August 2023).

Hempcrete

1. https://app.2050-materials.com/media/certificates/01-EPD-DOCUMENT1_EN_NOV22_Eng_Jd4ocDD_b5JAVSM.pdf (accessed 27 October 2023)
2. Ibid.
3. 'Roadmap plan to boost UK industrial hemp production and processing', University of York, 25 August 2021, https://www.york.ac.uk/news-and-events/news/2021/research/boosting-uk-hemp-production (accessed 29 August 2023).
4. Florentin Y, Pearlmutter D, Givoni B and Gal E, 'A life-cycle energy and carbon analysis of hemp-lime bio-composite building materials', *Energy and Buildings* 156, 2017.
5. Adapted from: https://www.researchgate.net/figure/Propagation-of-thermal-wave-Figure-9-Water-content-of-a-25cm-element-in-a-lime-and-hemp_fig4_266467019 (accessed 29 August 2023). Evrard A and de Herde A, 'Bioclimatic envelopes made of lime and hemp concrete', CISBAT 2005, *Renewables in a Changing Climate, Innovation in Building Envelopes and Environmental Systems*, Lausanne, 2005.
6. IPCC, 'The evidence is clear: The time for action is now', 4 April 2022, https://www.ipcc.ch/2022/04/04/ipcc-ar6-wgiii-pressrelease (accessed 10 September 2023).

Insulation

1. Climate Change Committee, Progress in reducing UK emissions 2023 Report to Parliament. Table 2 page 23. https://HYPERLINK "http://www.theccc.org.uk/wp-content/uploads/2023/06/Progress-in-"www.theccc.org.uk/wp-content/uploads/2023/06/Progress-in- reducing-UK-emissions-2023-Report-to-Parliament-1.pdf
2. AMA Research Building Insulation Products Market Report – UK 2021-25; https://www.amaresearch.co.uk/report/building-insulation-products-market-report-uk-2021-2025/ (accessed 30 November 2023)
3. Mintel Group Ltd, UK Thermal Insulation Market Report 2022 https://store.mintel.com/report/uk-thermal-insulation-market-report
4. Ibid.
5. ASBP Natural Fibre Insulation Group.
6. Retrofitting the UK's housing stock to reach net zero, 2021, https://energysavingtrust.org.uk/retrofitting-the-uks-housing-stock-to-reach-net-zero/ (accessed 30 November 2023)
7. AMA Research Building Insulation Products Market Report – UK 2021-25

Mycelium

1. Hawksworth DL and Lücking R, 'Fungal diversity revisited: 2.2 to 3.8 million species', *Microbiology Spectrum* **5(4),** July 2017.
2. Blackwell M, Fernando E and Vega C, 'Lives within lives: Hidden fungal biodiversity and the importance of conservation', *Fungal Ecology*, 9 July 2018, published by Elsevier on behalf of the British Mycological Society.
3. Zeng Q, Morales A and Cottarel G, 'Fungi and humans: Closer than you think', *Trends in Genetics*, 2002.

4. Davis D, Fisher M and Money N, 'Fastest flights in nature: High-speed spore discharge mechanisms among fungi', *Public Library of Science*, 2008.
5. The term 'wood-wide web' first appeared on the front cover of the August 1997 edition of *Nature* in response to a paper published by the journal and written by a Canadian PhD student, Suzanne Simard. The editor of *Nature* asked David Read to write a commentary on Simard's paper. It was during his discussions with the editor that Read coined the phrase 'wood-wide web'.
6. Sheldrake M, *Entangled Life*, The Bodley Head, 2020.
7. Kalogeiton VS, Papadopoulos DP, Georgilas IP, Sirakoulis G Ch and Adamatzky AI, 'Biomimicry of crowd evacuation with a slime mould cellular automaton model computational intelligence', *Medicine and Biology* 600, 2015.
8. Sheldrake M, *Entangled Life*, p 52.
9. https://www.ecovative.com (accessed 31 August 2023).
10. Baker-Brown D, *The Re-Use Atlas: A Designer's Guide Towards a Circular Economy*, p 47, RIBA, 2017.
11. https://www.biohm.co.uk (accessed 31 August 2023).
12. Baker-Brown D, *The Re-Use Atlas*, p 121.
13. Ibid., p 118.
14. https://www.biohm.co.uk/mycelium (accessed 31 August 2023).
15. Pownall A, 'Pavilion grown from mycelium acts as pop-up performance space at Dutch Design Week', *Dezeen*, 29 October 2019, https://www.dezeen.com/2019/10/29/growing-pavilion-mycelium-dutch-design-week (accessed 31 August 2023).
16. Woollacott E, 'The fungus and bacteria tackling plastic waste', BBC News, 30 July 2021, https://www.bbc.co.uk/news/business-57733178 (accessed 7 September 2023).

Plastics

1. Statistics from Cullen JM, Drewniok MP and Cabrera Serrenho A, *The P Word*, Resource Efficiency Collective, University of Cambridge, 2020, p 11, https://www.refficiency.org/publications/the-p-word (accessed 7 July 2022).
2. United Nations Development Programme (UNDP), 'What do plastics have to do with climate change?', 15 November 2022, https://stories.undp.org/what-do-plastics-have-to-do-with-climate-change (accessed 31 August 2023).
3. Omer N, 'Friday briefing: Why recycling plastic may not be as good for the planet as we thought', *The Guardian*, 26 May 2023, https://www.theguardian.com/world/2023/may/26/friday-briefing-why-recycling-plastic-may-not-be-as-good-for-the-planet-as-we-thought (accessed 31 August 2023).
4. Alliance of Sustainable Building Products, ZAP Toolkit, p 1, https://asbp.org.uk/wp-content/uploads/2023/05/ZAP-Toolkit-v2.pdf (accessed 31 August 2023).
5. Cullen JM, Drewniok MP and Cabrera Serrenho A, *The P Word*, p 19.
6. Omer N, 'Friday briefing: Why recycling plastic may not be as good for the planet as we thought'.
7. Carrington D, 'After bronze and iron, welcome to the plastic age, say scientists', *The Guardian*, 4 September 2019, https://www.theguardian.com/environment/2019/sep/04/plastic-pollution-fossil-record (accessed 7 July 2022).

8. Plastics Europe, 'How plastics are made', https://plasticseurope.org/plastics-explained/how-plastics-are-made (accessed 7 July 2022).
9. UNDP, 'What do plastics have to do with climate change?'.
10. Plastics Europe, 'Plastics – the Facts', 2022, p 16, https://plasticseurope.org/knowledge-hub/plastics-the-facts-2022 (accessed 26 May 2023).
11. Center for International Environmental Law, 'Plastic and climate: The hidden cost of a plastic planet', May 2019, https://www.ciel.org/wp-content/uploads/2019/05/Plastic-and-Climate-Executive-Summary-2019.pdf (accessed 7 July 2022).
12. Ibid.
13. Cullen JM, Drewniok MP and Cabrera Serrenho A, *The P Word*.
14. Ibid.
15. Ibid.
16. Ibid.
17. Ellen MacArthur Foundation, 'A vision of a circular economy for plastic', https://ellenmacarthurfoundation.org/plastics-vision (accessed 7 July 2022).
18. Alliance of Sustainable Building Products, *Plastics in Construction: Introductory Guide*, ASBP, 2021, p 1, https://asbp.org.uk/wp-content/uploads/2015/12/Intro-guide-v2-April-21.pdf (accessed 7 July 2022).
19. Green Square Accord, 'Building a greener future with the UK's first plastic free homes', 11 August 2022, (https://www.greensquareaccord.co.uk/news/housing/building-a-greener-future-with-the-uk-s-first-plastic-free-homes (accessed 31 August 2023).
20. Interview by Cullinan Studio with Carl Taylor of Green Square Accord on 1 September 2022.
21. United Nations Development Programme (UNEP), 'What you need to know about the plastic pollution resolution', 2 March 2022, https://www.unep.org/news-and-stories/story/what-you-need-know-about-plastic-pollution-resolution (accessed 31 August 2023).

Steel

1. ArcelorMittal Europe, XCarb recycled and renewably produced structural steel sections EPD, EPD-ARC-20210132-CBB1-EN, 2021. British Steel Rails and Sections Environmental Product Declaration, British Steel, EPD-TS-2020-003, January 2020
2. https://worldsteel.org/steel-topics/steel-markets (accessed 11 September 2023).
3. https://worldsteel.org/steel-topics/statistics/world-steel-in-figures (accessed 11 September 2023).
4. https://www.steelconstruction.info/The_recycling_and_reuse_survey (accessed 11 September 2023).
5. https://worldsteel.org/steel-topics/steel-markets/buildings-and-infrastructure (accessed 31 August 2023).
6. Ibid.
7. https://worldsteel.org/steel-topics/statistics/world-steel-in-figures (accessed 11 September 2023).
8. Eurofer, 'European Steel in Figures', 2022.Eurofer, 'European Steel in Figures', 2022.
9. World Steel Association, 'World Steel in Figures', 2022, https://worldsteel.org/steel-topics/statistics/world-steel-in-figures-2022 (accessed 2 September 2023).
10. SCI P363 Steel building design: Design data.
11. SCI P358 Joints in steel construction: Simple joints to Eurocode 3.
12. SCI P398 Joints in steel construction: Moment-resisting joints to Eurocode 3.
13. https://worldsteel.org/wp-content/uploads/Fact-sheet-Steel-industry-co-products.pdf (accessed 11 September 2023).
14. SCI P428 Guidance on demountable composite construction systems for UK practice.
15. For guidance on reuse of structural steelwork, refer to Gower P et al., *Circular Economy and Reuse: Guidance for Designers*, IStructE, July 2023, https://www.istructe.org/resources/guidance/circular-economy (accessed 2 September 2023).
16. SCI P427 Structural steel reuse: Assessment, testing and design principles.
17. SCI P440 Reuse of pre-1970 steelwork – Supplement to SCI P427. The British Constructional Steelwork Association has also published a 'Model specification for the purchase of reclaimed steel sections', 2022. The Institution of Structural Engineers has published a guide on reuse and circular economy for structures (Gower P et al., *Circular Economy and Reuse*).
18. British Constructional Steelwork Association (BCSA), National Structural Steelwork Specification for Building Construction, Annex J – Sustainability Specification, revised Seventh Edition, 2023.
19. Ibid.
20. https://www.responsiblesteel.org (accessed 2 September 2023).
21. https://www.theclimategroup.org/steelzero (accessed 2 September 2023).
22. https://bcsa.org.uk/member-directories/sustainability-charter (accessed 2 September 2023).
23. BCSA, UK Structural Steelwork: 2050 Decarbonisation Roadmap, 2021, https://bcsa.org.uk/resources/sustainability/steelwork-decarbonisation-roadmap (accessed 2 September 2023).
24. https://www.statista.com/statistics/589945/iron-ore-production-gross-weight-worldwide (accessed 2 September 2023).
25. https://www.iea.org/data-and-statistics/charts/global-end-use-steel-demand-and-in-use-steel-stock-by-scenario-2000-2050 (accessed 2 September 2023).
26. https://population.un.org/wpp/Graphs/Probabilistic/POP/TOT/900 (accessed 2 September 2023).
27. https://worldsteel.org/media-centre/blog/2018/future-of-global-scrap-availability (accessed 2 September 2023).
28. https://www.ssab.com/en-gb/fossil-free-steel/insights/hybrit-a-new-revolutionary-steelmaking-technology (accessed 23 September 2023).

Stone

1. https://www2.bgs.ac.uk/mineralsuk/download/dmq/Directory_of_Mines_and_Quarries_2020.pdf (accessed 14 September 2023).
2. https://www.mineralproducts.org/Facts-and-Figures.aspx (accessed 14 September 2023).
3. Natural Stone Institute - Sustainability Resources, https://www.naturalstoneinstitute.org/programs/sustainability/sustainability-resources/ (accessed 27 October 2023)
4. Ibid.
5. https://www.mineralproducts.org/MPA/media/root/Publications/2021/Profile_of_the_UK_Mineral_Products_Industry_2020_Statistical_Background.xlsx
6. Natural Stone Institute - Sustainability Resources, https://www.naturalstoneinstitute.org/programs/sustainability/sustainability-resources/ (accessed 27 October 2023)

Straw

1. School of Natural Building, *Technical Guide: Straw Construction in the UK*, 2022, p 22, https://strawworks.co.uk/technical-guide-straw-construction-in-the-uk (accessed 2 September 2023).
2. Minke G and Krick B, *Straw Bale Construction Manual*, Birkhäuser Verlag, 2020.
3. Straw as insulation material - UK, EPD; https://www.environdec.com/library/epd3854 (accessed 27 October 2023)
4. School of Natural Building, Technical Guide.
5. Straw as insulation material - UK, EPD; https://www.environdec.com/library/epd3854 (accessed 27 October 2023)
6. School of Natural Building, Technical Guide.
7. https://www.youtube.com/watch?v=9LOlV-01cAI (accessed 3 September 2023).
8. https://www.modcell.com (accessed 3 September 2023).
9. https://strawworks.co.uk/projects/north-kesteven-council-houses (accessed 3 September 2023).
10. www.EcoCocon.eu (accessed 3 September 2023).
11. https://www.ThatchAdviceCentre.co.uk (accessed 3 September 2023).
12. John Butler, www.sustainablebuildconsultancy.com (accessed 3 September 2023).
13. See School of Natural Building, *Technical Guide*, p 22.
14. http://schoolofnaturalbuilding.co.uk/uk-straw-bale-house-register (accessed 3 September 2023). https://strawbuilding.eu/strawbale-houses-europe (accessed 3 September 2023).
15. School of Natural Building, *Technical Guide*.
16. Many of the larger straw buildings are in France, where straw construction is estimated to be at least 10 years ahead of the UK.
17. School of Natural Building, *Technical Guide*, p 64.
18. http://builddesk.co.uk/software/builddesk-u/thermal-mass (accessed 3 September 2023).
19. Minke G and Krick B, *Straw Bale Construction Manual*.
20. Led by Professor Peter Walker and his team in the engineering department.
21. School of Natural Building, Technical Guide, pp 34–35.
22. Ibid.
23. https://www.architectscan.org/post/environmental-building-
24. https://www.constructionpaille.fr/statistiques (accessed 3 September 2023).
25. Information from Gabriel Martinez, RFCP, https://www.rfcp.fr/le-rfcp (accessed 3 September 2023).
26. Information from Stephanie Ventre, CNCP, https://cncp-feuillette.fr (accessed 3 September 2023).
27. https://www.ekopolis.fr/la-demarche-quartiers-et-batiments-durables-franciliens (accessed 3 September 2023).
28. Jones B, *Building with Straw Bales*, Third Edition, Green Books, 2015.
29. https://etcbygg.se/bygg/vill-du-vara-med-och-utmana-byggbranschen-pa-riktigt (accessed 6 September 2023).

Timber

1. Based on data from *The Timber Industry Net Zero Roadmap*, Timber Development UK, 2022, https://timberdevelopment.uk/resources/net-zero-roadmap (accessed 12 September 2023).
2. Forest Research, 'Forestry Statistics 2022: A compendium of statistics about woodland, forestry and primary wood processing in the United Kingdom', 2022.
3. Building Transparency EC3 database for European Plywood and OSB panels, https://buildingtransparency.org/ec3/material-search (accessed 27 October 2023).
4. BS EN 16449:2014. Wood and wood-based products. Calculation of the biogenic carbon content of wood and conversion to carbon dioxide.
5. Wood Recyclers' Association, 'UK waste wood market goes from strength to strength, exceeding 4 million tonnes of processed wood', 2022, https://woodrecyclers.org/uk-waste-wood-market-goes-from-strength-to-strength-exceeding-4-million-tonnes-of-processed-wood (accessed 5 September 2023).
6. Building Transparency EC3 database for European Plywood and OSB panels, https://buildingtransparency.org/ec3/material-search (accessed 27 October 2023).
7. Forest Research, 'Forestry Statistics 2022: A compendium of statistics about woodland, forestry and primary wood processing in the United Kingdom', 2022.
8. Food and Agriculture Organization of the United Nations, Dataset 'Forest area as a proportion of total land area', 2020, https://www.fao.org/sustainable-development-goals-data-portal/data/indicators/indicator-1511---forest-area-as-a-percentage-of-total-land-area/en (accessed 12 September 2023).
9. Forest Research, 'Forestry Statistics 2022'.
10. Ibid.
11. Hart J and Pomponi F, 'More timber in construction: Unanswered questions and future challenges', *Sustainability* 12(8), 2020.
12. Based on data from *The Timber Industry Net Zero Roadmap*.
13. Ministerial Conference on the Protection of Forests in Europe – FOREST EUROPE, 'State of Europe's Forests 2020', Zvolen, 2020, p 31.
14. Food and Agriculture Organization of the United Nations, 'Forestry production and trade', https://www.fao.org/faostat/en/#data/FO (accessed 4 September 2023).

15. Forest Research, 'Forestry Statistics 2021: A compendium of statistics about woodland, forestry and primary wood processing in the United Kingdom', 2021.

16. UKWAS, United Kingdom Woodland Assurance Standard, Fourth Edition, 2018.

17. Forest Research, 'UK wood production and trade 2021 provisional figures'.

18. Moore N, 'Timber utilisation statistics 2015'.

19. Partner Forest Program, *Sustainable Wood for Cities – A Guide for City Projects and Policies on Sourcing Wood to Benefit Climate, Environment and Society*, https://www.partnerforests. org/ sustainable-wood-for-cities-guide (accessed 4 September 2023)

20. BS EN 16449:2014. Wood and wood-based products. Calculation of the biogenic carbon content of wood and conversion to carbon dioxide.

21. https://www.pik-potsdam.de/en/news/latest-news/living-in-timber-cities-could-avoid-emissions-2013-without-using-farmland-for-wood-production?searchterm=100+billion (accessed 4 September 2023).

22. *The Timber Industry Net Zero Roadmap.*

23. UK territorial and overseas emissions from production of sawn timber for local consumption; data from *The Timber Industry Net Zero Roadmap.*

24. *The Timber Industry Net Zero Roadmap.*

25. See also Ridley-Ellis D, Stapel P and Baño V, 'Strength grading of sawn timber in Europe: An explanation for engineers and researchers', *European Journal of Wood and Wood Products* 74, 2016, pp 291–306.

26. Low-density hardwoods like poplar and sweet chestnut can be graded to C classes.

27. *The Timber Industry Net Zero Roadmap.*

28. UK territorial and overseas emissions from production of chipboard, fibreboard, joinery, treated timber, timber frame systems, trussed rafters, other engineered wood products, pallets and packaging for local consumption; data from *The Timber Industry Net Zero Roadmap.*

29. Potential reuse of structural products would necessitate strength grading processes which differ from grading new timber and are not yet trialled or covered by standards. Reuse potential here refers to typically recovered sizes and conditions of recovered elements, as well as potential for remanufacturing. Recycling potential means timber cascading.

30. Steffen M, *Moisture and Wood-Frame Buildings*, Canadian Wood Council, 2000.

31. STA, 'STA Advice Note 14: Robustness of CLT structures', Structural Timber Association, 2017.

32. European Committee for Standardization, BS EN 350. Durability of wood and wood-based products – Testing and classification of the durability to biological agents of wood and wood-based materials, 2016, p 10.

33. Bergman R, *Wood Handbook*, Chapter 13, 'Drying and control of moisture content and dimensional changes', USDA, 2010, pp 13–15.

34. European Committee for Standardization, 21/30433972 DC BS EN 460. Durability of wood and wood-based products. Natural durability of solid wood. Guide to the durability requirements for wood to be used in hazard classes, 2021.

35. For further information, see Jones D and Brischke C, Performance of Bio-based Building Materials, Chapter 4, 'Protection of the bio-based material', Woodhead Publishing Series in Civil and Structural Engineering, 2017, p 204. Hill C, *Wood Modification: Chemical, Thermal and Other Processes*, John Wiley & Sons, 2006.

36. European Committee for Standardization, BS EN 15228:2009. Structural timber. Structural timber preservative treated against biological attack, 2009.

37. Wood Recyclers' Association, 'UK waste wood market goes from strength to strength'.

38. Hart J and Pomponi F, 'More timber in construction'.

39. Cramer M and Ridley-Ellis D, 'A case study of timber demolition recycling in the UK', *Proceedings of the 16th Annual Meeting of the Northern European Network for Wood Science and Engineering*, Helsinki, 2020, p 90.

40. Sandin Y et al., *Design of Timber Buildings for Deconstruction and Reuse – Three Methods and Five Case Studies*, RISE, 2022.

41. Wood Recyclers' Association, 'UK waste wood market goes from strength to strength'.

42. Jelušič P and Kravanja S, 'Optimal design and competitive spans of timber floor joists based on multi-parametric MINLP optimization', Materials 15(9), 2022.

43. Ibid.

44. Stauffer N, 'Using nature's structures in wooden buildings', 12 January 2022, https://energy.mit.edu/news/using-natures-structures-in-wooden-buildings (accessed 5 September 2023).

45. Confor, 'Biodiversity, forestry and wood', 2020, https://www.confor.org.uk/media/247794/confor-biodiversity-forestry-report.pdf (accessed 5 September 2023).

46. Partner Forest Program, *Sustainable Wood for Cities.*

47. Hart J and Pomponi F, 'More timber in construction'.

48. Rose CM and Stegemann JA, 'Feasibility of cross-laminated secondary timber', Fifth International Conference on *Sustainable Construction Materials and Technologies*, Coventry, 2019, pp 495–507. Llana DF et al., 'Cross-laminated timber (CLT) manufactured with European oak recovered from demolition: Structural properties and non-destructive evaluation', Construction and Building Materials 339, 2022.

49. Zhang J et al., 'An easy-coating, versatile and strong soy flour adhesive via a biomineralised structure combined with a biomimetic brush-like polymer', *Chemical Engineering Journal* 450, 2022.

50. Beran R et al., 'Wood adhesives from waste-free recycling depolymerisation of flexible polyurethane foams', *Journal of Cleaner Production* 305, 2021.

Zinc

1. https://www.greenbooklive.com/search/scheme.jsp?id=346 / https://www.rheinzink.co.uk/epd-sustainability-certificates/ / https://www.nedzink.com/en/downloads/ (accessed 27 October 2023)

2. https://www.statista.com/statistics/264878/world-production-of-zinc-metal (accessed 11 September 2023).

3. https://galvanizeit.org/hot-dip-galvanizing/what-is-zinc/zinc-recycling (accessed 11 September 2023).

4. https://www.zinc.org/circularity-2 (accessed 11 September 2023).

5. https://en.wikipedia.org/wiki/List_of_countries_by_zinc_production (accessed 11 September 2023).

Future Innovations and Trends

1. Lasley S, 'Titanium – the lighter, whiter metal', North of 60 Mining News, 22 June 2020, https://www.miningnewsnorth.com/story/2019/06/01/critical-minerals/titanium-the-lighter-whiter-metal/5764.html (accessed 5 September 2023).

2. Farjana SH, Huda N, Mahmud MP and Lang C, 'Towards sustainable TiO2 production: An investigation of environmental impacts of ilmenite and rutile processing routes in Australia', *Journal of Cleaner Production* 196, September 2018, https://www.sciencedirect.com/science/article/abs/pii/S0959652618318067#:~:text=The%20major%20environmental%20issues%20of,fragile%20coastal%20areas%20and%20deforestation (accessed 5 September 2023).

3. Fortune Business Insights, 'Titanium dioxide market size, share and COVID-19 impact analysis, by process, by application and regional forecast, 2020–2027', January 2021, https://www.fortunebusinessinsights.com/titanium-dioxide-tio2-market-102390 (accessed 5 September 2023).

4. Lasley S, 'Titanium – the lighter, whiter metal'.

5. Ask Nature, 'The beetles that scatter all the light – white scarab beetles', 2021, https://asknature.org/strategy/the-beetles-that-scatter-all-the-light (accessed 5 September 2023).

6. https://www.impossiblematerials.com (accessed 5 September 2023).

7. HMC Harmony Chemical, 'The total amount of global TiO2 market value will reach USD 31.19 billion in year 2022', https://www.ti-line.net/resources/industry-news/the-total-amount-of-global-tio2-market-value-will-reach-usd-31.19-billion-in-year-2022.html (accessed 5 September 2023).

8. Global Newswire, 'The global titanium dioxide market size is expected to reach USD 27.9 billion in 2021', 8 July 2021, https://www.globenewswire.com/news-release/2021/07/08/2259676/0/en/The-global-Titanium-Dioxide-market-size-is-estimated-to-be-USD-20-9-billion-in-2021-and-is-projected-to-reach-USD-27-9-billion-by-2026-at-a-CAGR-of-5-9-between-2021-and-2026.html (accessed 5 September 2023).

9. Fortune Business Insights, 'Paints and coatings market size, share and COVID-19 impact analysis', June 2023, https://www.fortunebusinessinsights.com/industry-reports/paints-and-coatings-market-101947 (accessed 5 September 2023).

10. Greenspec, 'Paint: Health and environment', https://www.greenspec.co.uk/building-design/paint (accessed 5 September 2023).

11. Ask Nature, 'Butterflies hack light waves to produce brilliant colour', https://asknature.org/strategy/wing-scales-cause-light-to-diffract-and-interfere (accessed 5 September 2023).

12. https://www.cyprismaterials.com (accessed 5 September 2023).

13. George S, 'Cement and concrete industry's emissions have doubled in 20 years', Edie, 24 June 2022, https://www.edie.net/cement-and-concrete-industrys-emissions-have-doubled-in-20-years/#:~:text=According%20to%20this%20analysis%2C%20these,global%20annual%20emissions%20in%202021 (accessed 5 September 2023).

14. Belton P, 'Building's hard problem – making concrete green', BBC News, 14 May 2021, https://www.bbc.co.uk/news/business-56716859 (accessed 5 September 2023).

15. Simpkins K, 'Cities of the future may be built with algae-grown limestone', CU Boulder Today, 23 June 2022, https://www.colorado.edu/today/2022/06/23/cities-future-may-be-built-algae-grown-limestone (accessed 5 September 2023).

16. Ibid.

17. https://www.recyclingtoday.com/news/worldsteel-co2-report (accessed 5 September 2023).

18. Trafton A, 'New lightweight material is stronger than steel', MIT, 2 February 2022, https://news.mit.edu/2022/polymer-lightweight-material-2d-0202 (accessed 5 September 2023).

19. University of Maine, 'First 100% bio-based 3D-printed home unveiled at the University of Maine', 21 November 2022, https://umaine.edu/news/blog/2022/11/21/first-100-bio-based-3d-printed-home-unveiled-at-the-university-of-maine (accessed 5 September 2023).

20. Food and Agriculture Organization of the United Nations, 'Food wastage footprints – impacts on natural resources', https://www.fao.org/3/i3347e/i3347e.pdf (accessed 5 September 2023).

21. http://fabulajp.com

22. Material District, 'An edible building material made of food waste', 2 August 2022, https://materialdistrict.com/article/an-edible-building-material-made-of-food-waste (accessed 5 September 2023)

23. Boren M, Chan V and Musso C, 'The path to improved returns in materials commercialisation', McKinsey & Company, 1 August 2012, https://www.mckinsey.com/capabilities/operations/our-insights/the-path-to-improved-returns-in-materials-commercialization (accessed 5 September 2023).

Conclusion

1. cities4forests.com (accessed 20 August 2023).

2. architectscan.org/post/acan-futurebuild-2023 (accessed 17 August 2023).

How to read an EPDs

1. Construction LCA, https://constructionlca.co.uk (accessed 14 September 2023).

FURTHER READING

General

Berge, B, *The Ecology of Building Materials*, 2nd ed. Routledge, 2015.

Gauzin-Müller D, Vissac A, *TerraFibra*, Pavillon de l'Arsenal, 2021.

Halligan, C, Denison, J, A *Guide to Building Materials and the Environment*, 3rd ed. SGP, 2023 https://www.stephengeorge.co.uk/wp-content/uploads/2023/08/Guide-to-Building-Materials-and-the-Environment-v3.pdf

King, B, *The New Carbon Architecture: Building to Cool the Climate*, New Society Publishers, 2017.

Material Cultures, *Material Cultures: Material Reform, Building for a Post-Carbon Future*, MACK, 2022.

Aluminium

AFA, *Finishing Aluminium; a guide for architects*, AFA, 1999.

ALFED, *The Proprieties of Aluminium and its Alloys*, ALFED, 2014.

Carlisle S, Frielander E and Faircloth B, *Aluminium and Life Cycle Thinking: Towards Sustainable Cities*, Cwningen Press, 2015.

Stacey M, *Aluminium and Durability: Towards Sustainable Cities*, Second Edition, Cwningen Press, 2015.

Stacey M, *Aluminium Recyclability and Recycling: Towards Sustainable Cities*, Cwningen Press, 2015.

Stacey M, *Aluminium Flexible and Light: Towards Sustainable Cities*, Cwningen Press, 2016.

Stacey M, *Aluminium: Sympathetic and Powerful*, Riverside Architectural Press, 2020.

Stacey M, *Aluminium: a studio design guide*, RIBA Publishing, 2023.

Bamboo

Recommended resources

Bamboo U produces building guides and provides training: https://bamboou.com

The resources section of the International Bamboo and Rattan Organisation (INBAR) contains many valuable resources: https://www.inbar.int/resources

INBAR bamboo and rattan trade database: https://trade.inbar.int:10444

INBAR bamboo species selection tool: https://speciestool.inbar.int/bamboo

ISO Technical Committee 165 publishes standards for the structural use of bamboo: https://www.iso.org/committee/53584.html. Some of these are British Standards

General

Archila HF, Kaminski S, Trujillo D, Zea Escamilla E and Harries K, 'Bamboo reinforced concrete: A critical review', *Materials and Structures* 51, 2018, https://doi.org/10.1617/s11527-018-1228-6doi:10.1617/s11527-018-1228-6

Archila H, 'Bamboo hybrids can bolster UK timber supplies', *Materials World*, 2019, https://www.iom3archive.org.uk/materials-world-magazine/feature/2019/jul/01/bamboo-hybrids-can-bolster-uk-timber-supplies

Hidalgo-López O, *Bamboo: The Gift of Gods*, D'vinni Ltda, 2003

Kaminski S, Harries KA, Lopez LF, Trujillo D and Archila H, 'Durability of whole culm bamboo: Facts, misconceptions and the new ISO 22156 framework', *NOCMAT 2022 – 18th International Conference on Non-conventional Materials and Technologies*, 22 May 2022, p 14, https://zenodo.org/record/6575090doi:10.5281/ZENODO.6575090

Kaminski S, Lawrence A and Trujillo D, 'Structural use of bamboo Part 1: Introduction to bamboo', *The Structural Engineer*, August 2016, pp 40-43.

Liu KW, Xu QF, Wang G, Chen FM and Leng YB, *Contemporary Bamboo Architecture in China*, PRC, 2019.

Sustainability and LCA

Göswein V, Arehart J, Phan-huy C, Pomponi F and Habert G, 'Barriers and opportunities of fast-growing bio-based material use in buildings', *Buildings and Cities* **3(1)**, 2022, pp 745-755, http://journal-buildingscities.org/articles/10.5334/bc.254

Trujillo D, Archila HF, *Engineered Bamboo and Bamboo Engineering*, TRADA, 2016

van der Lugt P and King C, 'Bamboo in the Circular Economy: The potential of bamboo in a zero-waste, low-carbon future', International Bamboo and Rattan Organisation (INBAR), 2019.

Zea Escamilla E, Habert G, Correal Daza J, Archila H, Echeverry Fernández J and Trujillo D, 'Industrial or traditional bamboo construction? Comparative life cycle assessment (LCA) of bamboo-based buildings', *Sustainability* **10**, 2018.

Standards

ISO (2018) BS ISO 19624:2018 Bamboo structures – Grading of bamboo culms – Basic principles and procedures. BSI Standards Publication

ISO (2019) BS ISO 22157:2019 Bamboo structures – Determination of physical and mechanical properties of bamboo culms – Test methods. BSI Standards Publication

ISO (2021) BS ISO 22156:2021 Bamboo structures – Bamboo culms – Structural design. BSI Standards Publication

ISO (2022) BS ISO 23478:2022, Bamboo structures – Engineered bamboo products – Test methods for determination of physical and mechanical properties. BSI Standards Publication, p 36

BSI (2022) BS ISO 21629-2:2022 Bamboo floorings. Part 2: Outdoor use, p 20

Brick

Anon, 'EU Concrete Block Market Report: Consumption, production, trade and forecast to 2030', 2022, https://www.globenewswire.com/en/news-release/2022/07/08/2476408/0/en/EU-Concrete-Block-Market-Report-Consumption-Production-Trade-and-Forecast-to-2030-IndexBox.html

Brick Development Association, 'The UK clay brickmaking process', 2023, https://www.brick.org.uk/uploads/downloads/09.-The-UK-Clay-Brickmaking-Process-General-Guide-2023.f1678701625.pdf

British Geological Survey, 'Mineral planning factsheet – Brick clay', 2007, https://www2.bgs.ac.uk/mineralsuk/planning/mineralPlanningFactsheets.html

Department for Transport, Local Government and Regions; British Geological Survey, 'Brick clay: Issues for planning', 2001, https://www2.bgs.ac.uk/mineralsuk/download/planning/brick_clay_issues_for_planning_cr01118n.pdf

Lucas R, 'The tax on bricks and tiles, 1784–1850: Its application to the country at large and, in particular, to the county of Norfolk', *Construction History*, **13**, 1997, pp 29–55

Key M, *Sustainable Masonry Construction*, IHS BRE Press, 2009

Concrete

Standards

BS EN 197-1:2011 Cement. Composition, specifications and conformity criteria for common cements

BS EN 197-5:2021 Cement. Portland-composite cement CEM II/C-M and Composite cement CEM VI

BS EN 206 Concrete. Specification, performance, production and conformity

BS 8500:2019 Concrete. Complementary British Standard to BS EN 206

BS 8204 Screeds, bases and in-situ flooring

PAS 8820:2016 Construction materials – Alkali-activated cementitious material (AACM) and concrete specification

BS EN 1992 Eurocode 2, Design of concrete structures

PAS 2080:2023 Carbon management for buildings and infrastructure

General

Buildings and Cities, 'Embodied carbon in concrete: Problems of mis-messaging', 31 May 2022, https://www.buildingsandcities.org/insights/news/embodied-carbon-concrete.html

Chatham House, 'Making concrete change: Innovation in low-carbon cement and concrete', 2018, https://www.chathamhouse.org/2018/06/making-concrete-change-innovation-low-carbon-cement-and-concrete

The Concrete Centre, *Specifying Sustainable Concrete*, https://www.concretecentre.com/Publications-Software/Publications/Specifying-Sustainable-Concrete.aspx

Get It Right Initiative, https://getitright.uk.com

Green Construction Board, *Low Carbon Concrete Routemap*, Institution of Civil Engineers, 2022, https://www.ice.org.uk/engineering-resources/briefing-sheets/low-carbon-concrete-routemap

Guidance document for PAS 2080, http://www.constructionleadershipcouncil.co.uk/wp-content/uploads/2019/06/Guidance-Document-for-PAS2080_vFinal.pdf

Institution of Structural Engineers, 'How can we reduce the embodied carbon of structural concrete?', 2021, https://www.istructe.org/IStructE/media/Public/TSE-Archive/2021/How-can-we-reduce-the-embodied-carbon-of-structural-concrete_2.pdf

Institution of Structural Engineers, 'Balancing embodied and operational carbon in building envelope design', 2022, https://www.istructe.org/journal/volumes/volume-100-(2022)/issue-3/balancing-embodied-and-operational-carbon-facades

Institution of Structural Engineers, *IStructE Sustainability Resource Map*, 2022, https://www.istructe.org/IStructE/media/Public/Resources/IStructE-Sustainability-Resource-Map.pdf

Institution of Structural Engineers, *Structural Plan of Work Sustainability Checklist*, 2022, https://www.istructe.org/resources/climate-emergency/spow-sustainability-checklist

The National Structural Concrete Specification, Fourth Edition, http://www.engineeringsurveyor.com/software/NSCS-Edition-4.pdf

UK Concrete and Cement Industry Roadmap to Beyond Net Zero, MPA, 2023, https://www.thisisukconcrete.co.uk/Resources/UK-Concrete-and-Cement-Roadmap-to-Beyond-Net-Zero.aspx

Copper

European Copper Institute, https://copperalliance.org/regional-hubs/europe

Copper Development Association, https://www.copper.org, including their detailed design handbook, https://copper.org/applications/architecture/arch_dhb

International Copper Alliance has resources relating to copper production, https://copperalliance.org/policy-focus/health-safety/copper-mark

The Construction Material Pyramid allows for comparison of broader sustainability issues such as ecosystem impacts, www.materialepyramiden.dk

Cork

Amorim Isolamentos, 'Environmental Product Declaration (EPD) for expanded insulation corkboard', 2016, https://daphabitat.pt/assets/Uploads/dap/pdfs/e78a17a79c/Dap_ICB_EN_06-10-2016.pdf

Amorim Isolamentos, 'Expanded insulation corkboard datasheet', 2021, https://www.amorimcorkinsulation.com/xms/files/FICHAS_TECNICAS_2021/FT_Corkboard_EN_2021.pdf

Amorim Isolamentos, 'MD façade datasheet', 2021, https://www.amorimcorkinsulation.com/xms/files/FICHAS_TECNICAS_2021/FT_MDFacade_EN_2021.pdf

APCOR (Portuguese Cork Association), 'Cork yearbook 2020', 2020, https://www.apcor.pt/en/portfolio-posts/apcor-year-book-2020

ERFMI (European Resilient Flooring Manufacturers' Institute), 'Environmental Product Declaration (EPD) for cork floor tiles', 2019, https://erfmi.com/wp-content/uploads/EPD-9-Cork-floor-tiles-according-to-EN-12104-6Cork-floor-tiles-according-to-EN-12104.pdf

Knapic S, *et al.*, 'Cork as a building material: A review', *European Journal of Wood and Wood Products* **74(6)**, 2016, pp 775–791

Permanent Delegation of Portugal to UNESCO, 'Montado, cultural landscape', 2017, https://whc.unesco.org/en/tentativelists/6210

Smith JT, 'Process of treating cork', United States Patent Number US 456068 A, 1891

Ty-Mawr Lime, 'Expanded cork insulation system – external wall', 2022, https://www.lime.org.uk/applications/retrofit-insulation-systems-for-old-buildings/external-wall-insulation-system/expanded-cork-insulation-system-7711.html

Wilton O and Barnett Howland M, 'Cork: An historical overview of its use in building construction', *Construction History: Journal of the Construction History Society* **35(1)**, 2020, pp 1–22.

Wilton O and Barnett Howland M, 'Cork construction kit', The *Journal of Architecture* **25(2)**, 2020, pp 138–165.

Wilton O and Barnett Howland M, *Cork Construction*, Bartlett Design Research Folios: The Bartlett School of Architecture, 2022, https://issuu.com/bartlettarchucl/docs/design-research-wilton-barnett-howland-cork-constr

Earth

Heringer A, Blair Howe L and Rauch M, *Upscaling Earth: Material Process Catalyst*, gta Verlag, 2019

Houben H and Guillaud H, *Earth Construction: A Comprehensive Guide*, ITDG publishing, 2006

Maskell D *et al.*, 'Determination of optimal plaster thickness for moisture buffering of indoor air', *Building and Environment* **130**, 2018, pp 143–150.

ECOSEE ICBBM 2017 Bio based plaster for improved indoor air quality D.Maskell *et al.*

McLean W and Silver P, *Materials Technology*, RIBA Publishing, 2021

Minke G, *Building with Earth: Design and Technology of a Sustainable Architecture*, Birkhäuser, 2022

Volhard F, *Light Earth Building: A Handbook for Building with Wood and Earth*, Birkhäuser, 2016

Walker P and Keable R, *Rammed Earth: Design and Construction Guidelines*, BRE Bookshop, 2005

Training in earth building, https://ecvetearth.hypotheses.org

UNESCO Earth Building Network, https://terra.hypotheses.org

Earth Building UK and Ireland, http://ebuki.co/index.htm

Earth building organisations are now active in most European countries, notably AsTerre in France, DVL in Germany and Ebuki in the UK

Glass

ARUP, 'Re-thinking the life-cycle of architectural glass', 2018, https://www.arup.com/perspectives/publications/research/section/re-thinking-the-life-cycle-of-architectural-glass

Deloitte, 'Deloitte sustainability: Circular economy potential for climate change mitigation', 2016, https://www2.deloitte.com/content/dam/Deloitte/my/Documents/risk/my-risk-sdg13-circular-economy-potential-for-climate-change-mitigation.pdf

Hartwell R, Coult G and Overend M, 'Mapping the flat glass value-chain: A material flow analysis and energy balance of UK production', https://www.researchgate.net/publication/359074054_Mapping_the_flat_glass_value-chain_A_material_flow_analysis_and_energy_balance_of_UK_production

IGS Magazine Spring 2022: Decarbonising the Glass Industry, 2022, https://issuu.com/intelligentpublications/docs/igs_spring2022_hi-res/20

O'Regan C, *Structural Use of Glass in Buildings*, Second Edition, Institution of Structural Engineers, 2014, https://www.istructe.org/resources/guidance/structural-use-glass-buildings

Hempcrete

Websites

The Hemp Block Company, https://hempblock.co.uk

That Hempcrete Guy, William Stanwix, https://thathempcreteguy.com

Hemp-Lime Spray Ltd, http://hemplimespray.co.uk

Publications

Bevan R and Woolley T, *Hemp Lime Construction: A Guide to Building with Hemp Lime Composites*, BRE Press, 2008

Stanwix W and Sparrow A, *The Hempcrete Book: Designing and Building with Hemp-lime*, Green Books, 2014

Woolley T, *Building Materials, Health and Indoor Air Quality: No Breathing Space?* Routledge, 2017

Woolley T, *Natural Building Techniques: A Guide to Ecological Methods and Materials*, The Crowood Press, 2022

Insulation

Alliance for Sustainable Building Products, https://asbp.org.uk

Association for Environment Conscious Building, https://aecb.net

British Board of Agrément, https://www.bbacerts.co.uk

Building Research Establishment, https://bregroup.com

Construction Products Association, https://www.constructionproducts.org.uk/our-expertise/sustainability/decarbonisation-and-net-zero

Cradle to Cradle Products Innovation Institute, https://c2ccertified.org/resources

Environdec, https://www.environdec.com/home

The Green Register of Construction Professionals, https://www.greenregister.org.uk

Insulation Manufacturers Association, https://insulationmanufacturers.org.uk

Mineral Wool Insulation Manufacturers Association, http://www.mima.info

National Blown Bead Association, https://www.nbba.org.uk

National Insulation Association, https://www.nia-uk.org

Natureplus, https://www.natureplus.org

NBS, https://www.thenbs.com (the NBS provides detailed guidance on specification of materials, with a library of technical product descriptions and manufacturers' data)

One Click LCA, https://www.oneclicklca.com/library/articles

Part Z, https://part-z.uk

Retrofit Works, https://retrofitworks.co.uk

The Sustainable Traditional Buildings Alliance, https://stbauk.org

UK Green Building Council, https://www.ukgbc.org

Mycelium

Biohm (UK) https://www.biohm.co.uk/
Ecovative (USA) https://www.ecovative.com/
The Living, Architects of 'Hy-Fi' and 'Bionic Partition' http://www.thelivingnewyork.com/
Kazior J P, 'This is not a mushroom' *Icarus Complex Magazine*, Issue 4, 2022, p.26
Macfarlane R, *Underland: A deep time journey*, Penguin, 2020
Sheldrake M, *Entangled Life: How fungi make our worlds, change our minds and shape our futures*, Random House, 2020

Plastics

Adams K, Tate L, Harrison J, John H, Costello C, Allen G, Potter M L, Potter, J K, Sindall M, *ZAP Toolkit, Zero Avoidable Packaging Waste in Construction*, https://asbp.org.uk/wp-content/uploads/2023/05/ZAP-Toolkit-v2.pdf
Alliance of Sustainable Building Products, *Plastics in Construction: Introductory Q&A Guide*, https://asbp.org.uk/wp-content/uploads/2015/12/Intro-guide-v2-April-21.pdf
Cullen J, Drewniok M, Cabrera Serrenho A, THE 'P' WORD – *Plastic in the UK: practical and pervasive ... but problematic*, The Resource Efficiency Collective, 2020
The Alliance for Sustainable Building Products, Case Study: CHARM virtually plastic-free homes, https://asbp.org.uk/case-studies/charm-virtually-plastic-free-homes

Steel

SteelConstruction.info https://www.steelconstruction.info/Building_design_using_steel_-_a_summary_for_architects https://www.steelconstruction.info/Sustainability
The UK sector decarbonisation roadmap https://bcsa.org.uk/wp-content/uploads/2022/01/BCSA-2050-Decarbonisation-Roadmap.pdf
An article about achieving net zero in the steel sector https://accelerator.chathamhouse.org/article/achieving-net-zero-steel
SCI P167 https://www.steelconstruction.info/File:SCI_P167.pdf
AISC guidance and download Guide for Architects | American Institute of Steel Construction (aisc.org)
Architect guide to stainless steel SCI P179 https://portal.steel-sci.com/shop.html
Constructalia – guidance for architects Architects (arcelormittal.com)
Worldsteel Association Home page - worldsteel.org
Eurofer Home (eurofer.eu)

Stone

Albion Stone, https://www.albionstone.com (Portland, UK, stone quarry and supplier)
Ateliers Romeo, https://www.ateliers-romeo.com (Italian stonemasons specialising in composite stone products)
Historic England, 'Building stones of England', 2023, https://historicengland.org.uk/advice/technical-advice/buildings/building-stones-england

Mineral Products Association, https://www.mineralproducts.org (UK trade association for minerals, stone, etc.)
Polycor, https://www.polycor.com (worldwide stone suppliers offering a wide range of types)
Stone Federation Great Britain, https://www.stonefed.org.uk
The Stonemasonry Company, https://www.thestonemasonrycompany.co.uk (UK-based stonemasons specialising in reinforced and tensioned stone)
Szerelmey, https://www.szerelmey.com (UK-based stonemasons)

Straw

Building Limes Forum (BLF): Encourages expertise and understanding in the appropriate use of building limes and education in the standards of production, preparation, application and aftercare, https://www.buildinglimesforum.org.uk; https://www.buildinglimesforum.org.uk/lime-publications
EcoCocon: Slovakian company making MMC prefabricated timber and straw panels since 2008, https://ecococon.eu/gbEPD (Environmental Product Declaration) for UK straw, https://environdec.com/library/epd3854
Jones B, *Building with Straw Bales* – A Practical Manual for Self-builders and Architects, 2015 edition, UIT Cambridge, http://schoolofnaturalbuilding.co.uk/barbaras-book
School of Natural Building (SNaB): Offers theoretical and practical training in the use of natural materials and cement-free foundations, http://schoolofnaturalbuilding.co.uk
School of Natural Building, Technical Guide: Straw Construction in the UK, 2022, https://strawworks.co.uk/technical-guide-straw-construction-in-the-uk Scottish Lime Centre Trust: Scottish training centre and information resource, https://scotlime.org
Straw Works: Information and directory website, https://strawworks.co.uk
Strawbale Building UK (SBUK): Cooperative, membership organisation, https://strawbalebuildinguk.com

Timber

General

Built by Nature, an organisation dedicated to accelerating timber construction in the UK through the creation of networks and grant-funding, https://builtbn.org
Hart J and Pomponi F, 'More timber in construction: Unanswered questions and future challenges', *Sustainability* **12(8)**, 2020
Timber Development UK, The Timber Industry *Net Zero Roadmap*, 2022, https://timberdevelopment.uk/resources/net-zero-roadmap
Timber Development UK, 'Timber typologies – understanding options for timber construction', 7 June 2023, https://timberdevelopment.uk/resources/timber-typologies

Forestry certification

UKWAS, United Kingdom Woodland Assurance Standard, Fourth Edition, Edinburgh, 2018
Sustainable procurement of forest products, https://sustainableforestproducts.org/Introduction
Sustainable wood for cities, https://www.citywoodguide.com

Strength

gradingRidley-Ellis D, Stapel P and Baño V, 'Strength grading of sawn timber in Europe: An explanation for engineers and researchers', *European Journal of Wood and Wood Products* **74**, 2016, pp 291–306

Engineered wood products (EWPs)

Hairstans R, *Off-site and Industrialised Timber Construction*, Second Edition, BMTRADA, 2019

Longevity

Steffen M, *Moisture and Wood-Frame Buildings*, Canadian Wood Council, 2000

STA, 'STA Advice Note 14: Robustness of CLT Structures', Structural Timber Association, 2017

Jones D and Brischke C, *Performance of Bio-based Building Materials*, Chapter 4, 'Protection of the bio-based material', Woodhead Publishing Series in Civil and Structural Engineering, 2017, p 204

Hill C, *Wood Modification: Chemical, Thermal and Other Processes*, John Wiley & Sons, 2006

Fire

Fire Safety: Wood in Construction, https://timberfiresafety.org

Design for deconstruction and reuse

Cristescu C *et al.*, 'Design for deconstruction and reuse of timber structures – state of the art review', RISE Report, 2020.

Rose CM and Stegemann JA, 'Feasibility of cross-laminated secondary timber', *Fifth International Conference on Sustainable Construction Materials and Technologies,* Coventry, 2019, pp 495–507

Sandin Y et al., *Design of Timber Buildings for Deconstruction and Reuse — Three Methods and Five Case Studies*, RISE, 2022

Zinc

General

ZINC. International Zinc Association, www.zinc.org

Zinc and galvanising

Galvanizers Association, https://www.galvanizing.org.uk

Design assistance

VMZINC UK, https://www.vmzinc.com/en-gb/technical-support/design-assistance

Standard details

Rheinzink, https://www.rheinzink.co.uk/architects-designers/services/standard-details/

INDEX

IMAGE CREDITS

Figure 0.1 Rory Gardiner; 0.2, 1.1, 1.2 LETI; 0.3 Pelle Munch-Petersen; 0.4, 1.6, 1.7, 2.1, 2.2, 4.0, 4.1, 4.2, 4.4, 4.5, 5.1a, 5.1b, 5.2, 5.3, 7.1, 10.5, 11.1, 11.2, 11.4, 13.1, 14.1, 17.1, 17.2, 17.4, 17.7 FCBStudios; 1.3 Bioregional; 1.4 Buro Happold; 1.5 ©IStructE; 2.0 Michael Stacey; 2.3 Rural Studio / Nigel Rigden; 2.4 Wilkinson Eyre / James Brittain; 3.0 Tommaso Riva; 3.1 Hector F Archila; 3.2 Mauricio Cardenas; 3.3 SeARCH; 3.4a, 3.4b dEEP Architects; 4.3 Brick Development Association; 4.6, 15.8 Webb Yates / Agnese Sanvito; 4.7 Daniel Hopkinson for Historic England; 4.8 Carmody Groarke / Filippo Bolognese; 5.0 Richard Battye for FCBStudios; 5.4 © Rob Parrish; 5.5 Hugh Broughton Architects / James Brittain; 5.6 ETH Zurich, Block Research Group – photograph by Juney Lee; 6.0 C.F. Møller Architects / Adam Moerk; 6.1 BuckleyGreyYeoman / Ugne Pouwell; 7.0 Polysmiths / Lorenzo Zandri; 7.2 Oliver Wilton; 7.3 David Grandorge; 8.0, 8.3 Jonathan Tuckey Design; 8.1 Timothy Soar; 8.2 Bennetts Associates; 8.4 Hattie Hartman; 8.5 Philippe Madec; 9.0 Simon Kennedy; 9.1, 13.2 Hufton+Crow; 9.2 Tom Mcveigh of Feilden Clegg Bradley Studios; 9.3, 9.4, 9.5, 9.7a, 9.7b Eckersley O'Callaghan; 9.6 Patriarche Architectes ©Nicolas Grosmond; 10.0 © Frances Greenhalgh; 10.1, 17.10a, 17.10b White Arkitekter; 10.2 Material Cultures; 10.3 Graham Durrant; 10.4 Barrault Pressacco © Giaime Meloni; 11.0, 11.3 Thermafleece; 12.0 Biohm; 12.1 Photo by Amy Barkow, courtesy of The Living; 12.2 The Growing Pavilion (2019) by Biobased Creations. Image by Eric Melander; 13.0 Copyright Bernhard Lux | Dreamstime.com; 13.3 CAN / Jim Stephenson; 13.4 GreenSquareAccord; 14.0, 15.0 Elliot Wood / Rob Parrish; 14.2, 14.3 FCBStudios / Eurofer; 14.4 Buckley Gray Yeoman; 14.5a, 14.5b Heyne Tillett Steel / Grosvenor; 14.6 BCSA; 15.1, 15.2, 15.3, 15.5, 15.6, 15.7, 15.9 FCBStudios / Webb Yates; 15.4 RPBW / Michel Denancé; 15.10 Pierre Bidaud; 16.0 EcoCocon®; 16.1 FCBStudios / John Butler; 16.2 Grand Huit Architects / Clara Simay; 16.3a, 16.3b Henning Larsen A/S; 16.4 NZI Architects / Juan Sepúlveda Grazioli; 17.0 Jack Hobhouse; 17.3 Studio Weave; 17.5 Waugh Thistleton; 17.6 Studio Rhe / Dirk Lindner; 17.8 Suma Arquitectura / Jesus Granada; 17.9 Henley HaleBrown / Nick Kane; 18.0 Delvendahl Martin Architects © Tim Crocker; 18.1 Kirsty Maguire Architects / David Barbour; 19.1 Wikimedia Commons / NASA/Rick Guidice; 19.2 Wikimedia Commons / Olimpia1lli; 19.3 Wikimedia Commons / Emiliania huxleyi; 20.1 5th Studio / Timothy Soar; 20.2 ACAN Climate Literacy Group; 20.3 Atelier LUMA / Adrian Deweerdt

AUTHOR BIOS

EDITORS

Hattie Hartman Journalist and author Hattie Hartman is sustainability editor at the Architects' Journal and hosts the AJ's Climate Champions podcast. Her books include *Energy | People | Buildings* (co-author, RIBA Publishing, 2021) and *London 2012 Sustainable Design* (Wiley, 2012). Hattie is an Honorary Professor of Architecture at the University of Nottingham. A long-standing champion of sustainable design, she participates widely in industry juries, chairs events and lectures on design for climate emergency.

Joe Jack Williams A partner at Feilden Clegg Bradley Studios, Joe co-leads the sustainable design and research by the practice. He is a Passivhaus consultant and led the development of FCBS CARBON, an award-winning, free, whole-life carbon tool. Joe is a strong advocate for sustainable design, working with industry groups such as LETI and the UK Net Zero Building Standard, as well as lecturing extensively at universities across the UK.

CONTRIBUTORS

Aluminium
Michael Stacey
Professor of Architecture and Tectonics
Bartlett School of Architecture
University College London
www.ucl.ac.uk/bartlett/architecture
Educator, author and architect Michael Stacey combines practice, teaching, research and writing in his work. He is the author of numerous books, including five on aluminium. His most recent book is *Aluminium: A Studio Design Guide* (RIBA Publishing, 2023).

Bamboo
Hector Archila
Senior Lecturer, University of the West of England
Founder, Amphibia BASE Ltd
https://people.uwe.ac.uk/Person/HectorArchila
Hector is an architect, materials researcher and entrepreneurial academic, driving innovation on engineered bio-based structures for high-performance buildings. Hector works at the interface between industry and academia, teaching the next generation and leading businesses, raising funds and carrying out scientific and applied research on bio-based materials.

David Trujillo
Assistant Professor, Coventry University
Head of Bio-based Materials, Atelier One
https://pureportal.coventry.ac.uk/en/persons/david-trujillo
David has over two decades of experience of researching and designing with bamboo. He has authored or co-authored more than 40 publications in the field, including three ISO standards on the structural use of bamboo. He is currently the UK's designated expert to ISO Technical Committee 165 Working Group 12: Bamboo structures.

Edwin Zea Escamilla
Senior Assistant, Chair for Sustainable Construction ETH, Zurich
Chairperson, Task Force on Bamboo Construction, INBAR
https://sc.ibi.ethz.ch/en/people/scientific-assistants/dr--edwin-zea-escamilla.html
Edwin's work addresses the sustainability of construction materials and buildings, especially the use of bio-based materials in post-disaster reconstruction and affordable housing. Edwin develops simplified methodologies for life-cycle assessment of materials and buildings. He is leading the development of the first life-cycle inventories for bamboo forestry and bamboo-based construction materials on the Ecoinvent database.

Brick
David Watson
Technical Director, AKT II
https://www.akt-uk.com
At AKT II, David co-leads on all practice matters related to climate action. He is a structural engineer by training, with a particular focus on adaptive reuse, design efficiency, material innovation and production systems research.

Concrete
Eva MacNamara
Associate Director, Expedition Engineering
https://expedition.uk.com
Eva is a structural and civil engineer and an innovation consultant. She is pushing industry boundaries for regenerative outcomes and is currently leading R&D by the Institution of Civil Engineers (ICE) on the construction industry's impact on 'embodied biodiversity'.

Helen McGarry
Senior Engineer, Expedition Engineering
https://expedition.uk.com
Helen is a structural engineer motivated by design that enhances the built environment and society. She is actively involved in research and projects related to transitioning the concrete industry to a low-carbon future.

Bruce Martin
Associate Director, Expedition Engineering
https://expedition.uk.com
Bruce is a structural engineer who is passionate about minimising and reversing our contributions to the climate and biodiversity emergencies. As a member of the Low Carbon Concrete Group, he is working to change industry practice.

Copper, Lead, Zinc
Nick Hodges
Architect
Nick Hodges is an architect with more than 20 years' experience leading and delivering award-winning buildings for Feilden Clegg Bradley Studios. He has a long-standing interest in construction, materials and sustainability; and how buildings and their materials can be responsibly used to deliver longevity and quality.

Cork
Oliver Wilton
Associate Professor of Environmental Design
Director of Design Technology
The Bartlett School of Architecture
University College London
https://www.ucl.ac.uk/bartlett/architecture/people/oliver-wilton
Oliver is an educator at UCL and a director at architecture practice WW Studio. At the Bartlett, he works with colleagues to develop the school's design technology strategy. His design, research and teaching cover housing, environmental performance, material technology and new forms of construction.

Earth
Rowland Keable
CEO Rammed Earth Consulting
CEO EBUKI, Earth Building UK and Ireland
www.ebuki.co
Rowland has been building with earth since 1985 both in the UK and abroad. He has pioneered work on earth-building standards and guidelines and undertaken research and policy development on earth building in the UK, Europe and Africa.

Glass
Graham Coult
Technical Director, Eckersley O'Callaghan
Graham is a structural engineer with a passion for structural glass. With more than 25 years' experience in glass design, he is currently involved in industry-leading research and lobbying for the circular economy of glass. Graham regularly participates in safety committees and code bodies to improve the design and adoption of glass as a structural material.

Stéphany Le Rhun
Principal Engineer, Setec
Stéphany Le Rhun is an engineer and architect and a graduate of the École Centrale Lille and the Politecnico di Milano. She worked as a façade engineer before taking on the role of Global Sustainability Lead at Eckersley O'Callaghan. In 2023, she joined engineering and project management consultancy Setec to focus on the circular economy.

Hempcrete
Graham Durrant
Director, Hemp-Lime Spray Ltd
www.hemplimespray.co.uk
Graham Durrant holds an MSc Architecture in Advanced Environmental and Energy Studies and has been working as a hempcrete contractor since 2012. Through his company, Hemp-Lime Spray Limited, he has installed hundreds of cubic metres of hempcrete into a range of buildings. He jointly runs the Hempcrete Forum, which is dedicated to providing honest, practical guidance on building with hempcrete.

Insulation
Mark Lynn
Managing Director, Eden Renewable Innovations Ltd
Mark Lynn is managing director of sustainable insulation firm Eden Renewable Innovations Ltd and is vice-chair of the Alliance for Sustainable Building Products. An expert in building insulation and an ardent believer in health and sustainability, Mark takes a particular interest in building physics and the chemistry of building materials. Mark is a graduate chemist with a Masters in business from the University of Edinburgh.

Mycelium
Duncan Baker-Brown
Founder of BakerBrown Studio, author, environmental activist and Climate Literacy Champion at the University of Brighton
Duncan is a practising architect, academic and environmental activist. Author of *The Re-Use Atlas* (RIBA Publishing, 2017), Climate Literacy Champion at the University of Brighton and co-chair of the RIBA's Climate Action Expert Advisory Group, he has practised, researched and taught around issues of sustainable and closed-looped systems for more than 25 years.

Plastics
Carol Costello
Practice Leader, Cullinan Studio
https://www.cullinanstudio.com
Carol is a director of Cullinan Studio, where she has worked for more than 25 years. She believes in the power of the collective to find solutions to difficult problems and has led the design on numerous major cultural, education, workplace and infrastructure projects. Carol is a member of the ASBP Reducing Plastics in Construction group.

Steel

Will Arnold

Will is a Fellow and staff member of The Institution of Structural Engineers, responsible for bringing sustainability into all aspects of the institution's work. He is well known for his work on embodied carbon policy in the UK, including on Part Z and the UK Net Zero Carbon Buildings Standard.

Ana Girão-Coelho

Ana is the Director of Engineering at the British Constructional Steelwork Association (BCSA). She manages several BCSA committees, develops practical design and fabrication guides, and advises BCSA members on the design and fabrication of steel structures. Ana is an expert in steel construction with over 20 years of experience in Portugal, the Netherlands, and the UK.

Michael Sansom

Dr Michel Sansom is a civil engineer and chartered environmentalist and is sustainability manager at the British Constructional Steelwork Association (BCSA). He has 25 years' experience of environmental and sustainability work in consultancy, research and research management roles in the steel and construction sectors.

Stone

Pierre Bidaud

Creative Director, The Stonemasonry Company
https://www.thestonemasonrycompany.co.uk
Pierre is a stonemason with a background in heritage. He believes stone should be a commodity rather than a luxury. At The Stonemasonry Company, he works with a team of stone enthusiasts, pioneering the development of load-bearing structures and staircases. He advocates pivoting the stone industry towards manufacturing of preassembled elements for more sustainable and efficient construction.

Alex Lynes

Associate Director, Webb Yates Engineers
www.webbyates.com
Alex is a structural engineer with specialist knowledge of stone and other non-standard materials. With over a decade's experience working on complex projects, Alex advocates structural designs that combine beautiful architecture with low-carbon design. He is currently leading on an £80-million multidisciplinary project which utilises load-bearing stone in innovative ways.

Steve Webb

Director, Webb Yates Engineers
www.webbyates.com
Steve is a structural engineer and founding director of Webb Yates Engineers. A long-time advocate of low embodied carbon structures, Steve has pioneered new approaches to reinforced stone and timber design. In 2019, he was awarded the Milne Medal primarily for work in post-tensioned stone.

Straw

Barbara Jones

Director, The School of Natural Building
http://schoolofnaturalbuilding.co.uk
A trailblazer in natural building, Barbara is an entrepreneur and consultant, as well as a qualified carpenter. She is director of The School of Natural Building and a technical consultant to EcoCocon. She is the author of *Building with Straw Bales* (Green Books, 2009) and co-author of the open-source Technical Guide: Straw Construction in the UK (2022).

Timber

Marlene Cramer

Research Assistant, Edinburgh Napier University
Marlene is a researcher at the Centre for Wood Science and Technology at Edinburgh Napier University. She advocates more efficient use of timber by reusing recovered wood and underused species. UK hardwood properties and novel grading approaches are the focus of her current research.

Gabriele Tamagnone

Research Fellow, Centre for Advanced Timber Technology (CATT), New Model Institute for Technology and Engineering (NMITE), Hereford
Gabriele is a researcher and author of numerous papers in timber engineering, with a special interest in numerical modelling and testing of CLT structures. He is currently working on the implementation of a 'living lab approach' of open-innovation ecosystems within CATT.

Future Innovations and Trends

Sophie Thomas

Director of Circular Design, Useful Simple Trust
Founding Partner, etsaW Ventures
www.usefulprojects.co.uk
www.etsaw.com
As a communicator, campaigner and designer, Sophie Thomas is a UK pioneer of circular design practice. She works with businesses and the public sector to design new systems and materials, with a particular interest in the potential of mixed waste streams. Her latest venture, etsaW, is a venture studio focused on scaling circular economy innovation.